Multicivilizational Exchanges in the Making of Modern Science

Arun Bala · Raymond W. K. Lau ·
Jianjun Mei
Editors

Multicivilizational Exchanges in the Making of Modern Science

Needham's Dialogical Vision

palgrave
macmillan

Editors
Arun Bala
IHPST, University of Toronto
Toronto, Canada

Raymond W. K. Lau
Hong Kong Metropolitan University
Lam Tin, Hong Kong

Jianjun Mei
Needham Research Institute
University of Cambridge
Cambridge, UK

ISBN 978-981-97-3540-2   ISBN 978-981-97-3541-9 (eBook)
https://doi.org/10.1007/978-981-97-3541-9

This Palgrave Macmillan imprint is published by the registered company Springer Nature Singapore Pte Ltd.
The registered company address is: 152 Beach Road, #21-01/04 Gateway East, Singapore 189721, Singapore

Foreword

As a relatively young industrial chemist in the 1970s, I was privileged to work with Dr Joseph Needham in the East Asian History of Science Library in Cambridge University from 1975–77. I was exploring with his guidance ancient Chinese applications of Pigments and Dyes for Section 34 of Volume 5 of the *Science and Civilisation in China* (SCC) series. I vividly recollect the numerous occasions, during our regular late afternoon tea breaks, when we engaged in active discussions. One important concern was what would follow next once the SCC series was completed. Needham would always answer without hesitation that much remained to be done for similar studies of East Asian, South Asian, Persian, and Islamic civilizations. He always expressed strongly his view that modern science has multicivilizational origins, and that no single civilization can claim monopoly for contributing to the rise of modern science.

These conversations with Needham, and his understanding of modern science as the outcome of exchanges between civilizations were in my thoughts when I, as Chairman since 2014, steered the Joseph Needham Foundation for Science and Civilisation (JNFSC) to support Needham

Conference 2023 titled “Needham’s Dialogical Vision: Understanding Science as a Multi-Civilizational Outcome.” JNFSC jointly organized this international conference with the Hong Kong University of Science and Technology, the Needham Research Institute, the Faculty of Asian and Middle Eastern Studies of the University of Cambridge, and the Jao Tsung-I Petite Ecole of the University of Hong Kong.

The chapters of this book are compiled from selected and revised papers presented in the conference. I hope this study would be a first step in realizing the wider global vision of science articulated by Joseph Needham, and will serve not only as a must-read reference about the multicultural origins of modern science, but also provide a teaching resource for university-level courses.

Finally, I would like to express my gratitude and heartfelt appreciation for the most generous financial support from Bai Shan Tang Foundation, Jao Tsung-I Petite Ecole of Hong Kong University, the Croucher Foundation and the Hong Kong University of Science and Technology, and above all, to the distinguished speakers and many volunteers who made the conference a success and this book possible.

Peter L. Lee, PhD
Chairman
Joseph Needham Foundation for
Science and Civilization

Contents

Notes on Contributors

Arun Bala has an MSc in Physics and PhD in Philosophy. He is the author of *The Dialogue of Civilizations in the Birth of Modern Science* and *Complementarity Beyond Physics: Niels Bohr's Parallels*, and edited *Asia, Europe and the Emergence of Modern Science*, and *The Bright Dark Ages: Comparative and Connective Perspectives*. He held teaching and research appointments with various universities, including National University of Singapore, University of Toronto, and Western University. He is Director (Research) with Joseph Needham Foundation for Science and Civilization, and Visiting Scholar, Institute for the History and Philosophy of Science and Technology, University of Toronto.

H. Floris Cohen is Emeritus Professor of Comparative History of Science at Utrecht University (Netherlands). His active concern with the history of science in China and in Islamic civilization dates from the mid-1980s. He has a lively remembrance of his sense of excitement when, in the summer of 1987, he immersed himself in Joseph Needham's essay collections and corresponding chapters in *Science and Civilisation in China*.

Prasenjit Duara is a historian and holds the Oscar Tang Chair of East Asian Studies at Duke University. He has worked on rural China, nationalism and imperialism as well as environmental history. He was Professor and Chair of History at the University of Chicago (1991-2008) and Raffles Professor and Director of Asia Research Institute at the National University of Singapore. He was awarded the *doctor philosophiae honoris causa* from the University of Oslo in 2017 following the publication *The Crisis of Global Modernity: Asian Traditions and a Sustainable Future* (Cambridge 2015).

Nader El-Bizri is the Dean of the College of Arts, Humanities, and Social Sciences at the University of Sharjah, and is also affiliated with the Department of History and Philosophy of Science at the University of Cambridge. Prior to that, he was a tenured Professor in Philosophy and Civilization Studies at the American University of Beirut, and a Leverhulme Visiting Professor of History of Science at Durham University. He obtained his PhD in Philosophy in 1999 from the New School of Social Research in New York and the M.Arch-II in Architecture in 1994 from the Graduate School of Design at Harvard University.

Imre Galambos is a specialist of Chinese manuscripts. After receiving his PhD from Berkeley in 2002, he worked for the International Dunhuang Project (IDP) at the British Library for ten years. Then he moved on to teach at the University of Cambridge, from where he left in 2023, taking up a professorship at the School of Literature, Zhejiang University. His books include *Orthography of Early Chinese Writing* (2006), *Manuscripts and Travellers* (2012, co-authored with Sam van Schaik), *Translating Chinese Tradition and Teaching Tangut Culture* (2015), and *Dunhuang Manuscript Culture* (2020).

Roger Hart is a leading scholar of science and technology in twenty-first-century China, focusing on the Second Quantum Revolution. He is Professor of History at Texas Southern University. Previous appointments include Wilson Center, Seoul National Univ., UT Austin, UChicago, Institute for Advanced Study, Stanford, UC Berkeley, and Harvard. Previous awards include ACLS, NEH, Mellon, and Fulbright. Publications include two research monographs published by Johns Hopkins UP.

He earned his B.S. from MIT, M.S. from Stanford, and Ph.D. in Chinese history and history of science from UCLA. He has lived in China for seven years; his Chinese is near-native. Website: http://www.rhart.org/. Email: rhart@rhart.org.

George Gheverghese Joseph University of Manchester, is a mathematician and historian of mathematics. Born in Kerala, he grew up in Kenya, and completed his degrees in England. He has held university appointments across the world, including East and Central Africa, India, Papua New Guinea, New Zealand, and South East Asia. He has authored many books including *The Crest of the Peacock: Non-European Roots of Mathematics*, *A Passage to Infinity: Medieval Indian Mathematics from Kerala and Its Impact*, and *Indian Mathematics: Engaging with the World from Ancient to Modern Times*. He was also Editor of *International Seminar and Colloquium on 1500 Years of Aryabhateeyam* and *Knowledge and Cultures: Crossing Boundaries in History*.

Tung-Yi Kho is interested in comparative approaches to the study of civilizations and cultures. His research interests are transdisciplinary and focus on issues fundamental to human livelihood and well-being, especially under the conditions of coloniality/modernity. The fields implicated by such interests cut across the human-cultural and natural sciences, encompassing the disciplines of Anthropology, Cultural Studies, Political Economy, Global Development Studies, Sociology, History, Philosophy, and Quantum Physics.

Wen-Hua Kuo teaches Social Studies of Medicine at National Yang Ming Chiao Tung University, Taiwan. A licensed physician and acupuncturist, his work revolves around pharmaceutical regulation and controversies in East Asian medicines attempting to be modernized and used globally. His scholarly publications appear in a range of journals crossing several disciplines, including one on Needham that appears in Isis. In addition to his current research on care and caring professions, he has served as one of the Associate Editors of Social Studies of Science since 2023 and will serve as the President of Society for the Social Studies of Science from 2025 to 2027.

Raymond W. K. Lau was Professor of Sociology, Hong Kong Metropolitan University, until retirement and currently Visiting Senior Research Fellow, Hong Kong Shue Yan University. Originally trained as a sociologist focusing on theory, in the latter 2000s, he began researching into the development of ancient Chinese thoughts in a comparative context. He has published primary research on the philosophies of Xunzi, Laozi, and Zhuangzi in Chinese and in English. In 2020, he published in English a monograph titled *Intellectual Developments in Greece and China: Contingency, Institutionalization and Path Dependency*, which has led to his current work on the Needham Question.

Keekok Lee is a philosopher trained in the analytical tradition of "Anglo-Saxon philosophy." She taught at the National University of Singapore before moving to the University of Manchester, UK. Since taking early retirement in 1999, she has continued to research and publish in several philosophical domains, including the Philosophy of Biology/Genetics and the Philosophy of Medicine. The latter domain has investigated both Biomedicine and Classical Chinese Medicine. Of late, she has also been pre-occupied with the notion of Coloniality, both Exogenous and Endogenous with regard to non-European systems of knowledges. (For details, see http://www.keekoklee.org/)

Jongtae Lim is a professor, Department of Science Studies, Seoul National University, South Korea. He earned his PhD at Seoul National University in 2003, with his dissertation titled 'Chinese and Korean Scholars' Understanding of the Jesuits' Geographical Knowledge in the 17th and 18th Centuries.' Since 2006, he has been teaching the history of science in Korea in the Department of Science Studies at Seoul National University. His research interests cover topics in the history of early modern Korean science, such as Western learning, Sino-Korean scientific exchanges, and science in the state bureaucracy.

Gordon McOuat is Professor and former Director of the History of Science and Technology Program at the University of King's College/ Dalhousie University and Director of the "Cosmopolitanism and the Local in Nature and Science, East and West" project. His research focuses on the history and philosophy of logic, classification, and the origins of "natural kinds", including works on "Bentham's Logic," "The Origins of Natural Kinds," "Cataloguing Power," and "Species, Rules and Meaning." Recently, his research has turned to the study of the situatedness of knowledge and the international circulation of cosmopolitan science, co-editing *Circulation of Knowledge between England, India and China* (Brill 2013), *Narratives of Nature and Science, East and West* (Routledge, 2015), and *Spaces of Enlightenment Science* (Brill, 2021).

Jianjun Mei is Director of the Needham Research Institute in Cambridge, UK. He holds concurrent positions as Director of Research at the McDonald Institute for Archaeological Research, University of Cambridge; Visiting Professor of the University of Science and Technology Beijing; Editor-in-Chief of the Editorial Board for Advances in Archaeomaterials. His primary research interests are the history of metallurgical technology in China and the history of scientific and technological exchanges between China and other countries. He has published two monographs and more than one hundred academic papers in both Chinese and English.

Hyunhee Park is Professor of History at John Jay College and the Graduate Center of the City University of New York. She specializes in the history of cross-cultural contacts in East Asia and the Islamic world, particularly Sino-Islamic contacts, focusing on information/knowledge transfers, including transfers of geographical knowledge, foodways, and distillation technologies. She is the author of *Mapping the Chinese and Islamic Worlds: Cross-Cultural Exchange in Pre-Modern Asia* (2012), and *Soju: A Global History* (2021), and has published thirty academic articles and book chapters.

Hans Pols is Professor at the School of History and Philosophy of Science at the University of Sydney. He is interested in the history of colonial medicine and the transformation medical research and practice underwent during the process of decolonization. His research has focused on the Dutch East Indies and Indonesia, and on psychiatry and mental health. His book *Nurturing Indonesia: Medicine and Decolonisation in the Dutch East Indies* was published by Cambridge University Press in 2018.

Sundar Sarukkai has held positions of professor of philosophy at the National Institute of Advanced Studies, Founder-Director of the Manipal Centre for Philosophy and Humanities, and Visiting Faculty at the Centre for Society and Policy, Indian Institute of Science. He is the author of the following books—*Translating the World: Science and Language, Philosophy of Symmetry, Indian Philosophy and Philosophy of Science, What is Science?, JRD Tata and the Ethics of Philanthropy, Philosophy for Children: Thinking, Reading, Writing, The Social Life of Democracy*, and two books co-authored with Gopal Guru—*The Cracked Mirror: An Indian Debate on Experience and Theory and Experience, Caste and the Everyday Social*. His latest book is a novel, *Following a Prayer*. He is the Series Editor of Routledge's *Science and Technology Studies*, as well as the Co-Chief Editor of the *Springer Handbook of Logical Thought in India*.

John Steele is Wilbour Professor of Egyptology and Assyriology at Brown University. A historian of astronomy, he has research interests in Babylonian astral science, the circulation of astronomical knowledge, and the reception of ancient science from the seventeenth to the twentieth century. He is the author of several books including *The Babylonian Astronomical Compendium MUL.APIN* (Routledge 2019; co-authored with Hermann Hunger) and *The Allure of the Ancient: Receptions of the Ancient Middle East, ca. 1600–1800* (Brill, co-edited with Margaret Geoga).

List of Figures

Chapter 11

Chapter 12

Chapter 13

Chapter 14

Chapter 16

Chapter 17

List of Tables

1

Introduction

Arun Bala, Raymond W. K. Lau, and Jianjun Mei

1 Needham's Dual Legacy

This volume brings together scholars who have articulated or contributed to Needham's dialogical vision of science as the outgrowth of multicivilizational interactions. They span not only diverse civilizations but also diverse disciplines. The organization of the chapter themes of the book has drawn inspiration from the way Needham framed his monumental project into seven volumes each reflecting a disciplinary theme. Although Needham focused his studies on Chinese science as these volumes testify,

A. Bala (✉)
Institute for the History and Philosophy of Science and Technology, University of Toronto, Toronto, Canada
e-mail: arun.bala@yahoo.com

R. W. K. Lau
Hong Kong Shue Yan University, North Point, Hong Kong

J. Mei
Needham Research Institute, University of Cambridge, Cambridge, UK

A. Bala et al. (eds.), *Multicivilizational Exchanges in the Making of Modern Science*,
https://doi.org/10.1007/978-981-97-3541-9_1

he supported the dialogical turn when he reflected in his final summing up volume in 2004 on the broader implications of his achievement:

> Modern science is indeed composed of contributions from all the peoples of the Old World, and each contribution has flowed continuously in it, whether from Greek and Roman antiquity, or from the Islamic world or from the cultures of China and India. (Needham 2004, p. 25)[1]

What is significant is that Needham has an ecumenical image of modern science as growing through contributions from many civilizations in both its technological and, to a certain extent, theoretical development in terms of techniques and ideas. This is also evident in his 'rivers and ocean view' of modern science articulated later in the same volume:

> I have pictured modern science as being like an ocean into which the rivers of all the world's civilizations have poured their waters. (Needham 2004, p. 201)

In the later stages of his work, Needham had already begun making forays in such a multi-civilizational direction. In 1979 he published his reflections in his book *Within the Four Seas: The Dialogue of East and West*. Later in his inaugural lecture in 1985, that began what is now called the Needham Memorial Lecture Series, he titled his talk 'Gunpowder as the Fourth Power, East and West.' This reflected his view that exchanges between China, the West and other civilizations shaped modern society today.[2]

---

[1] Although Volume 7 was published in 2004, the text cited here was based on his Presidential Address delivered to Section X (General) on 31 August 1967 at the Leeds Meeting of the British Association (Needham 2004, p. 24, note 1).

[2] Needham writes: 'For three thousand years a dialogue has been going on between the two ends of the Old World. Greatly have they influenced each other, and very different are the cultures they have produced. We have now good reason to think that the problems of the world will never be solved so long as they are considered only from a European point of view. It is necessary to see Europe from the outside, to see European history, and European failure no less than European achievement, through the eyes of that larger part of humanity, the peoples of Asia (and indeed also of Africa)' (Needham 1979, p. 11). See also Needham (1985)

Joseph Needham developed two questions in the course of his study of the history of Chinese science. The first concerned why modern science emerged in Europe but not in China? This is his famous comparative question that has engaged historians, philosophers and sociologists ever since he framed it. It has parallels with the question why modern industrial society and capitalism emerged in Europe but not elsewhere that has engaged attention since it was framed by Max Weber.

The second is the question that Needham also raised and that can be seen as the dialogical question—how and possibly why China was able to contribute scientific and technological ideas that nurtured and enriched modern science. Like the Weberian question this has been extended to other civilizations—especially those such as the Middle Eastern and Indian that had achieved as high, or even higher levels, of achievement in the areas of technology and science than Europe before the modern era. But since Europe was about as developed as South East Asia at the time of Charlemagne, at least that part where modern science and capitalism first took of others have extended the question to ask 'Why did modern science emerge in Europe rather than South East Asia?'[3]

There is a third question to which attention has been paid even less than the dialogical question embodied in Needham's rivers-into-oceans metaphor. This concerns whether modern science has ignored dimensions of nature that could be rectified by learning from pre-modern cultures. This other aspect of Needham's view, albeit developing later in his project's life, also surprisingly emerges in quite a number of contributions to this volume. They mirror notions that in areas such as medicine and environmental natural knowledge modern science could profit by learning from pre-modern cultures.[4]

---

[3] South East Asia rarely features in the addressing of the Needham Question but there are exceptions. See Bala (2016) and Low (1999).

[4] A pioneering study in this area is Nandy (1989) See also Goonatilake (1999). Interestingly a leading contemporary philosopher of science Hasok Chang has argued that history of science, that offers understanding and appreciation of displaced theories in science could offer a reservoir of ideas and techniques that may turn out to be useful in constructing future science. Basically he sees history of science as complementing science in enriching its context for discovery. His argument could also be extended to scientific ideas from cultures outside the West, and in the light of dialogical histories of science (both ancient and modern). See Chang (2016).

Today there is a great deal of interest that includes essays on exchanges in Africa and the Atlantic Americas that is ignored in this volume. This is because the book focuses on the issue of why and how Needham's dialogical question has been marginalized in contrast to his comparative question. The comparative debates have to date centered on Chinese, Islamic, Indian, Babylonian, Japanese and Korean sciences. Hence the chapters are also confined to these areas. But the issues raised in the volume could in future be extended to include a wider global context.

A vast body of literature has emerged since Needham's pioneering work, in part inspired by his groundbreaking efforts and inspiration, which showed that what was important was to understand how interconnections of civilizations over the last three thousand years, and the exchanges of cosmological, mathematical, geographical, physical, biological and medical technologies, techniques, practices and knowledge had been woven together to make modern science possible. Although Needham himself had on numerous occasions stressed the importance of the connected history of science and technology, and the circulations of know-how and knowledge with associated dialogical exchanges, this crucial dimension of science history has hitherto been marginalized by the overemphasis on his comparative question.

2 Structure of the Book Themes

The organization of the themes in this volume follow closely the way Needham structured his volumes for *Science and Civilisation in China*.[5] Needham had framed his project into 7 volumes—many of them were later divided into sub-volumes. The book has five themes that are drawn from Needham's categorization of his volumes. The first volume is a history of China and the last volume on The Social Background has two parts—one deals with Language and Logic in China

[5] This structure also follows the way the panels were defined for Needham Conference 2023. The conference titled 'Needham's Dialogical Vision: Understanding Science as a Multicivilizational Outcome' was organized by the Joseph Needham Foundation for Science and Civilization, Needham Research Institute, Hong Kong University of Science and Technology, and the Faculty of Asian and Middle Eastern Studies in Cambridge University.

and the other on General Conclusions and Reflections concerns with addressing Needham's Grand Comparative Question concerning why modern science emerged in Europe and not China. The first volume and part one of the last volume have set the tone for the thematic issue of Historical Sociologies in Dialogue that is the theme of the first section of the book. It has three chapters by Prasenjit Duara, Raymond Lau and Tung-yi Kho.

Needham's second volume was entitled *History of Scientific Thought*. It largely focused on the philosophical background of Chinese science, especially on the way Chinese philosophies—Taoism, Confucianism, Mohism, Legalism and to an extent Buddhism—came to shape and influence Chinese science. The last theme of this volume follows this epistemological and methodological orientation of Needham and is titled *Modes of Inquiry in Dialogue*. We have four chapters by Sarukkai, Bala, McOuat and Cohen that deal with these generally philosophical questions.

Indeed, there has been little appreciation that three of Needham's seven volumes address the history (volume 1), philosophy (volume 2) and sociology (volume 7) of Chinese science. This framing of the current volume helps to rectify this neglect, and even stress the important significance Needham attached, long before it became fashionable, to connect together the history, philosophy and sociology of any tradition of science in order to understand its growth and evolution.

In between his historical, philosophical and sociological volumes Needham frames the four volumes that focus on specific scientific disciplines. These are titled *Mathematics and the Sciences of the Heavens and Earth (volume 3), Physics and Physical Technology* (v.4), *Chemistry and Chemical Technology* (v.5), *and Biology and Biological Technology* (v.6). Much of Needhamian scholarship has focused on widening his explorations of Chinese science, but ignored the wider science studies context in which he framed his project.

In acknowledgment of his studies in the specific sciences, the present book has three themes. The second on *Cosmologies in Dialogue* includes contributions by John Steele, Jongtae Lim and Hyunhee Park. The third section *Natural Sciences in Dialogue* includes the chapters of Joseph,

Hart, El-Bizri and Galambos. In the penultimate section *Medical Traditions in Dialogue* carries contributions by Hans Pols, Wen-Hua Kuo and Keekok Lee.

Moreover, the areas of cosmology, natural sciences of physics and chemistry, and biomedical sciences have largely been approached through a comparative perspective with the view to addressing Needham's question 'Why did modern science develop in Europe but not in China or any other major Eurasian civilization?' By contrast this volume focuses upon Needham's dialogical question 'How and why were China, and other major Asian civilizations, able to articulate knowledge that subsequently came to enrich modern science?' or carry knowledge that can enrich science in future. The essays take Needham's project forward in two directions—advancing his dialogical vision of the growth of science, and his multicivilizational perspective of modern science.

It is, of course, possible to ask whether framing the studies of Chinese science into categories of modern science is the right way to approach the issue. Are we not guilty of the sins of anachronism and teleology whichever direction we look at the historical connections. The renowned classical scholar Geoffrey Lloyd who has also researched into comparative studies of Chinese science has recently emphasized this by writing:

> The multiple volumes of SCC are truly monumental. Yet the plan and execution are, as we all know, very much of their time, and this has certainly contributed to its relative lack of influence in history and philosophy of science circles in the West – in contrast to the continuing mass of attention it still attracts in China itself. Joseph decided to organise the work according to Western categories: astronomy, mathematics, physics, engineering, medicine and so on. He was convinced not just that science is now universal but that those disciplinary boundaries can be used in relation to the science of much earlier times… (Lloyd 2020, p. 2)

But Needham himself had recognized this concern and made an answer to it four decades earlier when he wrote:

> I suppose we all generally agree that there is only one unitary science of nature, approached more or less closely and built up more or less successfully and continuously, even if very slowly, by the several groups

> of mankind from age to age. This means that we could expect to trace an absolute continuity between the first beginnings of astronomy and medicine in ancient Babylonia or ancient Egypt, through the advancing natural knowledge of mediaeval China, India, Islam and the classical Western world, to the break-through of late Renaissance Europe when, as has been said, the most effective method of discovery was itself discovered. (Needham 1978, p. 110)

Nevertheless, dialogical studies will continually face this problem of crossing boundaries—how do we select the conceptual frames for understanding knowledge in one frame in order to see its relevance for transmission to another frame. We will in this volume stay with Needham's choice even if it remains a problematic epistemological concern.

3 Explorations in Needham's Dialogical Vision

The chapters in this volume can be seen as exploring from multiple perspectives Needham's vision of science as evolving through multicultural engagements. They are defined by the different volumes of his grand project to both document and understand the intellectual, sociocultural and historical evolution of not only Chinese science in particular, but also its relevance for global science. It begins with the section Historical Sociologies in Dialogue. These are three chapters by Prasenjit Duara, Raymond WK Lau and Tung-yi Kho that explore the scope and limits of Needham's position.

The second chapter by Prasenjit Duara attempts to go beyond, and yet build upon, Needham's conception of pre-modern sciences from various civilizations as rivers flowing into the ocean of modern science. By extending and deepening the metaphor of the ocean Duara offers a more nuanced and rich perspective of historical exchanges of knowledge including that of science and technology. He offers a picture of history in terms of oceanic flows where single events ripple or crash across multiple boundaries, join other currents and eddies, or fall into thermohaline depths to upwell elsewhere. His chapter outlines a metaphor of knowledge exchanges as involving circulatory processes that interact and transform in patterns of stability and change that oceanic flows display rather than Needham's simplistic notion of separate rivers merging into a single ocean. He argues that such a circulatory history—as he terms it—does not privilege ownership of historical origins of scientific ideas and discoveries. He illustrates this by showing how a methodology of scientific rationality circulated through successive Eurasian religious traditions without any recognition of roots or pedigree.

In the following chapter Raymond Lau begins by acknowledging that, largely inspired by Needham, a strong case for dialogism or the dialogical perspective had been established by the 2000s. He notes that this has led some Eurocentric scholars to resort to neo-Eurocentric arguments that no longer claim science to have arisen from Europe's 'autochthonous' development. Instead, they argue that Europe was able to make use of non-European inputs to breakthrough into modern science due to certain 'unique' European 'qualities.' Such neo-Eurocentrism signals a connatural shortcoming of dialogism: While dialogism is able to show that intellectual inputs into Latin Europe from non-European civilizations constituted indispensable pre-conditions for the breakthrough into modern science, it is unable to explain why Latin Europe was able to make the breakthrough itself on the basis of such inputs. Lau proposes that it is necessary to transcend dialogism to provide an adequate non-Eurocentric account for the rise of modern science. He offers an analytical framework that takes into account scholarly traditions and the transformation of late Renaissance scholarly agency in the post-1453 era due to the conquest of Constantinople to transcend dialogism and account for modern science in fully non-essentialist terms.[6]

In contrast to the above views Tung-yi Kho's historical sociology presents a decolonial perspective which repudiates the notion of accommodating multicivilizational knowledges into modern science. He maintains that as modernity was emerging and being celebrated by the West, it was accompanied by a darker side—coloniality—in non-Western societies. Coloniality is implicated not only in its nurture of an exploitative

[6] For a more detailed and thoroughgoing account of this non-essentialist analytical framework see Lau (2020).

political economy, but also its repression of cultural and knowledge production outside Europe. This makes Needham's perspective problematic because of two premises undergirding his Grand Question why modern science first arose in Europe but not elsewhere. Firstly, it makes the questionable assumption that modern science represents the pinnacle of knowledge; and second, that because modern science was first 'discovered' in the West, the latter represents the ideal of civilization. Using Chinese examples, Kho shows why Needham's line of inquiry is misdirected, and offers in its stead a more expansive understanding of science and civilization which does not embrace Needham's exclusive vision of modern science as the sole and ultimate arbiter of natural knowledge.

The next section cosmologies in dialogue includes discussions on the dialogical exchanges that shaped scientific exchanges in three areas—in the astral sciences of diverse civilizations by John Steele, in astronomy, mathematics and geography between Qing China and Korea by Jongtae Lim, exchanges of geographical knowledge between China and the world of Islam by Hyunhee Park, ending with Joseph's account of the growth of mathematics as a result of knowledge circulations across many Afro-Eurasian cultures.

Steele examines how elements of Babylonian astral science circulated over wide areas including the eastern Mediterranean, Iran and, in particular, he explores its adoption, adaptation and assimilation into different cultures and languages. His study raises questions such as why and how did such circulations occur, what were the processes of transformation that shaped it adaptations into local cultures and traditions, and why these changes were such that it becomes no more useful to refer to the knowledge that traveled as 'Babylonian.' Its spread has made this knowledge truly multicivilizational. He uses three case studies of the assimilation of aspects of Babylonian astral science by Greek, Indian and Jewish scholars to illustrate two key features of the circulation of knowledge in general, and scientific knowledge in particular. One is the appropriation of perceived useful Babylonian knowledge, and the second is the adaptation of such material within existing knowledge systems. His key notions of appropriation and adaptation contribute to a wider understanding of the way knowledge travels across civilizations, instead of being locked into civilizational silos of science as is widely assumed.

Lim explores how, during the early years of the eighteenth century, with the consolidation and stabilization of Qing imperial rule in China, it became possible for scholars from Korea and China to exchange knowledge and practices. This was mediated through government projects, especially in astronomy and geography. His chapter is an attempt to explore Needham's 'dialogic perspective' in the history of science by delineating the political and cultural dynamics involved in these exchanges where the Sino-European science of the Qing was transferred to the Korean tributary state. He finds that in spite of being the weaker party, the Korean actors also played a leading role in these exchanges, by adroitly sidestepping restrictions by Qing agents designed to safeguard their political and epistemic privileges. They did this by not only permitting, but also creating boundaries for knowledge crossings. Korean scholars thereby used the exchanges to build a local science of their own that conformed to its own scholarly agenda and cultural identity. In effect, instead of being passive importers of metropolitan science, they also acted as knowledge producers.

Hyunhee Park observes that the 1959 publication of *SCC* Volume III, which includes novel discussions about Chinese geography and cartography, has led to new methods of geographical study that have made conspicuous lesser known but more practical and even more 'accurate' forms of geographical knowledge than those found in well-studied European societies thanks to flourishing networks of cross-cultural interactions. She demonstrates that geographers and cartographers in both Chinese and Islamic cultural worlds in fact enjoyed expansive and accurate views of the world at large, relying on more accurate and up-to-date literary and cartographical data accumulated through multicivilizational connections and contacts. By developing new approaches that utilize greater information from abundant geographic and cartographic works developed in Chinese and Islamic worlds, her chapter suggests how we can re-evaluate the relative importance of practical knowledge in the development of geography in world history. Her study shows how the evolution of geographical knowledge and cartography can enrich our understanding of the global history of science.

The third section Natural Sciences in Dialogue has four essays—two by George Gheverghese Joseph and Roger Hart on the circulation of mathematical knowledge, followed by two pieces on exchanges in the theory of optics and the transmissions of book forms.

Joseph contests the view that histories of mathematics can be adequately understood by tracing their growth within civilizations in isolation from each other. Such tunnel histories of science fail to show those historical processes and epistemological reorientations that are often engendered by such inter-civilizational exchanges. But from the Bronze Age about 5,000 years ago mathematical ideas have circulated across the Afro-Eurasian region between the agrarian civilizations that emerged in the river valleys of the Tigris-Euphrates, the Nile, the Indus and the Hwang-Ho. Over time these interactions became more intensive and extensive as the connections between these civilizations expanded through empires, religions and commerce. His chapter explores such exchanges between Babylonian, Egyptian, Greek, Indian, Islamic and Chinese civilizations. By taking a global perspective of the circulations of mathematics, Joseph explores how the flows of ideas across Afro-Eurasia have crucially influenced the growth of mathematics. In effect his chapter is an attempt to 'liberate mathematics from civilizations' and constitutes an approach that could be extended to other disciplines as well.

The chapter following also looks at the knowledge exchanges across civilizations that expanded mathematical ideas. Roger Hart notes that the best currently available evidence for 'understanding science as a multicivilizational outcome' is the early history of linear algebra. Linear algebra is a core field of modern mathematics and one of the mathematical foundations of quantum mechanics and quantum information sciences. He argues that it first developed in China before circulating into Europe by the late Middle Ages. He shows that many of the key techniques and methodology of the discipline can be found in Chinese treatises dating back to the first century CE and continuing into the early-modern era. These include what are now seen as distinctive discoveries in Europe in the early-modern era associated with Gaussian elimination—augmented matrices, row operations, row reductions, leading to upper-triangular form which can be found in the earliest transmitted Chinese mathematical treatise, *The Nine Chapters on the Mathematical Art*, from the second century CE. Hart concludes that the history of linear algebra shows how modern science has always been 'dialogical.'

In the next tenth chapter Nader El-Bizri focuses on the science of optics of the eleventh-century Arab polymath Ibn al-Haytham (known in Latin as 'Alhazen,' 965-1041 CE). He critically engaged the ancient Greco-Roman principles of visual perception of figures such as Plato, Aristotle, Ptolemy, Euclid, Galen. El-Bizri shows how Ibn al-Haytham's monumental *Optics* (*Kitāb al-Manāẓir*; *Book of Optics*) became the locus of dialogical multicivilizational confluences in science through its seminal influence in medieval Europe on its science and arts. A specific emphasis is placed in this regard on the adaptive assimilation of his theories of vision and light within the European *perspectiva* traditions in science, and its impact on European aesthetics and architecture—an influence that stretched from the High Middle Ages, to the Italian Renaissance and early-modern period. El-Bizri concludes that the case of Ibn al-Haytham's *Optics* illustrates how European contributions to modern science have been underpinned by drawing upon intellectual resources from Arabic civilization, Egypt, Mesopotamia, Persia, India and China. It illustrates the multicivilizational exchanges that went into the making of modern science.

Imre Galambos' chapter is unique in this volume because it is not about dialogical exchanges of ideas but about what we can learn about them through the study of material exchanges of book forms. He argues that the history of the Chinese book has largely been based on transmitted sources until a series of large-scale manuscript discoveries of the twentieth century. Most of these discoveries happened along the peripheries of the Sinitic cultural sphere, at sites along the Silk Road such as Dunhuang, Turfan and Khara-khoto. They require us to recognize the presence early pre-modern manuscript cultures with similar manuscripts that survive in other regions beyond the Sinitic cultural center, including Japan, Korea and the regions of so-called ethnic minorities. Galambos argues that this shows that while some book forms have disappeared without a trace from the central areas of Sinitic culture, they survive—or may even continue to be used—along the peripheries. Such survivals are evidence of patterns of material transmissions from other areas of research that corroborate, albeit indirectly, multicivilizational exchanges of knowledge in the history of science and technology.

The following section on medical traditions in dialogue has three contributions that complement each other—Han Pols is concerned with the transmissions of traditional therapies and practices and their accommodations with modern medicine, Wen-Hua Kuo looks at how modern epistemological and methodological practices have been deployed to integrate traditional acupuncture and its theoretical orientation (where possible) into modern medicine, while Keekok Lee's chapter looks at how modern epidemiological approaches and theoretical notions are now being re-discovered in modern medicine. All of these chapters problematize how one system of traditional knowledge finds accommodation into another system of knowledge associated with modernity.

Pols describes the Dutch East Indies and Indonesia as a meeting place of a wide range of practitioners and suppliers of medicines, cures and approaches from multiple traditions including Indonesian, Chinese and Western. These traditions were selectively drawn from, combined and transformed by the sick, their carers and healers they sought help from. This diverse medical marketplace is explored by Pols by studying the offerings of pharmacists and the magazines through which they advertised their products as pills, elixirs, potions and technologies of treatment. They ranged from those designed to alleviate illness, enhance beauty or promote sexual vigor. Pols argues that in order to understand this eclectic mix we should go beyond the influential texts based on Western, Ayurvedic, Chinese and Arabic traditions. Instead only by focusing on the 'material' elements of the marketplace can we appreciate the global circulation of medicinal products within the Indonesian context. Even the marketplace as an important arena for the exchange of medical ideas from diverse traditions requires deeper appreciation of the material circulations of medicinal products and technologies it purveyed.

Wen-Hua Kuo describes attempts to make acupuncture a standardized therapeutic for study and for clinical practice. It has led to institutional efforts to set a scientific foundation for acupuncture by laying down scientific standards for it. Kuo's chapter updates Needham's approach to acupuncture that treats it as both experience-based and clinically oriented by looking at the career and work of Jaung-Geng Lin, the first physician of Chinese medicine elected as an academician at Academia Sinica. It looks at Lin and his achievement in a larger context of his attempts to validate this non-biomedical tradition with scientific experiments and standards. Tracing and evaluating Taiwanese Professor Lin's research career, Kuo not only allows us to see the changing meaning of this living tradition by researchers and experts to make it comparable to biomedicine, but also how it offers an opportunity to refashion it, as Needham expected, by making it a part of ecumenical medicine in future. Kuo's study exemplifies how methodological practices from modern medicine and techniques and theoretical insights from non-biomedical traditions can be combined to advance scientific knowledge.

In her chapter, Lee Keekok maintains that comparative and dialogical approaches to answering the Needham Question are interconnected. She argues that Needham's comparative approach itself may have been limited because of his limited conception of scientific medicine. Even as we take a dialogical orientation, we also need to deploy a comparative one to enrich our understanding. She chooses Chinese medicine to illustrate her point. According to her ignoring epidemiology in modern medicine, and only attending to clinical medicine, led Needham, despite his historical ecumenism, to not only travel the road of coloniality but also misrepresent Chinese philosophy-cosmology as well. But modern medicine, especially in epidemiology is shifting from a Newtonian to a post-Newtonian paradigm that is bringing it into closer alignment with classical Chinese medicine. This aligns with the classical Chinese medical view which acknowledges as key notions personalized and preventive medicine which operate within the framework of a non-linear model of causality. Lee argues that a comparative study of modern trends in biomedicine and classical Chinese medical approaches and ideas can generate a fruitful dialogue to enrich medicine in future.

The final section on Modes of Inquiry in Dialogue has four chapters that explore the methodological and epistemological dimensions of intercultural exchanges for the growth of scientific knowledge. It begins with Sundar Sarrukai's account of the epistemological orientation of Indian mathematics and how this may enrich our understanding of mathematics in general. It is followed by Arun Bala's adoption of the notions of three modes of nature knowledge drawn from Arabic optics, Indian mathematics and Chinese technologies to explain how their interactions in Europe led to the birth of modern science. Gordon McOuat points to how the quite different notion of genealogy arising from Japanese culture leads to a novel conception of dialogue and exchange between civilizational knowledge systems. In the last chapter, Floris Cohen argues that the comparative and dialogical approaches to civilizational traditions of nature knowledge are by no means exclusive, but could be mutually fructifying.

The section begins with Sundar Sarukkai's observation that the link between mathematics and philosophy was very tenuous in Indian thought, and that it led to the absence of any well-established Platonic orientation in Indian mathematics. Instead, Indian mathematics was inspired by linguistics. It resulted in a mathematical practice that was not only more pragmatic in character, but also placed emphasis on algorithms rather than axioms as in Euclidean geometry. It leads him to reject the universalism of mathematical method rooted in Platonic philosophy, and defend a pluralistic approach for the foundations of both the ontology and epistemology of mathematics. He proposes that this would allow us to better accommodate other traditions of mathematics, and moreover such a dialogue between the Western and Indian approaches can also be fructifying for both. He concludes that instead of beginning with the assumption of universalism we should accept the foundational pluralism of mathematics. In that case other mathematical traditions are accepted as different independent ways of doing mathematics. We would not use dialogue across these different mathematical traditions to reduce one system to another, but retain plural perspectives as valid in their own terms. However, it is possible to wonder whether such dialogue may not also amplify and transform mathematical foundations.

Arun Bala explores the comparative approach of Floris Cohen's explanation for the rise of modern science. Cohen argues that three modes of nature knowledge—realist mathematical, kinetic corpuscular and fact-finding experimental—combined to produce modern science. Moreover, Cohen sees these as transformations of earlier traditions in ancient Greek thought and a medieval Renaissance approach to the study of nature. Bala argues that although the notion of these three 'modes of nature knowledge' is very illuminating for understanding the origin and growth of modern science, their roots have to be traced beyond the European context in which Cohen locates them. According to Bala they are more illuminatingly seen as outcomes of three outside influences on Renaissance Europe—the flow of mechanical technologies from China, the computational techniques associated with the Indian number system, and the Arabic optics of Ibn al-Haytham. He concludes that in order to understand the origins of modern science we need to go beyond the comparative approach within which Cohen confines his analysis of modes of nature knowledge, toward a dialogical perspective which links them into the global context of Renaissance Europe.

In the following chapter, Gordon McOuat looks at an alternative model for genetics developed by a Japanese biologist in the early twentieth century. It emerged out of a deep encounter between Eastern and Western science on the issue of how the relationships between living beings can be conceived. He notes that this 'Dynamic System,' as it was labeled by its proponent Bunzo Hayata, offered a way out of the interminable disputes over 'origins,' 'genealogies,' 'nature' and 'inheritance' associated with discovering a uniquely true unilinear phylogenetically 'natural system.' It offered an image of a new perspectival and mulitidimensional network of relationships between living beings that side-stepped the problems associated with unilinearity. McOuat's chapter explores the rise of the 'dynamic system,' its spread across the West, and how it might offer up a new model of dialogue and exchange between knowledge systems. In particular, McOuat concludes that some Western biologists are attempting to go beyond the search for unique phylogenetic trees inspired by neo-Darwinism, and beginning to recognize the significance for biology of hybridity, interrelated sets and perspectivism of systems theory, in ways reminiscent of Hayata's theory.

The final chapter by Floris Cohen contests the notion that the differences between dialogical and comparative approaches to the cross-cultural analysis of the history of science render them mutually exclusive. He does this not only by comparing them, but also offering a reconstruction of Needham's own thought on the matter. Then he builds on this analysis to set forth how he conceives of their difference, and why he sees them as mutually compatible. He advocates that cross-cultural historians should stop dichotomizing narratives of 'dialogue believers' and 'dialogue sceptics' but instead combine the virtues of comparative and dialogical approaches for more accurate and richer narratives of the growth of scientific knowledge. He also disputes Arun Bala's characterization of him as a comparativist as a caricature of his position, and disputes that the alternative inter-civilizational dialogical vision to the account he presents can be coherently defended. Cohen's chapter is a cogent reminder that notions of comparativism and dialogism are highly contested issues, and that any attempt to understand the multicivilizational exchanges that made modern science must negotiate such differences carefully.

4 Toward a Science and Civilizations in Dialogue Project

The chapters in this volume that explore Needham's dialogical vision strongly suggest that we can draw on the template and methodology of his *Science and Civilisation in China (SCC) Project* to articulate a wider *Science and Civilisations in Dialogue (SCD) Project.* Indeed, the chapters in this volume bring together both a set of themes and body of scholars that could furnish both opportunity and ideas to articulate this project. Needham's work turned the historiography of science and its origins from an insular study focusing only on Europe to one that located Europe within the wider context of other major civilizations. It could be seen as globalizing the historiography of science. If the former approach may be seen as focused on a tunnel history of science, Needham lent credibility to the notion that there were other traditions of science that could be studied on their own terms too. Indeed, the success of his achievement has led to the contemporary movement of 'Asia as Method' that calls for Chinese and other traditions to be studied on their own terms.[7]

However, what was revolutionary about Needham's work was that it opened the door into two directions that went beyond the tunnel history approaches that can be termed 'comparative' and 'dialogical.' Comparative studies of science involve comparing the similarities and differences of nature knowledge in two or more civilizations. They raise questions

[7] For some recent attempts to rethink the Needham legacy see Hsia and Schäfer (2019) and Mei (2020).

such as why scientific knowledge developed in differentiated ways in the civilizations being compared. Dialogical studies concern how flows of knowledge across civilizations shape the growth of knowledge within them. In important respects they may be seen as building upon the success of comparative histories of civilizational sciences. But they also bring into light the new questions such as demarcation, incommensurability, translation, transmission, teleology and anachronism.[8]

The essays in this volume suggest that we can consider reorienting the Needham project, in light of his dialogical vision, by embracing his vision of science as a multicivilizational outcome. This would also require expanding it beyond the Chinese science context in which it began into a wider multicivilizational framework. Indeed Needham's studies *Science and Civilisation in China* (SCC) could provide the framework for a new *Science and Civilisations in Dialogue* (SCD) project of taking the Needham legacy to its next level. In this expansion his layout of the volumes, however problematic it may appear to some historians, can provide a template for going from a Science and Civilization in China project (SCC) to a broader Science and Civilizations in Dialogue (SCD) project. It would be a natural development from what Needham achieved in his SCC project, and his vision of a universal history of science of multiple rivers entering one ocean, however problematic this may be.

Notwithstanding the difficulties of such a venture here is a quick outline of what could be achieved. It could begin, as he did, with an *Introductory Orientations* (SCD vol.1) volume surveying the languages, geographies, histories and scientific contacts across world civilizations that led to the growth of knowledge through multicivilizational exchanges. This could be followed by a *History of Scientific Thought* (SCD vol.2) that would consider this history, the philosophies of nature that shaped it, and its development within the context of traditions of worldviews in various civilizations shaped by multicivilizational exchanges. The next three volumes would follow the SCC structure but informed by Needham's dialogical vision—*Mathematics and the Sciences of the Heavens and Earth* (SCD vol.3), concerned with the growth of

[8] For attempts to take the Needham Question in a dialogical direction, or address its significance, see Bala (2006), O'Brien (2013), Raj (2007), Renn (2012) and Schaffer et al. (2009).

knowledge in geography, astronomy and mathematics through multicivilizational exchanges; *Physics and Physical Technology* (SCD vol. 4) dealing with physics per se, as well as mechanical, civil and nautical engineering and how multicivilization exchanges shaped them; *Chemistry and Chemical Technology* (SCD vol. 5) looking at the evolution of chemistry and its associated technologies through circulations of knowledge across civilizations (especially in alchemy); and *Biology and Biological Technology* (SCD vol.6) that would study the circulations of knowledge in botany and agriculture, zoology, medicine and pharmaceuticals across civilizations. Finally, the dialogical series would end with a volume on *The Social Background* (SCD vol.7) that would examine the social contexts of trade, empires and religions in scientific exchanges, and the problems of linguistic, cultural, political and geographical factors that can answer Needham's dialogical question: How did a dialogue of multicivilizations shape the growth of science in history?

Articulating Needham's dialogical legacy presents a number of problems that have to be addressed. The first is that it offers a more complex problem than confronted Needham and others who followed him on his comparative project. Here the main concern was with the binary comparisons of Chinese and Western civilizations, later extended to the binary comparisons of Arabic and Indian cultures, among others, to the Western. But looking at advances in any scientific discipline as a multicivilizational outcome presents a more complex picture of considering the polyvalent roots of any discipline.

One way out is not to have a few authors for each volume as with the SCC Project, but multiple authors for each of the SCD volumes. What would tie a particular volume together is that it is looking at an individual discipline, say astronomy or medicine, and the way inter-civilizational crossings shaped it.

But the Needham comparative question also would not be excluded. Instead, it would become more nuanced. Seeing science as a developing enterprise across cultures, of which modern science is a further development, opens the question of why science at different times and places underwent dramatic transformations. For example, why did Chinese, Indian and Greek sciences mutate into new distinct traditions in the Axial Age? Why did further scientific and technological transformation take place in Europe in the modern era? These relate to Needham's comparative question. But taking into account the dialogical exchanges that nurtured science over millennia also changes the comparative question we have to now ask: Why did a *dialogue of civilizations* in particular areas and times produce distinct and marked changes in science in some places but not in others? How were these places able to draw upon such dialogue to advance science in new directions?

Finally, we can also ask if the comparative and dialogical approaches may not enrich each other? The comparison of civilizational sciences lays the basis for a deeper understanding of how it permitted fructifying exchanges across them. Also, dialogical exchanges can illuminate the growth of science within particular civilizations. In fact, it is the interweaving of dialogical and internal influences that may better explain the evolution of scientific traditions both across space and time in history. One could say that comparativism is lame without dialogism, and dialogism is blind without comparativism.

Thus, including Needham's comparative and dialogical approaches, albeit extending their scope beyond what Needham accomplished, raises new problems for science studies and paves the way for exciting new prospects. Here the chapters of this volume can be significant. It can draw upon the resource of scholarly expertise it has brought together to explore the scope, relevance and feasibility of the proposed *Science and Civilisations in Dialogue Project* to carry Needham's legacy forward.

5 Bridging Needhamian and Modern Science Studies

Apart from opening up the possibility of reorienting Needham's project in a new globalized dialogical direction, the essays in this volume also point toward a new bridging of Needhamian and modern science studies. Despite the great homage paid to Needham in documenting Chinese science, and paving the way to multiple studies of the history of Indian and Islamic sciences, and now Japanese, Korean, African and other traditions, mainstream science studies of modern science has largely seen his work as marginal to its project. This has been bemoaned by Mark Elvin in his introductory chapter to Needham's final volume:

> What is hard to come to terms with, almost half a century after the appearance of the first volume in 1954, is the limited assimilation of Needham's work into the bloodstream of the history of science in general; that is, outside the half-occluded universe of East Asian specialists and a handful of experts sensitive to the decisive contributions of comparisons. For these to be useful, there has to be enough in common between the two domains to make comparisons and contrasts relevant, and enough differences to make such juxtapositions reveal critically distinctive aspects of one or the other. (Elvin 2004, p. xxv)

What Elvin said more than one and a half decades ago continues to hold true today. Two decades later the situation has really not changed very much. In a recent review of Needham's legacy the pioneer classicist and comparative scholar Geoffrey Lloyd writes:

> The multiple volumes of SCC are truly monumental. Yet the plan and execution are, as we all know, very much of their time, and this has certainly contributed to its relative lack of influence in history and philosophy of science circles in the West – in contrast to the continuing mass of attention it still attracts in China itself. (Lloyd 2020, p. 2)

The essays in this volume, by going beyond the comparative question of why non-European civilizations did not develop modern science, but exploring how they exchanged and produced modern science, have made Needhamian-inspired dialogical studies relevant to modern science.

References

Bala, Arun. 2006. *The Dialogue of Civilizations in the Birth of Modern Science*. New York: Palgrave Macmillan.

Bala, Arun. 2016. The Needham Question and Southeast Asia: Comparative and Connective Perspectives. In *The Bright Dark Ages: Comparative and Connective Perspectives*, eds. Arun Bala and Prasenjit Duara. Leiden: Brill.

Chang, Hasok. 2016. Who Cares About the History of Science? 2015 Wilkins-Bernal-Medawar Lectures. *Notes and Records* 71: 91–107. Published online by The Royal Society, 19 October 2016, https://doi.org/10.1098/rsnr.2016.0042

Elvin, Mark. 2004. Vale Atque Ave. In *Science and Civilisation in China*, ed. J. Needham, vol. VII: 2, xxiv-xliii. ed. K.G. Robinson. New York: Cambridge University Press.

Goonatilake, Susantha. 1999. *Toward a Global Science: Mining Civilizational Knowledge*. Indiana University Press.

Hsia, Florence, and Dagmer Schäfer. 2019. History of Science, Technology, and Medicine: A Second Look at Joseph Needham. *ISIS* 110 (1): 94–99.

Lau, Raymond W.K. 2020. *Intellectual Developments in Greece and China: Contingency, Institutionalization and Path Dependency*. Cambridge Scholars Publishing.

Lloyd, Geoffrey. 2020. After Joseph Needham: The Legacy Reviewed, the Agenda Revised—Some Personal Reflections. *Cultures of Science* 3 (1): 11–20. https://doi.org/10.1177/2096608320917579.

Low, Morris F., ed. 1999. *Beyond Joseph Needham: Science, Technology, and Medicine in East and Southeast Asia*. Chicago: Chicago University Press.

Mei, Jianjun, ed. 2020. 'Joseph Needham's Intellectual Heritage', *Cultures of Science* 3 (1).

Nandy, Ashis, ed. 1989. *Science*. Hegemony and Violence: A Requiem for Modernity. Oxford University Press.

Needham, Joseph. 1978. Address to the Opening Session of the 15th International Congress of the History of Science, Edinburgh, 11 August 1977. *The British Journal for the History of Science* II (38): 103–113.

Needham, Joseph. 1979. *Within the Four Seas: The Dialogue of East and West*. University of Toronto.

Needham, Joseph. 1985. *Gunpowder as the Fourth Power, East and West. First East Asian History of Science Foundation Lecture in 1983, and Inaugural Needham Memorial Lecture*. Hong Kong University Press.

Needham, Joseph. 2004. General Conclusions and Reflections. In *Science and Civilisation in China*, ed. Kenneth Robinson, vol. 7, part 2. Cambridge: Cambridge University Press.

O'Brien, Patrick. 2013. Historical Foundations for a Global Perspective on the Emergence of a Western European Regime for the Discovery, Development, and Diffusion of Useful and Reliable Knowledge. *Journal of Global History* 8 (1): 1–24.

Raj, Kapil. 2007. *Relocating Modern Science: Circulation and the Construction of Knowledge in South Asia and Europe, 1650–1900*. Palgrave Macmillan.

Renn, Jürgen, ed. 2012. *The Globalization of Knowledge in History*. Max Planck Institute for the History of Science, online version at http://edition-open-access.de/studies/1/.

Schaffer, Simon, Lissa Roberts, Kapil Raj, and James Delbourgo, eds. 2009. *The Brokered World: Go-Betweens and Global Intelligence, 1770–1820*. Uppsala Studies in History of Science 35. Sagamore Beach, MA: Watson Publishing International.

Part I

Historical Sociology in Dialogue

2

An Oceanic Paradigm of Historical Flows

Prasenjit Duara

Needham famously depicted the scientific history of the world until the modern scientific revolution as one of different civilizational rivers flowing into the vast ocean of science. While Needham's metaphor of the ocean of science—which suggests currents, flows and circulations at different levels and spaces—is acute and apposite, I wish to show that it applies equally to the first part of the metaphor, to the period he refers to as one of different civilizational rivers. From what we now know about the circulatory flows of historical, including scientific and technological processes, since at least the Bronze Age, we may need to re-think not only the idea of 'civilizational rivers' which suggests a relatively bounded one-way flow, but perhaps also about a single moment in the emergence of modern science.

Thanks are due to Arun Bala and Raymond Lau for their very helpful comments on the original paper.

P. Duara (✉)
Department of History, Duke University, Durham, UK
e-mail: pd77@duke.edu

A. Bala et al. (eds.), *Multicivilizational Exchanges in the Making of Modern Science*,
https://doi.org/10.1007/978-981-97-3541-9_2

In what was perhaps his last set of writings between 1992 and 1994, Needham (2004) disagreed with 'those who call modern science 'Western', for though it began in Western Europe it has long ceased to be exclusively Western. Indeed, I have pictured modern science as being like an ocean into which the rivers from all the world's civilisations have poured their waters' (Needham 2004, p. 201). Not only did he think modern science was not exclusively Western, he often called it modern 'international science.' At the same time, however, he held a linear, evolutionary view of science. He distinguished stages of science: (1) the primitive stage…or proto-science; (2) the medieval stage when techniques were differentiated and general principles were formulated, often related to some all-embracing theory. Stages 1 and 2 were to be distinguished from (3) Renaissance Science, (4) nineteenth-century 'modern science' and (5) twentieth-century attempts to establish a canon of accepted Unity of Science (p. 202).

He subsequently goes on to 'make a sharp distinction between modern science and that which preceded it, namely ancient and medieval science.... (which) consists in two things, mathematization of hypotheses about Nature on the one hand, and on the other continuous and relentless experimentation. In my opinion *there is no science outside of modern international science* (emphasis added)' (p. 203). Around the time he acknowledged that cultures other than medieval China, including Sri Lanka, India and Japan may have had other scientific achievements closer to those of Western Europe in some respects, he also inverts the 'why not China' question to ask why it is that modern science should have arisen only in Western Europe. To this question he adduces various familiar factors linked to Greek logic, capitalism, the emergence of the bourgeoisie through the history of self-governing cities and their interest in precise hypotheses and measurement for commercial profit, etc. He neglects to mention what would later become commonly accepted, the role of 'war capitalism' and resource control by W Europeans across the new and old world as drivers and incentives for the development of precise knowledge.

I agree with the idea that certain ideas and practices that gather in a cluster of cities in a region may have catalyzed a range, even a burst, of technical and scientific practices (e.g., in W Europe). But in the deep view of history, there were long epochs of slow development interrupted by sudden bursts in innovation from at least the Axial Age of the sixth century BCE and long before as seen in the archaeological record (Kolodny et al. 2015). However, in following the evolutionary path, Needham comes to hold a teleological view of science. Note this passage:

> (I)t is the rise of the bourgeoisie in Western Europe from the 15th century onwards which decided that Europe would not continue trying to emulate its classical past, nor evolve in Chinese fashion, improving its technology and making small advances in various sciences without actually coming to grips with scientific method (emphasis added), but would strike out in a new direction, developing capitalism, scientific method and the industrial revolution one after the other. (Needham 2004, p. 229)

Today, the idea that there is a single 'scientific method' would probably not be acceptable to most of the scientific community, even though hypotheses, experimentation and verifiable accuracy remain important especially in the non-theoretical sciences. The idea of a universal truth of science is a leftover idea of the last century or earlier that was associated with a linear evolutionary view. Post-war studies of scientific pluralism have challenged the idea of a unified science reducible to microphysical theories. Some appeal to a 'disorder of things' or a 'dappled world,' 'that is, a world both of nomologically ordered systems and unruly ones.' They also 'present a different view of progress, not as unification, but rather through the emergence of new fields that proliferate scientific concepts, methods and theories' (Ludwig and Ruphy 2021).

I try to develop a methodology of historical sociology that I call circulatory histories. Empires and nations and smaller polities have not developed historically, contra the claims of contemporary national and civilizational histories, as originary events, forces and processes that progress in a linear, tunneled manner to fulfill their telos. Linear histories are written by advocates of nations and civilizations in the modern period and frequently gloss over cross-cultural and circulatory processes that contradict their unimpeded story of awakening, struggle and progress (Duara 1995). Rather, histories take a *circulatory* form: emerging in place

A, they move to place B, C or E where, interacting with other local processes and structures, they take on a different form and circulate on. On occasion, this circulatory process *may* return to A, whether or not they are recognized or acknowledged as having emerged from there, i.e., circle back.

This kind of circulatory history does not privilege notions of *ownership* of historical processes or the issue of historical origins which often become foundational claims of *sovereignty*. Rather, this history may be better understood through a paradigm of oceanic flows where an event ripples or crashes across multiple boundaries, circulates into other currents or eddies, or sinks into the thermohaline depths which can upswell in another space. The effort in this chapter is to outline the different modes which circulatory processes undergo and transform. Oceans display spatio-temporal patterns that display both stability as well as interactive change (Duara 2021).

The ocean metaphor differs from Needham's view of civilizational rivers, each developing its sciences indigenously within their banks, as it were, in suggesting that there was always circulation and exchange before European expansion. In other words, there had long been an ocean of scientific knowledge where currents emerging from different locations interacted or merged. Some of these currents sank into the deeper ocean where their heavier and slower movement shaped surface developments imperceptibly, much as the case described by Robert Beckwith, discussed below, of the unremembered transformations of educational institutions and modes of argument across Eurasia over a millennium and which represented a significant source of modern science.

1 Circulatory Histories

Jack Goody, David Pingree and others have demonstrated repeatedly that Eurasia has been basically unified since the onset of the Bronze Age. Pingree has declared, as 'a simple historical fact, scientific ideas have been transmitted for millennia from culture to culture, and transformed by each recipient culture into something new' (Pingree 1992, p. 563). The early technological revolution enabled intensive agriculture that permitted surplus accumulation, class stratification, literacy, bureaucracy, urbanization and inter-linked city-based civilizations across the zone. Consequent upon these factors and depending on the changing circumstances of access in a particular region, the zone was connected by long distance trade, warfare, technology, ideas and organizational and institutional practices (including mass industrial production). Indeed, a recent interactive map indicating genetic mixing shows how human races have been migrating and mixing for over four thousand years (Hellenthal et. al. 2014).

In some of his late works, Goody shows the ways in which European scholars have striven to dis-embed 'Western' history from its actual, interactive historical context starting from Antiquity to the present. Carthage, Phoenician traders, Persia, Egypt, India and most of all, China and the Islamic world have played a much greater role in Western histories than even the most radical analysts have acknowledged. Thus, while it cannot be gainsaid that the industrial revolution of the late eighteenth century was a northern European achievement, however, this revolution cannot be understood without the background of 'industrialization, mechanization and mass-production developed...in China with textiles, ceramics, and paper, in India with cotton, later taken up in Europe and the Near East...'. Goody also discusses the realm of institutions, particularly of knowledge production and values, including mathematics, medicine, humanism, democracy, individualism and romantic love. In none of these cases, does he find European societies to be the exclusive proprietors of these institutions and values. Rather the Europeans owed these traditions in a significant way to other, most notably, Islamic societies. (Goody 2006, p. 210)

At different times, different cities or regions in Eurasia—whether joined by trading, religious and technological networks or separated by empires, disease, political instability, piracy or climate change—created different nodes of absorption, innovation or isolation in this gradually expanding zone over the millennia. Seen in this wider canvas of changing zones of innovation, the capacity of some European social formations to institutionalize secular knowledge production and sustained economic growth over the last two centuries represents a very important but not unique achievement.

China during the Song dynasty which was open to multiple influences from the Mongol empire to the north and to the vast maritime trade in cities like Quanzhou and its hinterland in the south which was such a center catalyzing innovation over several centuries. The circulation of goods, ideas and technologies in and with China and the Abbasid empire is now well studied. During the Song and Mongol period, Quanzhou may have been the largest port city in the world and through its extensive trading and transshipment activities represented an emporium in the 'medieval international world trade system' (Schottenhammer 2001). Zhao Rugua (1170–1231), the customs official at Quanzhou wrote about the major trading cities and goods from Srivijaya (modern Indonesia) to Malabar, Cairo and Baghdad. Quanzhou had populations of Arab, Persian, Thai and Tamil trading communities among others and their monuments, temples and legacies which continue to be displayed to this day. It was in this cosmopolitan city that the Chinese developed the most sophisticated ship-building technology and also experimented with the use of paper money albeit with ultimately negative results.

Drawing from underwater archaeological artifacts from a ninth-century Belitung shipwreck, among others, Amanda Respess (2020) reveals the vast realm of trade and exchange between the Chinese and the Islamic world before European trade and knowledge influenced this world. Respess also demonstrates that the principles of Chinese medicine were absorbed into Islamic medicine during the Mongol period. Medieval Islamic medicine was not simply a development of the Galenic tradition, but a 'globalized movement and interaction of multiple centers, articulated by multiple minds' (Respess, 226). Baghdad, the Abbasid capital beginning in the 8th CE, was another such center particularly with regard to the intellectual ambience, attracting ambitious men from all over the vast empire, especially from Central Asia. A great number of Sanskrit scientific texts in astronomy, mathematics, medicine, alchemy were translated into Arabic, followed by a vast influx of Greek learning, particularly in the ninth century (Beckwith 80–85). In different periods, various societies were poised to learn the lessons from other parts of the Afro-Eurasian world quickly enough—given the timescale of history—and we need not identify the distinctive qualities of a region or nation as reaching back into the mists of time, that is, unless that society has a competitive agenda.

2 An Oceanic Metaphor

Returning to Needham's idea of rivers of science flowing into the ocean of the Western scientific revolution, I believe the discussion above shows that rather there have always been something like oceanic currents that may flow separately but often also interact with different currents. These interactions are not only sequential or lateral, they are also shaped vertically by different layers or scales in the ocean and atmosphere. In this section, I think with the circulating waters of oceans to capture an understanding of the circulations, interactions and volume of historical processes. Whereas national and other linear histories distort the real processes that intersect and shape 'domestic' institutional processes, the oceanic metaphor demonstrates how human history, including the study of history, overlaps with and often merges with natural history where they lead in directions different from or beyond the human intention. The oceanic metaphor is particularly useful to demonstrate the gap between the intentional purpose of a historical project (what I call historiography) and its unintended consequences.

Oceans reveal circulatory currents that are of different *temporalities*, layers and effects as they go through diverse conditions. Surface currents are faster moving because they carry heat and are shaped by winds; gyres tend to move more slowly than eddies which churn up smaller spaces. Oceans and seas are realms where spaces and temporal processes interact at varying scales. Between the surface and the deep water, there are intermediate levels with a kind of *middling temporality* such as the Levantine Intermediate Water which plays an important role in the formation of all deep Mediterranean waters but its vortices also spread warmer, saltier waters into larger Mediterranean basins (Shultz 2015).

Deep currents in the ocean are heavier because water becomes colder at the poles and is pulled down by salinity and gravity. But deep currents also flow across the various oceans and cycle through roughly once every thousand years. Over the last several decades, this deep-water flow across the oceans has been understood as a 'conveyor belt' reflecting the relatively stable pathway through which the warm waters pushed to the poles, overturned and made its way across the oceanic world. Recently, scientists have 'deconstructed' the model of the conveyer belt suggesting that while overturning and coursing remain true, the conveyor belt idea was too simple. Rather several different pathways have appeared, formed interactively by surface eddies and wind-fields (Lozier 2010). This suggests that while its temporality—the rate of flow and the types of activity it produces—has been maintained and the idea of the conveyer belt remains, it is also interactive with other geo-atmospheric forces. Even the deepest level is affected and affects the rest.

Social formations—communities, empires or nations—are also shaped by different temporalities. In recent centuries, the most slowly changing temporality shaping most contemporary nations are European Enlightenment values with its ideal of the human mastery of nature principally through modern science and the nation form. At an intermediate level our institutions, including civic, economic, statist, ethnic, religious are sustained for longer periods than the most surface level which may involve policy changes, campaigns, networks and social groupings which change more rapidly over time. A major event, such as a war or a discovery or climate disaster, may well bring about changes at all three levels, resembling the periodic 'upwellings' and 'downwellings' from and toward the deep oceans bringing about changes in temperatures and lives of oceanic beings.

The flow of water patterns times not only cyclically and sequentially, but also by differentiating spaces and through interactive scales. The ocean has many different spaces, such as the relatively still Sargasso Sea or the volatile Bay of Bengal. The boundaries where waters meet are similar to *ecotones* occupying the productive and heterogeneous transition space between different ecological zones, say between wetlands and rainforests or places like Baghdad where Central Asian and South Asian intellectual contributions encounter Chinese technology and medicine and Greek thought.

History also often followed the oceanic and related natural patterns. One such example is the historical Asian coastal maritime trade stretching from the China coast to East Africa (now called the Maritime Silk Road) traceable to the early centuries of the first millennium which hewed closely to the monsoon patterns. From the 1400s, the routes were held together most importantly at the cosmopolitan port city of Malacca to which the southwest monsoon winds brought Indian, Persian and Arab traders. There they waited for the northeasterly monsoon winds to carry in the trade from China and eastern regions before returning with the same winds. Until the nineteenth century, the European traders also followed these patterns, albeit with more militarized vessels (Lughod and Lippman 1989). Geographers have recently developed the notion of 'wet ontologies' to challenge the flat ideas of territoriality by looking to the oceans to open up a three-dimensional vertical world of volume. The oceans, not only challenge the modern state-based territorial conception of power by fluid boundary-crossings of all kinds, but also by 'posing the problem of reach, instability, resistance, mixing and separation in the voluminous sphere of Earth' (Steinberg and Peters 2015, p. 251).

Finally, to return to the theme of Needham's Dialogical Vision, it is undoubtedly the case that different societies have always circulated knowledge and technologies. This is now well accepted even in several STS circles (Schaffer et al. 2009). On the issue of commensurability of different traditions, Galison has discovered that there is no need for a common language to exchange goods, ideas, or technologies. In what he calls 'trading zones' he writes, 'it is possible to share a local understanding of an entity without sharing the full apparatus of meanings, symbols and values in which each of us might embed it. Images, symbol systems, calculational and diagrammatic schemes—even complex objects—could be part of a generalized notion of language that is far from 'just words.' Indeed, language, as I want to use it, is a regular yet flexible apparatus that may take many forms, from the recognized, everyday 'natural world languages' to the myriad, systematic registers in which we communicate.' (Galison 2010, p. 44) As we move under the radar of the high sciences, it becomes easier for us to understand how practices, techniques, technologies as well as ideas have always moved into unfamiliar spaces, much as a eddies, streamlines, pathlines and streaklines in the ocean could merge with other wave formations.

3 A Methodological Current across Eurasia

In circulatory histories, we need to consider the unconsidered chain that an action generates involving other material objects and natural organisms that are engaged in the effects as 'actants.' Historical events, processes and products impact, churn, disperse, parts die and other parts transform and even transmute, but their continuity may still be recognizable. The oceanic flow can be seen as the vital metaphor or paradigm of this continuum.

I will follow a case study of the way in which a historical formation continues and is adapted as it circulates over time in different societies. This case is taken from Christopher Beckwith, professor of Eurasian Studies at Indiana University. His book is titled, *Warriors of the Cloisters: The Central Asian Origins of Science in the Medieval World*

Beckwith demonstrates through textual, archaeological and architectural studies that a principal, and even indispensable, scientific methodology of late medieval and Renaissance science emerged from what he dubs the '*recursive argument method*' (RAM) which originated among Buddhist scholars in Central Asia in places like Khotan, Balkh, Tokharistan, Gandhara among others. Quoting Edward Grant, he says, RAM 'was the 'the basic vehicle for the analysis of problems in natural philosophy and theology' from the medieval intellectual revolution to the scientific revolution' (Grant 2007, p. 188; Beckwith 2012, p. 1).

Furthermore, he suggests that this methodology was sustained by the institution of the Buddhist college or *vihara*, and together they were adapted by the later Islamic rulers in the eighth century. Thus, did emerge the RAM mode of inquiry in the Islamic sciences, particularly for jurisprudential and theological inquiries, while the *vihara* became the *madrasa* center of learning. These developments took place before the influx of Greek learning in the ninth century (Beckwith 2012, p. 84). Finally, with the Crusades, and particularly through Islamic Spain in the twelfth and thirteenth centuries, classical Arabic civilization became widely absorbed into European knowledge.

While it is well acknowledged that Greek and particularly Aristotelean science represented a significant portion of Islamic knowledge that went into Europe thus enabling the Renaissance and scientific development, Beckwith argues that Arab scientific contributions cannot be understood without its earlier foundations in pre- and non-Grecian synthesis in Abbasid Islamic thought. This included RAM and the *madrasa* inherited from Central Asian Muslims and Buddhists, the Indian place-system numerals and mathematics adapted by the Central Asian Muslim scholar al-Khwarizm (Algorithmus), as well as works on alchemy, astronomy, mathematics, metaphysics, medicine, physics (especially optics), psychology and many other fields developed by Islamic scholars (see also Bala 2006, ch 7).

Furthermore, it is not simply the ideas or logic that circulates, the physical and aesthetic architecture of the vihara together is also recognizable across these greatly different religio-political regimes. The first European colleges such as in Bologna, Paris and Oxford were modeled along the four-cornered cloisters, the arcaded vaulted passages, the courtyard and the semi-dome structure of the *madrasa* and *vihara*. Moreover, as in Oxford and other universities in Europe, the *madrasa* and *vihara* were funded by endowments of land grants or village revenues. We may think of these as the non-human actants that enable or sustain the circulation of ideas between different historical groups in space and time.

Beckwith's distinctive contribution is the early scientific mode of argument that is RAM. Unlike the 'scholastic method' or the 'disputed question' of medieval Europe, RAM is distinguished by its recursive structure; Beckwith claims that the former methods leave out the recursive structure which 'represents thinking about one's own thinking, including arguing about one's own arguments, a variety of what is known as linguistic recursion' (Beckwith 2012, p. 25) He outlines this mode of argument in a variety of scholars from Central Asian Buddhists like Vasubandhu of Gandhara, Abu Sina (Avicenna) and Thomas Aquinas and others like Nagarjuna in second century in South India.

The method (1) states the argument; (2) lists (often numbered) a string of Subarguments_1 for and against argument 1. This is followed by (3) a second string of Subarguments_2 for and against Subarguments_1. The author establishes his own point of view through this mode of recursive argument where the arguments in the second list dispute the arguments in the first list and the arguments in the first dispute the main argument. The general idea of the method is to break each topic down into analyzable parts and exhaustively debate each one from every possible direction. (Beckwith 2012, p. 24) Although RAM was abandoned as old-fashioned in the seventeenth century, Beckwith argues that the logic of the methodology was replaced by the role of 'hypothesis and experimentation' in modern science which was structurally related to the medieval recursive argument method.

While declaring that there is no longer a single 'scientific method,' Beckwith argues that even to this day, one can find instances of an 'ideal' experimental method. First, precise hypotheses are explicitly stated; second, the scientist describes how the hypotheses were tested and gives the results of the tests. Finally, the scientist confronts the hypotheses one by one with the results and draws a conclusion about the validity of each hypothesis, in order. This, he argues, was the logic of RAM as it was utilized even until the seventeenth century, only presumably the experiment was not a practical affirmation or negation, but a theoretical or empirical observation. He goes on to furnish instances of the parallel from the experimental methods of modern psychological reports. He argues 'the method encourages logical investigation of problems and opens the door to possibilities that might not have been imagined without the demands of the method, with its exhaustive lists of arguments and counterarguments, author's arguments, and refutations' (Beckwith 2012, p. 165).

To be sure, Beckwith is absolutely clear that only Western Europe possessed a 'full scientific culture' which ultimately led to modern science in Europe. Thus, while, Central Asia, Tibet, India, the Islamic world may have used the method and its sustaining institutions, none had the 'passion for science—a passion to learn about all aspects of the world, to record what they learned, and to improve upon earlier knowledge, which marks them from their very first voyages of discovery…' (p. 145). The principal thrust of his argument is to insist on the *origins* of the scientific *method* which is to be found among the Central Asian Buddhist scholars and the institution of *vihara*. He argues that the first major *vihara* was built in Taxila (in Pakistan, today) after the Kushans (ca. 50 BCE to ca. AD 225), a nomadic people from C. Asia, established an empire from Sogdiana to the Ganges valley. The *vihara*, so characteristic of Buddhist South Asia in the first millennium, only spread there after the establishment of the Kushan empire and its patronage of Buddhism.

Aside from his great concerns to demonstrate the origins of RAM and the *vihara*, Beckwith reveals several important dimensions of transmission and circulation. Political and ideological power have much to do with it. From the perspective of circulatory history, it is the power of the Kushans that appears to re-energize Buddhism in India. In Islamic empires, political and religious powers appear to have negated or obscured the theological and religious dimensions of Buddhism but retained important intellectual and institutional forms, without necessarily acknowledging or perhaps even thinking of it as an inheritance. Finally, in the European case, the absorption of these forms together with the wider Islamic contributions were buried for centuries under an entire philosophy of history which declared histories to be civilizationally bounded and teleological. Most of all, his work reveals the deep historical continuities across layers of historical cultures and spaces, that to my mind, resemble the temporality of the deep ocean, unnoticed but often conditioning the flows of surface currents through feedback loops.

Whether or not one can agree with the idea that RAM is the prototype of the experimental method of modern science, the burgeoning literature on scientific pluralism includes many different methods and 'styles of scientific reasoning' (Hacking 2012). One is reminded of Needham's 'ocean into which the rivers from all the world's civilizations have poured their waters' with the fore-mentioned qualification that the methods and styles of scientific reasoning had already been circulating and developing through currents in the sea. Developing Philip Kitcher's model of a 'well-ordered' science (Ludwig and Ruphy 2021, p. 20), Jonardon Ganeri explores another method of scientific reasoning but in doing so, seeks to turn Needham's argument on its head. 'In answer to Joseph Needham, it was not 'modern science' which failed to develop in India or China but rather non-well-ordered science, science as unconstrained by social value and democratic consent. What I argue is that this is not a deficit in the civilisational histories of these countries, but a virtue.'

Ganeri (2013) argues for a polycentric understanding of the history of scientific modernity and urges a focus on the cultural traffic between the nodes, recognizing that their reciprocal and transformational relationships often take place under asymmetrical power relations. He explores two historical accounts of the 'ideal conversation whose deliverances seem to govern the aspirations of scientific inquiry: the synthetic pluralism (*syādvāda*) advanced by Jaina thinkers in the second half of the first millennium, and a Buddhist account of balanced deliberation (*vādayutti*)' which was first compiled from the reign of Ashoka. (268-232 BCE), the great patron of Buddhism. The Jaina style of reasoning seeks to develop the perspective on a proposition from all possible human points of view. The method encoded seven levels of specification with each level representing a dialogical system with its own rules. Ganeri adds that this system also contains a warranty of agreement to disagree; there is no rejection of the law of contradiction, but rather toleration, which according to Kitcher is the habitat for a 'well-ordered' procedure of investigation that counts as science. At the same time, Ganeri finds the dialogical positions to be fixed and inflexible and looks to Buddhist resources for a more dynamic engagement.

In the *Kathāvatthu* or Points of Controversy the principal concern is with balance and fairness in the conduct of dialogue by guaranteeing a consideration and weightage given to all parties involved. In some ways, the dialogue is structured like RAM, divided into sub-dialogues involving refutation, reformulation and conclusion. The highly structured mode of exchange involves several steps described below by Ganeri,

> In the way forward, the proponent solicits from the respondent the endorsement of a thesis and then tries to argue against it. In the way back, the respondent turns the tables, soliciting from the proponent the endorsement of the counterthesis and then trying to argue against it. In the refutation, the respondent, continuing, seeks to refute the argument that the proponent had advanced against the thesis (Ganeri 2013, p. 357).[1]

[1] See also a more detailed explanation of the mode of argumentation in Ganeri (2001, pp. 487–492).

The *Kathāvatthu* shows the form that any counter-argument should take, joining a positive thesis and a defensive repost. According to Ganeri, this procedure clarifies the implied or veiled commitments of the participants and reveals the normative framework of a dialogical exchange. (Ganeri 2001, p. 490) As such, it constitutes a 'well-ordered' science which represents all human points of view in its investigation through engagement. The *Kathāvatthu* 'sets out an ideal of scientific inquiry by the light of which one can judge any actual investigative process' (Ganeri 2013, p. 358).

The metaphor of the ocean of science may be particularly suited to our contemporary 'dappled' world of proliferating theories, methods and conceptions. I will mention two themes which the metaphor may illuminate. The first is the theme of intentionality which is also related to the question of contingency taken up by Raymond W.K. Lau in this volume. The second is the theme of potentialities of future developments in the ocean of science. The case I describe here is Arun Bala's interpretation of Needham's understanding of Chinese science as organic.

4 Intentionality, Contingency and Potential in the Oceanic Metaphor

The oceanic paradigm of history urges us to look not only to the cross-boundary, multi-temporal and multi-scalar or volumetric dimensions of a historical event or process, but to the gap between purposive historical action by human agents and its unseen or obscured effects or consequences. I have distinguished the record of purposive historical action which I call Historiography from a wider natural—an oceanic understanding of history (Duara 2021).

A historical event, idea or artifact that was intended for one purpose may or may not result in its desired purpose, but it is often carried away by currents reflecting other assemblages. Here, let me give an example from one of the most important discoveries of twentieth-century oceanography, itself. Until the middle of the century, there was no certain knowledge of the deep ocean currents or the conveyor belt. The answer came in the form of an incidental scientific discovery. Before

World War I, there were no viable sub-surface ocean-going vessels. Attack submarines were used during World War II, and during the Cold War, submarine technology became one of the most important areas of naval development for both the US and the Soviet Union. In *Science as a Mission*, Naomi Oreskes (2021) demonstrates the role of the US Navy's mission in the post-war era: not only to develop sophisticated submarines but to be able to detect enemy submarines where radio waves did not spread through water.

The US Navy invested a great deal of resources in the technology and the best ocean scientists of the time to research the levels of salinity and density in the thermocline, the layer of ocean below sunlight which separates surface from deep water, to investigate its effect on sonar. The goal was to detect sound channels that could lead to a submarine while hiding one's own in an 'acoustic shadow' (Oreskes 2021, p. 83). In the process of this research, Henry Stommel and Arnold Arons made one of the most important discoveries in twentieth-century oceanography: the thermohaline deep ocean current driven by density gradients and Coriolis effect, which would become indispensable for understanding contemporary climate change (Oreskes 2021, p. 72).

As Oreskes remarks, 'The operational problem was the effect of the thermocline on sonar; the breakthrough was the understanding of deep-ocean circulation' (Oreskes 58). Oreskes is principally interested in how military funding for scientific research results not only in epistogenic; i.e., knowledge producing, but principally in 'agnatogenic'; i.e., ignorance producing results. The latter refers to areas of scientific research that may be good for the world, but not necessarily for military or political interests and remain unfunded or even concealed. Yet this instance suggests how an assemblage designed for procuring one type of narrow knowledge goal, resulted in contingent knowledge formation for entirely different purposes. The military was engaged in a purposive, 'historiographic' project. What resulted was a historical event that emerged from this project and propagated many other projects.

Second, the oceanic metaphor also embeds a reservoir of knowledge that follows certain courses even though it is interactive at multiple levels. Arun Bala urges us to follow Needham's later writings on Chinese science. While classical Chinese science may have blocked the development of Newtonian mechanical science in China, Chinese organic materialism which was related to its correlative cosmology may be much more closely related to other developments in science. Bala cites the instance of the oak tree, which in an Aristotelean view is the outcome of the acorn—its essence, as it were, whereas the environment is treated merely as the medium in which it grows. In organic materialism, 'the oak became what is not only by virtue of what is innate in the acorn, but also by virtue of the environment in which the tree grew' (Bala 2020, p. 66). This is a relational argument where interdependencies, connections are of utmost importance and is increasingly popular, especially in the biological and environmental sciences. Needham, according to Nakayama Shigeru, was inclined to see 'late modern science as a synthesis of the Chinese tradition of organism and early modern Western mechanism' (Cited in Bala 2020, p. 70). An oceanic view that permits linkages across temporalities and currents that move in contingent directions may be a more suitable mode of understanding the state of scientific plurality today than a linear evolutionary view of the growth of modern Western science leading to a single revolutionary leap.

In many ways, as we have seen, Needham anticipated many of the insights of scientific pluralism through his deep understanding of Chinese science. Yet, much of his writing took place when a linear conception of history and civilization and the teleology of progress toward a unified science remained the dominant paradigm of the time. There is no need to deny the revolutionary upsurge in science that took place in Western Europe from the perspective of the oceanic paradigm. Upwellings, oceanic storms and tsunamis create turbulence and rapid transformations around them just as the coalescence of historical forces can catalyze a revolution in the terrestrial sphere. But the ocean also reveals circulatory and interactive patterns of currents and waves that generate new spaces, movements and patterns depending upon geo-atmospheric and hydrological factors, while also following predictable paths. Some of these new movements are the result of a tsunami or a tectonic shift, while others have the potential to contribute to new patterns just beginning to be seen.

In this context, I want to turn our attention to a radical transformation in the very role of the ocean as paradigm. The discovery of the thermohaline current or the conveyer belt and its subsequent observations also led scientists to a fuller knowledge of climate change. In the last decade or so, we are beginning to see the counter-finalities of the high levels of greenhouse gases that we have released into the atmosphere and in the oceans. As the polar caps melt, the salinity and density of the waters decline, the North Atlantic current of the conveyer belt does not tend to sink as it did before. The icier waters toward the surface have tended to slow or stop the warmer surface Gulf Stream. The Gulf Stream had warmed parts of N America and Europe during the winter months creating more temperate climates in these regions during the Holocene. In 2021, the Gulf Stream had reached the weakest point in 1600 years threatening icy cold winters in the N Atlantic shores, sea level rise and more volatile climactic conditions globally. The Earth systems balance between the oceansphere and the atmosphere is being thrown off. (Duara 2021, p. 158)

Oceans can perhaps no longer simply play the metaphorical or mirror role for historical activities that it once did. According to geologists and other scientists in the Anthropocene working group, a technosphere has been constructed by national capitalist states to keep humans alive in growing numbers on the planet including houses, factories, farms, mines, roads, airports and shipping ports, rockets, financial and computer systems, together with its discarded waste (Haff 2019, p. 137). The technosphere is estimated to be five orders of magnitude greater than the human biosphere it is designed to serve. It is the connector of the 'different natural and artificial systems and functions to extract high quality energy from the environment to sustain itself and its parts, including humans' (Haff 2014, p. 133). But it has a self-expanding life of its own, with no braking mechanism or negative feedback loops or the capacity to recycle its emissions. It is the most proximate agency of planetary destruction.

In this context, the ocean as metaphor has been transformed into the ocean as object, not only of knowledge, but also knowledge to control and exploit its deepest recesses, to treat it as capitalist nation-states have treated the rest of the planet. The oceanic hydrosphere and other spheres in the Earth system have been severely impacted by growing human demand for natural resources leading to many species extinction, the collapse of fish stock, ocean acidification, pollution, and not least, climate change. The exploitation of resources through technologies of verticality—whether below the ocean surface or high in the atmosphere (and beyond)—has met its match in the counter-finalities of the historiographic project. The greatest threat to human and much other life on earth today may well come from the oceans with rising sea levels and an uncontrollably volatile climate. The tsunami of the scientific revolution has ended up damaging the planet grievously; it remains to be seen whether some other means in the ocean of science can rescue it.

References

Bala, Arun. 2006. *The Dialogue of Civilizations in the Birth of Modern Science*. New York: Palgrave Macmillan.

Bala, Arun. 2020. Chinese Organic Materialism and Modern Science Studies: Rethinking Joseph Needham's Legac'. *Chinese Cultures of Science* 3 (1): 62–74.

Beckwith, Christopher L. 2012. *Warriors of the Cloisters: The Central Asian Origins of Science in the Medieval World*. Princeton, NJ: Princeton University Press.

Duara, Prasenjit. 1995. *Rescuing History from the Nation: Questioning Narratives of Modern China*. Chicago: University of Chicago Press.

Duara, Prasenjit. 2021. Oceans as the Paradigm of History. *Theory Culture and Society* 38 (7–8): 143–166.

Galison, Peter. 2010. Trading with the Enemy. In *Trading Zones and Interactional Expertise: Creating New Kinds of Collaboration*, ed. M.E. Gorman, 26–52. Cambridge: MIT Press.

Ganeri, Jonardon. 2001. Argumentation, Dialogue and the *Kathāvatthu*. *Journal of Indian Philosophy* 29 (4): 485–493.

Ganeri, Jonardon. 2013. Well-Ordered Science and Indian Epistemic Cultures: Toward a Polycentered History of Science. *Isis* 104 (2): 348–359.

Goody, Jack. 2006. *The Theft of History*. New York: Cambridge University Press.

Grant, Edward. 2007. *A History of Natural Philosophy: From the Ancient World to the Nineteenth Century*. Cambridge: Cambridge University Press.

Hacking, Ian. 2012. Language, Truth and Reason 30 Years Later. *Studies in History and Philosophy of Science* 43: 599–609.

Haff, P. 2014. Humans and Technology in the Anthropocene: Six Rules. *The Anthropocene Review* 1 (2): 126–136.

Haff, P. 2019. The Technosphere and Its Physical Stratigraphical Record. In *The Anthropocene as a Geological Time Unit: A Guide to the Scientific Evidence and Current Debate*, ed. J. Zalasiewicz, et al., 137–155. Cambridge: Cambridge University Press.

Hellenthal, Garrett, Busby, G. B. J., Band, G., Wilson, J., Capelli, C., Falush, D., and Myers, S. 2014. Supplementary Materials for A Genetic Atlas of Human Admixture History. *Science* 343, 747, published 14 Feb. https://doi.org/10.1126/science http://admixturemap.paintmychromosomes.com/.

Kolodny, Oren, Creanza, N., and Feldman, M. W. 2015. *Evolution in Leaps: The Punctuated Accumulation and Loss of Cultural Innovations*. PNAS Published online 23 Nov. https://www.pnas.org/doi/full/10.1073/pnas.1520492112

Lozier, M. Susan. 2010. Deconstructing the Conveyor Belt. *Science* 328 (5985): 1507–1511.

Ludwig, David, and Ruphy, S. 2021. Scientific Pluralism. In *The Stanford Encyclopedia of Philosophy*, ed. Edward N. Zalta (Winter 2021 edition). https://plato.stanford.edu/archives/win2021/entries/scientific-pluralism/

Lughod, Abu, and Lippman, Janet. 1989. *Before European Hegemony: The World System AD 1250–1350*. Oxford: Oxford University Press.

Needham, Joseph. 2004. General Conclusions and Reflections. In *idem.*, *Science and Civilization in Chinese History* vol. 7, part 2, 199–231. Cambridge: Cambridge University Press.

Oreskes, Naomi. 2021. *Science on a Mission: How Military Funding Shaped What We Do and Don't Know about the Ocean*. Chicago/London: University of Chicago Press.

Pingree, David. 1992. Hellenophilia Versus the History of Science. *Isis*, 83 (4).

Respess, Amanda. 2020. *The Abode of Water: Shipwreck Evidence and the Maritime Circulation of Medicine Between Iran and China in the 9th Through 14th Centuries*. PhD Dissertation, University of Michigan.

Schafer, Simon, Kapil Raj, et al. eds. 2009. *The Brokered World: Go-betweens and Global Intelligence, 1770–1820*. Sagamore Beach, MA: Watson Pub. Inc.

Shultz, D. 2015. Underwater Gliders Find Swirling Vortices of Warm, Salty Water. *Eos*, 96 (13 Oct).

Schottenhammer, Angela. ed. 2001. *The Emporium of the World: Maritime Quanzhou, 1000–1400.* Leiden/Boston: Brill.

Steinberg, P., and K. Peters. 2015. Wet Ontologies, Fluid Spaces: Giving Depth to Volume Through Oceanic Thinking. *Environment and Planning D: Society and Space* 33 (2): 247–264.

3

The Needham Question: A Non-Eurocentric Approach Transcending Dialogism

Raymond W. K. Lau

1 The Needham Question and Dialogism Versus Eurocentrism and Neo-Eurocentrism

The Needham Question (hereinafter NQ) is a double question, consisting of a positive one ('Why did modern science emerge in Europe?', hereinafter the positive NQ) and a negative one ('Why did it not emerge elsewhere?', hereinafter the negative NQ).[1] The late Angus Graham (1973 46, 51–2), the preeminent sinologist of modern times, observed that both the negative question and the NQ's dual nature are

R. W. K. Lau (✉)
Hong Kong Shue Yan University, North Point, Hong Kong
e-mail: raywklau1122@gmail.com

[1] The present chapter is revised from our paper titled 'Revisiting the Needham Question: Beyond transmission in critiquing Eurocentrism' presented at the 'Needham's dialogical vision: Understanding science as a multi-civilizational outcome' conference held on 23–24 March 2023. We are indebted to Arun Bala for his extended exchanges with us over the past several years, as well as to the conference participants for their feedback. We would also like to thank Dr. Peter Lee and Professor Christian Daniels for their support.

A. Bala et al. (eds.), *Multicivilizational Exchanges in the Making of Modern Science*,
https://doi.org/10.1007/978-981-97-3541-9_3

conceptually problematic. We agree, though space forbids an explanation of why, hence, the present chapter is concerned with the positive NQ only. Nonetheless, given the fact that, like Needham, Eurocentric scholars invariably raise the NQ in dual form, hence, in critiquing their arguments concerning the positive NQ, it is impossible to avoid critiquing at the same time their arguments concerning the negative NQ, since these latter arguments are meant to complement their arguments concerning the positive NQ. As we are more familiar with China, in doing so, we restrict ourselves to its case. Needham was, of course, not the first person to have pondered about the NQ. Cohen (1994) has provided an account of how different scholars, some pre-dating Needham, have addressed the positive NQ, which we can bypass for our purposes. Before Needham, the sociologist Max Weber, while not specifically referring to modern science,[2] had, in fact, provided an answer to the double NQ. Indeed, the accounts of some of the scholars examined by Cohen are, to varying degrees, inspired by Weber's Protestant-ethic thesis. Weber provides a good starting point, though not with reference to that particular thesis.

Weber's famous notion of rationalism (or rationality or rationalization) underwent a significant change in conceptualization in 1915, when contrary to that of the Protestant-ethic thesis, he re-defined rationality as 'logical or teleological consistency' in his new 'forms of rationality' thesis (Collins 1985, pp. 70–1). This thesis argues that different civilizations have developed different (sets of) 'forms of rationality' in a whole range of intellectual practices from science, political thought, art, architecture…to bookkeeping (Weber 1992, pp. xxviii ff). 'It is', Weber stated, 'hence our first concern to work out and to explain genetically the special peculiarity of Occidental rationalism', concerning which he said that 'The author admits that he is inclined to think the importance of biological heredity very great'. Weber admitted that science had not

[2] The notion of the scientific revolution occurring in the seventeenth century has been subject to dispute. On this, we adopt Lindberg's (2007, pp. 357–67) view that there is substantial continuity in important respects between pre-1600 and post-1600 developments, but it is still true that there is a qualitative breakthrough occurring around 1600 to justify speaking of post-1600 science as modern science. For convenience, we often speak of the breakthrough of ca. 1600.

yet been able to prove 'biological heredity', hence, in the meantime, the sociologist would have to make do with seeking an explanation of the 'peculiarity of Occidental rationalism' 'from the peculiarities of the social structure of the Occident' in comparison with that of other civilizations. Thus, for Weber, as part and parcel of his 'forms of rationality' thesis, modern science emerged in Europe but not in other civilizations (the double NQ) because of their different 'forms of rationality', which derive ultimately, Weber believed, from differences in 'biological heredity', and certainly from differences in their 'social structures'.

In Weber's conception, different civilizations are like different impervious water-tight compartments, with each *independently* and *autonomously* developing its own particular 'forms of rationality' due to *purely internal* factors. If inter-civilizational cultural exchanges are acknowledged within this conception, such exchanges are assumed to have no significant effect on each civilization's own independent and autonomous development of its particular 'forms of rationality'. We call such a conception, with reference to the 'West',[3] the *autochthonous 'Western' development (AWD[4]) paradigm.* To refer to modern science, the AWD paradigm believes that modern science arose out of the autochthonous development of the 'West' as a result of the working of purely endogenous 'Western' factors. These factors fall into several categories, what Weber called 'biological heredity', social factors (Weber's 'the peculiarities of the social structure'), and, adding to Weber's schema, cultural factors. We refer to the beliefs in 'biological heredity' *bio-psychological essentialism*, in social factors *social essentialism*, and in cultural factors *cultural essentialism*. In the above quotes, Weber was a believer in the former two, though he was certainly also a believer in the last as well. With racist theory now abandoned by most scholars, contemporary AWD paradigm supporters mainly rely on social and/or cultural essentialism.

The above beliefs are called essentialist because they speak of a purported 'essence', whether of a 'racial', social, or cultural kind. Thus,

---

[3] The reason why the word *West* is put in quotes (except in quotes from other scholars) will be explained in due course.

[4] With reference to note 3 above, since putting a letter in an acronym in quotes is odd and inconvenient, we use AWD instead of A'W'D.

whereas the term 'essentialism' can be used in several different though inter-related senses, in the present work, except indicated otherwise, *essentialism* refers to the view that a particular society's development is due to factors *unique to that society which have arisen in that society through purely endogenous dynamics*. If this definition seems clear enough, what essentialism does and does not imply or entail is often less than well understood, hence some clarification is necessary.

Firstly, in historical explanation of a particular civilization's intellectual development, factors purely internal to that civilization do, of course, figure, and taking them into account does not necessarily constitute essentialism, for instance, to refer to the most straightforward case, when they are simply given contingent facts. Similarly, it is completely legitimate to consider the effects of a particular cultural value of a certain civilization, if that value had arisen as a result of that civilization's interactions with other civilizations, then considering its effects would not be essentialist. Likewise for social explanation. For instance, we have argued elsewhere (Lau 2020) that China's early development of a sophisticated social division of labor led to what we call a divorce between mathematics and philosophy, with mathematics becoming the special vocation of hereditary lineages while philosophers were, without exception, untrained in mathematics beyond elementary level. This divorce had significant effects on China's intellectual development. In this case, a social institution (mathematics becoming the special vocation of hereditary lineages) is offered as an explanatory variable, but it cannot be considered as constituting any 'unique essence' that marks China off from other civilizations.

Because, in Weber's conception, each civilization's development is autochthonous, the 'forms of rationality' that are 'unique' to the 'West' must, therefore, be lacking in other civilizations. We call such a view, with reference to the non-'West', the *development-marked-by-lacks (DML) view*, i.e., the view that value-laden developments found in the 'West' are lacking in the non-'West'. It should be noted that the DML view logically follows from the AWD paradigm. The concept of *Eurocentrism* can be understood in a number of different though inter-related senses. In the present chapter, Eurocentrism is understood in the sense of the belief that the development of the 'West'[5] is due to 'uniquely' 'Western' factors. Hence, Eurocentrism is inherently essentialist, and the AWD paradigm and DML view are Eurocentric. When Needham first raised the NQ, the AWD paradigm and DML view were dominant. How about today? We address this issue shortly afterward.

Whereas this cannot be discussed here, the various answers to the NQ offered by Needham himself cannot stand up to scrutiny, while he remained under the heavy influence of Eurocentrism. Be that as it may, Needham's work has been of monumental importance in two respects. Firstly, by showing that China had a long and substantial scientific tradition, he pioneered the investigation into non-European scientific traditions. Further, Needham openly called for abandoning Eurocentrism: '[T]he river of Chinese science flowed, like all other such rivers, into the sea of modern science' (Needham 1981, p. 9); 'Modern science is indeed composed of contributions from all the peoples of the Old World, and each contribution has flowed continuously into it, whether from Greek and Roman antiquity, or from the Islamic world or from the cultures of China and India' (Needham et al. 2001, p. 25). As formulated in this manner, Needham's argument is untenable, it is nonetheless seminal in seeing modern science to have arisen as a result of the cross-fertilization between different civilizations.

Following Needham's lead, other scholars also began to recover and restore the development of science in other non-European civilizations besides China, thereby showing that science was a widespread phenomenon (Nasr 2003; Bose et al. 1971; the History of Cartography project edited by Harley and Woodward 1987, 1992, 1994, 1998). Further, some scholars also began to advance analyses that challenge Eurocentric accounts of the history of intellectual developments, including scientific developments, by pointing out the crucial role played by Europe's borrowing and learning from non-European scholarship (Bernal 1987, 1991, 2001; Joseph 2011[1991]; Hess 1995; Saliba 1994, 2007; Goody 2006; Bala 2006; Katz 2007; Hart 2013). Referring to the

[5] Europe in the term Eurocentrism refers not to Europe in the geographical sense, but to the 'West' (see Sect. 3).

rise of modern science, the perspective that modern science is a multi-civilizational outcome, in the sense that while the breakthrough of ca. 1600 occurred in Latin Europe, that breakthrough was made possible by what it had learnt from various non-European civilizations either directly or indirectly, is now generally referred to as *dialogism* (as opposed to Eurocentrism) or *the dialogical perspective*. This term came into use in the early 2010s (see shortly below), and, as far as we can tell, is due to the title of Bala's book of 2006, *The Dialogue of Civilizations in the Making of Modern Science*.

By the 2000s, a solid case had been established for dialogism, though the strength of the evidence supporting the claims of Latin Europe's debts to non-European civilizations varies from case to case. Under these circumstances, the AWD paradigm and the DML view can hardly be maintained in their original form. Yet, we can still find contemporary arguments that adhere to them in a manner almost identical to that of the early twentieth century. Thus, in a work that addresses the NQ with reference to mathematical astronomy, Chen (2009, p. 634) argues in conclusion that 'Western science…is the embodiment of the spirit of its [the 'West's'] entire civilization'. For Chen, this 'spirit' constituted the 'West's' 'essence' which was responsible for the rise of modern science in the 'West' but not elsewhere. This is the AWD paradigm in pure form based upon a mixture of bio-psychological and cultural essentialism, and it is all the more remarkable in view of the fact that Chen gives full acknowledgment to the role played by Latin Europe's debts to Arabic scholarship. We consider Chen's argument in Sect. 3. Similarly, in a collaborative project between the classist G.E.R. Lloyd and the sinologist Nathan Sivin that began in the early 1990s until Sivin's passing away in 2022 (Lloyd 1990, 1996, 2015, 2018; Sivin 1995, 2018; Lloyd and Sivin 2002), they argue, in a comparison between Greece and ancient China, that ancient China lacked the Greek search for ultimate foundations, the equivalent of Greek natural-philosophical element theories, the concept of nature which is purportedly a Greek 'invention', and so on. According to them, Greece possessed all these developments while China lacked them due, respectively, to the presence and absence of agora debate. This social-essentialist thesis is pure Weber all over again, and, with regards to China, the DML view in unadulterated form.

On the other hand, however, the challenge of dialogism has had a significant effect on some other scholars, who have responded to that challenge in two different ways which are complementary to one another. Firstly, they try to weaken the dialogical perspective by attempting to minimize the debts that Latin Europe owed to non-European civilizations in the making of modern science.[6] The second response, which is sometimes adopted by the same scholars who also adopt the just-mentioned response, involves a shift in the terrain of argument, a shift first noticed by the critic of Eurocentrism John Hobson. Hobson's concern is with the rise of capitalist modernity instead of modern science, but his observations are highly relevant to us. According to Hobson (2012, pp. 17, 24), 'global dialogism is in danger of being outflanked by the emergence of what I will call "neo-Eurocentrism" or "neo-Orientalism"' which admits that Europe had borrowed extensively from other civilizations, but argues that 'it was this "cosmopolitan" attitude that marked the exceptional quality of Europe and the West. Above all, this focus shifts the theoretical terrain away from pure European inventive genius towards the idea of European creative adaptability'. In other words, it is now argued that *what enabled Europe to make use of what it had borrowed from non-European cultures to make the breakthrough into capitalist modernity/ modern science are certain 'uniquely European 'qualities'* ('creative adaptability', attitude, trait, values, drives, and so on) absent in other civilizations. We follow Hobson to call this shift in approach *neo-Eurocentrism*, defined by the above-italicized sentence. Thus, hereinafter, the term 'Eurocentrism' and its cognates will continue to be used in the generic sense, but when reference is made specifically to the above new approach, the prefix 'neo' will be added.

Why does Hobson say that dialogism is in danger of being outflanked by neo-Eurocentrism? In our exchanges with Bala (see note 1), we raised the question with him as to why, on the basis of what it had learnt from

[6] This is done by two means, firstly, by adopting a legal definition of evidence (onus of proof, the presumption of innocence) concerning knowledge-transmission claims, which is contrary to the common criterion of the balance of probabilities or plausibility adopted in various historical disciplines; secondly, by claiming cultural barriers to knowledge transmission. On the former, see Lau (2023 section 3); the latter will be touched upon below in relation to the cultural-conservatism argument.

non-European civilizations, Latin Europe was able to make the breakthrough of ca. 1600. He replied that this remains, to quote from him, 'a black box'. In other words, while dialogism is able to show that a key pre-condition for the breakthrough of ca. 1600 is what Latin Europe had learnt from elsewhere, it is unable to explain why, on that basis, scholars such as Galileo were able to make that breakthrough. By contrast, neo-Eurocentrism does have a particular argument—the claim concerning certain 'uniquely European qualities'—to account for that. Though neo-Eurocentrism fails as badly as conventional Eurocentrism (see Sect. 4), its emergence does reveal a gaping shortcoming of dialogism.

2 A Theoretical-Methodological Framework for Analyzing Intellectual Development on a Non-essentialist Basis

The following framework is a special construction integrating a number of concepts, theories, and perspectives from various disciplines into a unitary whole that caters specifically to the analysis of intellectual development. It comprises of the following components: (a) sensitive dependence of historical development on starting point; (b) contingency in history as a systematically analyzable explanatory variable; (c) the integration of externalist and internalist explanations; (d) scholarly tradition as a self-reinforcing informal institution that can be made subject to non-essentialist analysis; (e) the distinction and relationship between the formal aspects and substantive knowledges of a scholarly tradition; (f) historical development as a path-dependent process; (g) process approach versus mechanistic approach; and (h) scholarly agency, one type of internal factors discussed under (c) above, as a non-essentialist explanatory variable. Not all of the above components will figure in the analysis below, but we still discuss them all so as to provide the overall theoretical context in which that analysis is undertaken.

According to chaos theory in physics (Ruelle 1991), a slight change in the initial conditions of a dynamical system will, in time, lead to a very different outcome, a phenomenon known technically as sensitive dependence on initial conditions (which we adapt as *sensitive dependence on starting point*) and popularly as the butterfly effect; since the slight change cannot be known precisely, the system becomes unpredictable beyond a certain time. Applying this to historical explanation, this implies that analyses that focus only on relatively late stages of different developments are ill-equipped to understand why such differences have arisen. Sensitive dependence on starting point becomes especially significant if the starting point itself is to a significant degree the outcome of contingency.

The role of *contingency* in history cannot be emphasized enough. Needham (Needham et al. 2001, p. 209) rejected 'pure chance' as emphatically as he did bio-psychological essentialism as a legitimate explanatory variable; in our view, this is because, like many others, he mistakenly believed that contingency cannot be systematically pinned down. Contingency can pertain to *cause* (a contingent factor or event causing a certain event to occur), as well as to *outcome*. The latter case is well-captured by the notion of unintended consequences in history, first formulated by Merton (1936), though he restricted it, unnecessarily, to the case of purposive social actions—hereinafter, we refer to it as *unforeseen consequences in history* to do away with Merton's tying of the notion to purpose. The history of intellectual developments is full of such contingent unforeseen consequences, one particular form of which we refer to as *ironies of history*, for instance, a practice that conforms to the scientific spirit might subsequently prove counter-productive, while a mystical idea might subsequently constitute the basis of a rational construction.

In the literature on intellectual development, a distinction is often made between *externalist* and *internalist* explanations. Social explanation, whether of the essentialist or non-essentialist variety, is a major type of externalist explanation. Some social explanations fall into the trap of *sociological reductionism*, i.e., seeing the substance of intellectual development (ideas, etc.) as a direct reflection of some social factor(s). What social explanation, even of the non-essentialist variety, can provide is to identify the social conditions that impact on intellectual development either positively or negatively or in other ways; the substance of intellectual development itself will have to be explained otherwise. Hence, social explanation is, by nature, inadequate, and needs to be combined with an account of the substance of intellectual development itself. Internalist arguments are concerned with the internal pre-conditions and conditions of intellectual development, which fall into two categories: (1) the intellectual resources from which a certain development draws upon, which can be subdivided into two types, namely, [a] its intellectual ancestry, and [b] other intellectual sources inputting into it; and (2) the agency of the scholars concerned. Intellectual ancestry refers to *scholarly tradition*; other intellectual sources refer to knowledges transmitted from other cultures. *Agency* refers to the active role played by social agents; with reference to intellectual development, it refers to the *scholarly* values, drives and so on of *the scholars concerned, not to society-wide* values and so on. Scholarly agency is absent in dialogical analyses. This is because *of the universal belief, shared by Eurocentric and non-Eurocentric scholars alike, that scholarly values, drives and so on can only be interpreted on an essentialist basis.* Section 4 shows how and why this is a huge mistake. From our distinction between external and internal factors, and the categorization of internal factors into different types and sub-types, *it is clear why the dialogical perspective is inadequate*, for it focuses only on 1b above.

Scholarly tradition is a potentially misleading issue because it is generally interpreted on an essentialist basis. Elsewhere (Lau 2020), we show that the founding, reproduction, and development of scholarly traditions can be satisfactorily explained only on a non-essentialist basis. Scholarly tradition is a *self-reinforcing social institution* in the sociological sense as analyzed in sociological neo-institutionalist theory (Powell and DiMaggio 1991; Scott 2001). From this theory's perspective, much of social action is non-instrumental, but *normative-dispositional* in nature: normative with regard to internalized goals and values; dispositional in the sense of automatically, as it were, following acquired routines, rules, and procedures on the basis of internalized taken-for-granted shared knowledges and beliefs. Scholars raised in a particular scholarly tradition do not only acquire its substantive knowledges, they also internalize its goals, values, and taken-for-granted knowledges and beliefs, adopt its practices (routines, rules, procedures), and act in a normative-dispositional way on that basis in their work. The same individual, if raised in a different scholarly tradition, would end up as a very different scholar. Scholars raised in a particular tradition reproduce that tradition by virtue of their daily work. Once a tradition is sufficiently well established, the factors or conditions that originally gave rise to it are no longer relevant to scholars raised in it, since from now on, what shape its members normatively and dispositionally in their daily work are no longer these factors or conditions, but the tradition's institutionalized features. Because a scholarly tradition is a self-reinforcing institution, scholars of a particular *entrenched* tradition do not abandon it other than in the most exceptional of circumstances. This is analogous to the tenacious grip that a scientific paradigm exercises over scientists raised in that paradigm analyzed by Kuhn (1962).

There are to two inter-related aspects of a scholarly tradition. As noted, scholars raised in a particular scholarly tradition do not only acquire its substantive knowledges, but also internalize its goals, values, beliefs, orientations, and approaches, as well as adopt its practices. The values, approaches, and so on constitute what we refer to as its *formal aspects*, these formal aspects are *embodied* in the tradition's *substantive knowledges* themselves—which is why we call these aspects formal, with the substantive knowledges constituting contents. Form and content are inextricably linked together. For instance, Greece developed what can be called a theoretical tradition in mathematics, the tradition's theoretical approach is embodied in the substantive knowledges comprising Euclid's *Elements*.

Scholarly tradition is non-culture-specific in the sense that, *under certain conditions*, a particular culture (such as *Islamic* society) can borrow a tradition developed by a very different culture (such as *polytheistic* Greece). Islamic scholars were able to borrow the Greek tradition because they had *not* hitherto developed any entrenched tradition of a significantly different nature. This also applied to the case of Latin Europe learning from Arabic scholarship beginning in the eleventh century. In contrast to this, where a society had *already* developed a deeply *entrenched* scholarly tradition of its own, its scholars will be highly resistant to the knowledges of a tradition of a very *different* nature, especially to substantive knowledges of the latter that embody its formal aspects.[7] Scholars lacking the perspective presently being analyzed invariably put this down to alleged cultural conservatism, a favorite form of cultural-essentialist argument today. That such resistance is *intellectual* and *not* cultural in nature can be seen from Kuhn's analysis of the resistance to paradigm change which occurs within the *same* culture. It can thus be seen that an understanding, on a non-essentialist basis, of scholarly tradition and its character and features enables a rebuttal of a popular form of cultural-essentialist argument.

In historical sociology, there is a theory which argues that historical outcomes are neither necessary nor predictable by theory, in which contingency features systematically as a variable. This is path dependency (PD) theory. Mahoney (2000) distinguishes between two types of PD processes. In the first kind, called self-reinforcing sequence, a contingent event or factor sets into motion a *self-reinforcing process* in which a particular pattern is *institutionalized* and hence reproduced over time. The self-reinforcing mechanisms are deterministic, but because the event or factor that kicks off the process is contingent, hence the outcome is unpredictable by theory. In this scenario, development is said to be *locked-in* along a specific trajectory as a result of the self-reinforcing mechanisms, which pose *obstacles to trajectory change*. These mechanisms include such factors as sunk costs, durable commitments, and learning curves (Aminzade 1992, p. 463). The self-reinforcing mechanisms ensure the *reproduction* of the institutionalized pattern, and the contingent event or factor that originally kicked off the process subsequently becomes irrelevant to that reproduction (Stinchcombe 1987, pp. 103–4). Once there is lock-in, breaking away from the institutionalized trajectory rarely occurs; if it does, it is not for self-generated reasons, but under huge *external* impact. It is clear that *the concept of self-reinforcing PD is eminently applicable to analyzing the foundation and development of scholarly traditions in a non-essentialist way.*

Mahoney's second kind of PD process is called *reactive-sequence* PD, according to which a certain event or factor A produces event B in reaction, which in turn produces event C in reaction, and so on, until the

[7] Some individual items of a tradition's substantive knowledges are less tied to that tradition's formal aspects (such as China's notion of the universe being an empty infinite space), and face much less resistance.

process terminates with, say, event H. While A's production of B is explicable *ex post facto*, A could have led to, say, X instead, depending on what intervening factors, whether or not contingent, are present; similarly for B to C, and so on. Hence, whereas A may or may not be contingent, each reactive step in the sequence is to varying degrees non-deterministic, hence H is neither necessary nor predictable by theory on the basis of the initial condition A. In short, it is the non-deterministic path that determines the historical outcome.

The *timing* of the intervention of a factor or event is a crucial variable, i.e., the same factor or event intervening at one point in time in the developmental process instead of at another may well have very different effects. This can clearly be seen in the cases of the adoption of the Greek scholarly tradition by Islamic scholars and of that of Arabic scholarship by Latin Europe; in both cases, this occurred when the borrower had not developed any entrenched scholarly tradition of a very different nature, hence facilitating the adoption.

We do not rule out deterministic theories in principle. Goldstone (1998) astutely observes that there is no one correct type of universally applicable theories to explain history, where similar patterns are observed across diverse societies, deterministic theories are likely to apply, where unique and exceptional outcomes are observed in a particular society, PD most likely applies. The breakthrough into modern science constitutes an exceptional and unique event, to account for which we believe our non-essentialist and non-deterministic framework is best suited.

Our framework adopts a *process approach*, which sees developmental outcomes as being produced by a long series of linked moments in a continuous process, in which some factors (such as scholarly tradition) may exert their effects throughout or through most of the process, while other factors appear at different stages of the process exerting their effects on that process at those particular stages. Most of these factors do not directly produce the final outcome and may seem far removed from the latter, but nonetheless contribute to the production of the final outcome. The process approach is in contrast to what we call the *mechanistic approach*, which is the approach taken by all essentialist arguments. Essentialist arguments are not concerned with process, but see intellectual outputs as being produced separately and individually in the same mechanical way, like billiard balls hitting one another, by the same purported 'essence'. Thus, Weber argued that *each and every one* of the 'unique' 'forms of rationality' found in, and allegedly only in, Europe had been produced, probably by the Europeans' 'biological heredity', and certainly by Europe's 'peculiar social structure'.

3 Greece and the Needham Question

Europe in Eurocentrism does not refer to Europe as a geographical area, but as a civilization, which is often referred to as 'Western' civilization. In both popular and scholarly discourses, the 'West' is assumed to be a '*trans-historical*' entity that began with Greece; hence Greece is part and parcel of the Europe in Eurocentrism. Thus, the key role played by the Greek tradition in the rise of modern science has always been interpreted in Eurocentric terms, especially because scholarly tradition itself has invariably been interpreted in essentialist terms. Put another way, as a result of (a) the essentialist interpretation of scholarly tradition and (b) the 'common sense' that Greece is the origins of the 'West', the role of the Greek tradition in the rise of modern science has always been taken by Eurocentric scholars to be supportive of Eurocentrism, because on the basis of (a) and (b), the development from Greece to Latin Europe can then be taken to be an '*internal* "Western" development' (how the issue of the role of Arabic science is dealt with in this 'internal "Western" development' view will be discussed in due course). It is also for this reason that the role of Greece has always been a source of trouble for non-Eurocentric scholars. We have already debunked (a); elsewhere (Lau 2020) we have debunked (b) by subjecting the notion of the 'trans-historical' 'West' to what Foucauldians call a genealogical analysis, i.e., to trace how such a notion has historically arisen. For lack of space, we are unable to provide even a brief summary of that analysis here, hence we can only make a few conclusive statements below.

Greek civilization and Latin European civilization were two different civilizations. Latin Europe and its successor did take up (initially through Arabic scholarship) the Greek tradition, but *adoption* does not constitute any '*internal*' or '*organic*' development from one stage to another of the *same* 'trans-historical' entity. Islamic society also took up the Greek tradition, but nobody would claim that the Abbasid caliphate and Greece constituted one and the same 'trans-historical' entity. One reason for this is that, unlike the Greeks and the Latin Europeans, who are both white, Arabs, Persians, etc. are not white.[8] Indeed, the construction of the 'West' as a fictitious 'trans-historical' entity occurred in the context of rapidly rising racism in Europe upon the success of the colonial enterprise. During this process of construction, as Latin Europe's scholarly development began to overtake Arabic scholarship which was in decline, Greece, previously regarded, like the Arabs, as constituting a 'them', was claimed to be Europe's 'childhood', while Europe's scholarly debt to Arabic scholarship came to be progressively denied by reducing the latter's role to one of merely 'preserving' Greek learning in its original state for ultimate 'transmitting back to the "West"'. It was on this basis that the AWD paradigm claimed that modern science had arisen from the 'autochthonous' development of the 'West' from Greece to seventeenth-century Europe.

Today, even with Europe's intellectual debts to Arabic scholarship (which took Greek learning to new heights) fully acknowledged, *given the role played by the Greek tradition in the rise of modern science, the AWD paradigm continues to survive in virtually pure form thanks to regarding Greece as the origins of the 'trans-historical' 'West'*. This can most clearly be seen in Chen (2009, p. 634, our emphases): '*Western* science is not only a branch of its [the '*West's*'] many scholarly fields, it is the embodiment of the *spirit* of its entire civilization'. In order to square this AWD-paradigmatic claim with his full acknowledgment of the role of Arabic scholarship, Chen invents the notion of '"the West" in the broad sense' (p. 631), whereby the Arabs *by virtue of their taking up of the Greek tradition* are transformed into 'Westerners' 'in the broad sense'. Besides Chen's blatant invention, subtler forms of surviving claims that the rise of modern science is a direct Greece-to-Europe (the 'trans-historical' 'West') 'internal' inheritance also exist, though space forbids discussing them here.

---

[8] Whites, blacks, and so on are valid categories in the classification of population groups but without any extra-complexion implications assumed in racist understanding (Malik 2008).

Hence, debunking the notion of the 'trans-historical' 'West', still almost universally held in both popular and scholarly discourses even in today's post-racism age, is critical for dismantling a key pillar of Eurocentrism. By the same token, showing that Greece constituted a different civilization from Latin Europe and its successor greatly strengthens the dialogical perspective.

4 A Non-Eurocentric Approach to the Needham Question Transcending Dialogism

For lack of space, we are unable to provide an overall analysis of the issue here, but must remain content with confining ourselves to one core issue to be explained presently. In Sect. 2, in mentioning the need to integrate external and internal factors, we categorize internal factors into two types, 1 and 2, and sub-divide 1 into 1a and 1b. Dialogism is inadequate and hence unable to explain why Latin Europe was able to make the breakthrough of ca. 1600 because it focuses only on 1b. On the other hand, neo-Eurocentrism does offer the semblance of an answer to that question simply by sticking to the familiar argument that this was due to some 'unique' European 'qualities', though with the modification that such 'qualities' were responsible not so much for inventing something anew *ex nihilo*, as for making use what it had learnt from other civilizations to break new ground. Such claimed 'uniquely European qualities' are all too familiar, namely, dynamism, restlessness, curiosity, aspiration to break with orthodoxy, and so on, which neo-Eurocentric scholars believe can only be interpreted on an essentialist basis. Dialogism shares this mistaken belief. In the following, we show that the 'black box' spoken of by Bala (see Sect. 1) can be cracked open, and neo-Eurocentrism rebutted, *only* by analyzing, on a *non-essentialist* basis, the scholarly (not society-wide) values, drives and so on—all *internal* factors in our distinction between internal and external factors—of late-Renaissance scholars such as Galileo.

The conquest of Constantinople in 1453—an *external* factor in relation to scholarly development—brought about various *(contingent) unforeseen consequences* relevant to our discussion. Firstly, between 1200 and 1450, Aristotelianism was without rival in Latin Europe (Grant 1996, p. 162). Byzantine scholars fleeing from the Ottoman's westward advance broke that monopoly which helped *undermine its authority* by bringing with them works on Hermeticism, Neoplatonism, atomism, Archimedean mathematics, and so on. Secondly, it made Europe determined to seek an alternative route to Asian spices and grab the spice trade by force from their hated enemy (Arab merchants) on whom they had for centuries depended; this was the driving force behind the so-called 'voyages of discovery' (Keay 2005). Reports from Portuguese and Spanish travelers revealed empirical facts that contradicted various Greek theoretically-derived postulations made by Aristotle and Ptolemy, which forced scholars to acknowledge the superiority of experience over rational speculation (Cohen 1994, pp. 354–7; Poskett 2023, Part I). This did not only further *undermine Aristotelianism's authority*, but also engendered what can be called *an empirical turn* among scholars such as Tycho Brahe. In 1517, Luther published his *Ninety Theses*, which, together with the progressive undermining of Aristotelianism's authority, engendered a greater willingness to *break with orthodoxy*. Latin Christianity's split is an *external* factor in relation to scholarly development which was endogenous to Europe, but it does not constitute any European 'essence' since religious schisms occur elsewhere as well. While Copernicus might possibly have been inspired by Hermetic sun worship (Kuhn 1957, pp. 128–31; Yates 1979[1957], pp. 153–4), his breaking with the orthodoxy of Aristotelian-Ptolemaic geocentrism occurred precisely in the context of this changing scholarly environment. Galileo has also often been hailed by Eurocentric scholars as embodying the aspiration and drive to break with orthodoxy—Ronan (1983, p. 186): '[in China] there was no aspiration to break the mould of existing orthodoxies as inspired men such as Galileo'. One of Galileo's breaking-with-orthodoxy concepts concerns the relativity of motion, it has been well established that he had taken this concept from the fourteenth-century scholastic Oresme (Lindberg 2007, p. 285). Oresme used a shipboard example also used by Galileo, but Oresme refused to accept the conclusion from that example that Galileo later endorsed. Why the difference? As Lindberg points out, Oresme wanted to preserve orthodoxy, whereas Galileo did not, and the only possible reason for this contrast is the different scholarly climates of the fourteenth and late sixteenth centuries. This is an illustration of the point about the critical importance of *timing* in *path-dependent* analysis—the same concept met with different fates at different points in time.

The 'voyages of discovery' and the imperial enterprise consequent upon them had another *(contingent) unforeseen consequence* on scholarship as well. When Bacon was 16 (ca. 1578), he disliked Aristotle because his teachings were not conducive to practical benefits; for Bacon, 'Truth, therefore, and utility, are here perfectly identical…that which is most useful in practice is most correct in theory' (Bajaj 1990, the quote from Bacon is on p. 44; cf. Hall 1962, p. 15). Bacon was the keenest spokesman of what may be called a *practical turn* among various late-Renaissance scholars. The empirical turn and the practical turn complement each other, hence one can speak of a *practical-empirical turn.* Galileo, commonly regarded as the central figure of the breakthrough of ca. 1600, was an embodiment of this combined turn. The conceptual-theoretical continuity between Galileo and the scholastics (much more than the shipboard example) with regard to motion has been pointed out by various science historians, where Galileo broke away from the scholastics was his application to practical-empirical problems (Crombie 1959; Grant 1996; Lindberg 2007); and this is what enabled him to make his breakthroughs. Most of his breakthroughs occurred from 1592 to 1610 when he was in Padua, where he collaborated closely with the Venetian navy. Renn and Valleriani (2001) point out that the two sciences that Galileo presented in his *Discourse* were the outcome of this collaboration.

The practical-empirical turn embodied a fundamental shift in *scholarly values* (putting the practical and the empirical above the purely theoretical), and the greater readiness to break with orthodoxy represented a fundamental shift in *scholarly drives*. This double shift in the scholarly climate among late Renaissance scholars—which pertains to the *internal* factor of *scholarly agency*—was engendered as *(contingent) unforeseen consequences* of *external* events and factors (1453 leading to the 'voyages of discovery', Latin Christianity's split, Latin Christianity's hatred of Islam and so on) as explained above.

Let us now compare our above analysis with Cohen's (2011, 2015) sophisticated thesis on the positive NQ. Cohen argues that the breakthrough of ca. 1600 consists of 'three revolutionary transformations'. Unfortunately, we are unable to discuss this part of his thesis here. What concerns us here is Cohen's answer to his own question of why these transformations all occurred in the early seventeenth century:

> The only viable answer resides in the unusually extraverted *propensities* of European civilization—its 'roving' *qualities* and its outer-directed religious orientation[9] (...)....Europe had by 1600 reached a point where the *will* toward innovation was furthered by an overall climate in which *daring* novelty stood at a certain premium (Cohen 2011, pp. 273, 277, our emphases).
>
> European society and culture [have been endowed] with an extraordinary *dynamism...restlessness* and *curiosity* instigated wide-ranging explorations culminating in the Voyages of Discovery (*ibid*., p. 135, our emphases).

This is a classic instance of cultural-essentialist neo-Eurocentric argument popular today. Let us bypass Cohen's repetition of the ideological narrative of the so-called 'voyages of discovery' as 'explorations'—to find this being repeated today in serious scholarship by eminent scholars is both surprising and unexpected. What we want to focus on are the 'qualities' emboldened in the above quotes. In our analysis of the role of agency in the rise of modern science, by this agency, we refer to the practical-empirical turn and the greater readiness to break with orthodoxy, and the newly emerging values and drives embodied in them, in a fundamental shift in the intellectual climate *among intellectuals*. The relationship between this agency among scholars such as Galileo and the breakthroughs achieved by them is *concretely* shown. In contrast, it can be seen that the 'qualities' that Cohen speaks of pertain to, to quote from him, 'European society and culture' in general. While we have

---

[9] This particular component of Cohen's argument is a perversion and distortion of Weber's Protestant-ethic thesis, though we cannot discuss it here.

no space to discuss the validity (or lack of it) of these claimed *society-wide* 'qualities', how they are related in *concrete* terms to the occurrence of his three 'revolutionary transformations' *among scholars* is left unexplained. In other words, this does not amount to any explanation at all. Rather, it is simply a statement of a pre-conceived cultural-essentialist view misconceived as an explanation.

We would like to end, perhaps fittingly, with Needham. In making his above cultural-essentialist argument, Cohen quotes from none other than Needham for support:

> the exceptional poverty of Europe in the precious metals, the continual desire of Western peoples for commodities which they themselves could not produce (...)...By contrast China was a coherent agrarian land-mass, a unified empire since the third century B.C. [sic] with an administrative tradition unmatched elsewhere till modern times, endowed with vast riches...Europe a culture of rovers [cf. Cohen's '"roving" qualities'], was always uneasy within her boundaries [cf. Cohen's 'restlessness'], nervously sending out probes in all directions [cf. Cohen's 'curiosity']...The greater population of China was self-sufficient...generally content with only occasional exploration, essentially incurious about those far parts of the world...It may well be that here, at this point of tension, lies some of the secret of the specific European creativeness when the time was ripe[10] (Needham 1969, quoted in Cohen 2010, p. 135).

To quote from someone who is supposed to be non-Eurocentric for support for the Eurocentric position seems ideal; the only trouble is that, firstly, as pointed out in Sect. 1, Needham remained under the heavy influence of Eurocentrism (the above quote is precisely an instance of this), and, secondly, while his knowledge of China's technological developments is formidable, his knowledge of other aspects of China leaves much room for desire, and most of his sweeping observations and generalizations are, unfortunately, derived from such a deficiency. The above quote is a typical example. Each and every claim made by Needham

[10] 'When the time was ripe'? As though this clause, all by itself, carries self-evident explanatory substance and meaning.

in the above quote about China is *empirically false*. We cannot, obviously, discuss this in full here, so let us illustrate with just one instance. According to Needham, Europe 'was always uneasy within her boundaries' whereas China was not ('a unified empire' since the third century), hence it 'nervously' sent out 'probes in all directions' whereas China was 'essentially incurious about' the rest of the world. As everyone with even an elementary knowledge of China's history can tell, exactly the reverse is the case.

From 1046 BCE with the founding of the Western Zhou Dynasty (the beginning of China as a state based upon what after Confucius came to be known as the Confucian ethic) until 1911 when dynastic China ended spans 2,957 years. We have calculated that out of this total, a unified state under Han-Chinese rule over the whole of China[11] existed only in 44.1% of the time (the Western Zhou, Qin, Han, Western Jin, Sui, Tang and Ming dynasties); in 43.6% of the time, China was either fractured, or had large parts of the country occupied by invading nomads, or both; and in 12.3% of the time (the Mongolian Yuan and the Manchurian Qing dynasties), the whole of China was ruled by invading nomads. Even when the country was unified under Han-Chinese rule, nomadic invasions were a *perennial* threat—to illustrate with the three most powerful Han-Chinese dynasties in China's history, the Marshall Emperor of Han was engaged in a fifty-year war with the invading Xiongnu (匈奴); the invading Tibetans once occupied the capital city of the Tang Dynasty when the latter was fighting an internal rebellion; the Ming overthrew the Mongolian Yuan Dynasty in 1368 and drove the Mongols back to the north, but the latter kept preparing to invade China again which they eventually did in 1449, taking the Ming emperor captive for four years which almost led to Ming's collapse, and after the Mongols came the Manchus who were to overthrow the Ming in 1644. The number of nomadic peoples who invaded and occupied (in part or in whole) China over the millennia number well over a dozen. China was, to borrow Needham's expression, 'uneasy within her boundaries' *throughout the ages*. As a result, China sent 'probes in all directions' from very early on. To

[11] China's borders varied over time, but for simplicity, we bypass this detail, which does not affect our argument.

mention only the earliest instances, the Silk Road was established by the Marshall Emperor of Han (reigned -141 to -87), none other than because he was looking for allies in central Asia against the invading Xiongnu. In addition, the Marshall Emperor also sent envoys by sea to the south, who reached today's SE Asia and Sri Lanka. These corridors of exchanges were the origins of the land-based Silk Road in the north and the maritime trade route in the south that were soon to flourish.

References

Aminzade, Ronald. 1992. Historical Sociology and Time. *Sociological Methods and Research* 20 (4): 456–480.

Bajaj, Jatinder K. 1990. Francis Bacon, The First Philosopher of Modern Science: A Non-Western View. In *Science, Hegemony and Violence: A Requiem for Modernity* (1990[1988]), ed. Ashis Nandy, 24–67. Oxford University Press.

Bala, Arun. 2006. *The Dialogue of Civilizations in the Birth of Modern Science*. Palgrave Macmillan.

Bernal, Martin. 1987. *Black Athena: The Afroasiatic Roots of Classical Civilization, Volume 1 The Fabrication of Ancient Greece 1785–1985*. New Brunswick: Rutgers University Press.

Bernal, Martin. 1991. *Black Athena: The Afroasiatic Roots of Classical Civilization Volume 2 The Archaeological and Documentary Evidence*. New Brunswick: Rutgers University Press.

Bernal, Martin. 2001. *Black Athena Writes Back: Martin Bernal Responds to His Critics*. Durham: Duke University Press.

Bose, D.M., Sen, S.N., and Subbarayappa, B.V. 1971. *A Concise History of Science in India*. New Delhi: Indian National Science Academy.

Chen Fangzheng. 2009. *Inheritance and Rebellion: Why Modern Science Emerged in the West*. 陳方正,《继承与叛逆:现代科学为何出现于西方》(北京:三联, 2009).

Cohen, H. 1994. *Floris*. The Scientific Revolution: A Historiographical Inquiry. University of Chicago Press.

Cohen, H. Floris. 2011. *How Modern Science Came into the World: Four Civilizations, One 17th-Century Breakthrough*. Amsterdam University Press.

Cohen, H. Floris. 2015. *The Rise of Modern Science Explained. A Comparative History*. Cambridge: Cambridge University Press.
Collins, Randall. 1985. *Max Weber*. London: Sage.
Crombie, A.C. 1959. The Significance of Medieval Discussion of Scientific Method for the Scientific Revolution. In *Critical Problems in the History of Science*, ed. Marshall Clagett, 79–101. University of Wisconsin Press.
Goody, Jack. 2006. *The Theft of History*. Polity.
Graham, Angus C. 1973. China, Europe, and the origins of modern science: Needham's The Grand Titration. In *Chinese Science*, ed. Shigeru Nakayam and Nathan Sivin, 45–69. Cambridge, Mass: MIT Press.
Grant, Edward. 1996. *The Foundations of Modern Science in the Middle Ages*. New York: Cambridge University Press.
Hall, A. Rupert. 1962. General Introduction. In *The Scientific Renaissance 1450–1630*, by Marie Boas. London: Collins.
Harley, J.B., and Woodward, David, eds. 1987. *Cartography in Prehistoric, Ancient, and Medieval Europe and the Mediterranean*. University of Chicago Press.
Harley, J.B., and Woodward, David, eds. 1992. *Cartography in the Traditional Islamic and South Asian Societies*. University of Chicago Press.
Harley, J.B., and Woodward, David, eds. 1994. *Cartography in the Traditional East and Southeast Asian Societies*. University of Chicago Press.
Harley, J.B., and David Woodward, eds. 1998. *Cartography in the Traditional African, American, Artic*. Australian and Pacific Societies: University of Chicago Press.
Hart, Roger. 2013. *Imagined Civilizations: China, the West, and Their First Encounter*. Baltimore: John Hopkins University Press.
Hess, David. 1995. *Science and Technology in a Multicultural World*. Columbia University Press.
Hobson, John. 2012. Global Dialogical History and the Challenge of Neo-Eurocentrism. In *Asia, Europe and the Emergence of Modern Science: Knowledge Crossing Boundaries*, ed. Arun Bala, 13–33. New York: Palgrave Macmillan.
Joseph, George G. 2011. *The Crest of the Peacock Non-European Roots of Mathematics*, 3rd ed. Princeton: Princeton University Press.
Katz, Victor, ed. 2007. *The Mathematics of Egypt, Mesopotamia, China, India, and Islam: A Sourcebook*. Princeton, NJ and Oxford: Princeton University Press.
Keay, John. 2005. *The Spice Route*. London: John Murray Publishers.
Kuhn, Thomas. 1957. *The Copernican Revolution*. Harvard University Press.

Kuhn, Thomas. 1962. *The Structure of Scientific Revolutions*. Chicago: The University of Chicago Press.

Lau, Raymond W.K.. 2020. *Intellectual Developments in Greece and China: Contingency*. Institutionalization and Path-Dependency: Cambridge Scholars Publishing.

Lau, Raymond W.K. 2023. Revisiting the Needham Question: Beyond Transmissionism in Critiquing Eurocentrism', Paper Presented at the 'Needham's Dialogical Vision: Understanding Science as a Multi-Civilizational Outcome' Conference Held on 23–24 March 2023. Available online on ResearchGate and Academia.edu.

Lindberg, David. 2007. *The Beginnings of Western Science*, 2nd ed. Chicago: University of Chicago Press.

Lloyd, G.E.R. 1973. *Greek Science After Aristotle*. London: W.W. Norton.

Lloyd, G.E.R. 1990. *Demystifying Mentalities*. Cambridge: Cambridge University Press.

Lloyd, G.E.R. 1996. *Adversaries and Authorities: Investigations into Ancient Greek and Chinese Science*. Cambridge: Cambridge University Press.

Lloyd, G.E.R. 2015. *Analogical Investigations: Historical and Cross-Cultural Perspectives on Human Reasoning*. Cambridge: Cambridge University Press.

Lloyd, G.E.R. 2018. Introduction: Methods, Problems and Prospects. In *Ancient Greece and China Compared*, ed. G.E.R. Lloydand, Jenny Jingyi Zhao, 1–29. Cambridge University Press.

Lloyd, G.E.R., and N. Sivin. 2002. *The Way and the Word: Science and Medicine in Early China and Greece*. New Haven: Yale University Press.

Mahoney, James. 2000. Path Dependence in Historical Sociology. *Theory and Society* 29 (4): 507–548.

Malik, Kenan. 2008. *Strange Fruit: Why Both Sides Are Wrong in the Race Debate*. Oxford: Oneworld.

Nasr, Seyyed Hossein. 2003. *Science and Civilization in Islam*, 2nd ed. Islamic Texts Society.

Needham, Joseph. 1969. *The Grant Titration*. Taylor and Francis.

Needham, Joseph. 1981. *Science in Traditional China*. Hong Kong: The Chinese University Press.

Needham, Joseph, with the collaboration of Robinson, Kenneth Girdwood and Huang, Ray (Huang Jen-yu), and including contributions by them, Introduction by Elvin, Mark, edited by Robinson, Kenneth G. 2001. *Science and Civilisation in China, Volume 7, Part II: General Conclusions and Reflections*. Cambridge University Press.

Poskett, James. 2023. *Horizons: A Global History of Science*. Penguin.

Powell, Walter, and P. DiMaggio. 1991. Introduction. In *The New Institutionalism in Organizational Analysis*, ed. Powell Walter and P. DiMaggio, 1–38. Chicago: The University of Chicago Press.

Renn, J., and Valleriani, M. 2001. Galileo and the Challenge of the Arsenal. Lecture Delivered in Florence on 21 March 2001, available online, accessed 23 May 2022.

Ronan, Colin. 1983. *The Cambridge Illustrated History of the World's Science*. Cambridge: Cambridge University Press.

Ruelle, David. 1991. *Chance and Chaos*. Princeton: Princeton University Press.

Saliba, George. 1994. *A History of Arabic Astronomy: Planetary Theories during the Golden Age of Islam*. New York and London: New York University Press.

Saliba, George. 2007. *Islamic Science and the Making of the European Renaissance*. Cambridge, Mass: MIT Press.

Scott, W. Richard. 2001. *Institutions and Organizations*, 2nd ed. Thousand Oaks: Sage.

Sivin, Nathan. 1995. *Medicine, Philosophy and Religion in Ancient China Researches and Reflections*. Variorum.

Sivin, Nathan. 2018. Why Some Comparisons Make More Difference Than Others. In *Ancient Greece and China Compared*, ed. G.E.R. Lloyd, and Jenny Jingyi Zhao, 33–9. Cambridge University Press.in *Ancient Greece and China Compared*. Cambridge University Press.

Stinchcombe, Arthur. 1987. *Constructing Social Theories*. Chicago: University of Chicago Press.

Weber, Max. 1992. Author's Introduction. In *The Protestant Ethic and the Spirit of Capitalism*, ed. Max Weber, trans. by Parsons, T., Introduction by Giddens, A, xxviii–xlii. Routledge.

Yates, Frances. 1979 [1957]. *Giordano Bruno and the Hermetic Tradition*. University of Chicago Press.

4

The Need to Extend Needham's Vision of Science: A Decolonial Perspective

Tung-Yi Kho

1 Introduction

Needham's view of modern science is celebratory and appears to be undergirded by the belief that it is both a superior and 'universal' form of knowledge. These two beliefs underlie the two questions he then proceeds to pose. The first question, which is comparative insofar as it seeks to draw comparisons between Europe and non-Europe, may be stated thus: Why did modern science emerge in Europe and not elsewhere? The second question, which is dialogical insofar as it strives to understand the contributions of the non-West to the modern scientific enterprise, may be cast as follows: How did non-Western civilizations develop modern knowledge and techniques that contributed to the development of modern science?

By way of his comparative and dialogical questions, Needham appears to regard modern science to have historically taken root in Europe even

T.-Y. Kho (✉)
Independent Scholar, Hong Kong SAR, China
e-mail: kho.tungyi@yahoo.com

A. Bala et al. (eds.), *Multicivilizational Exchanges in the Making of Modern Science*,
https://doi.org/10.1007/978-981-97-3541-9_4

as it had been nourished by the various contributions of non-Western civilizations. That is to say, modern science is celebrated because it is seen somewhat as an end-point of knowledge. This invokes a 'many-rivers-into-one-ocean' image of the development of modern science, whereby the knowledges and techniques of various non-Western civilizations flow into and converge in modern scientific knowledge.

Although Needham explicitly commits to an inter-cultural study of modern science and is aware of the contributions of non-European civilizations to it, the beliefs implicit in his two questions allude to a certain Western exceptionalism, howsoever unwitting.

Yet Western exceptionalism as a belief has been predominant in the modern era, so Needham can hardly be faulted for his predilections. After all, the evidence appears to vindicate such a position: not only was the West the first to have undergone a complete modern transformation, the improvement in material standards of living that such a transformation brought about was also conspicuous vis-à-vis pre-modern lifeways. As such, it would have been difficult not to feel exuberant about modernity, or modern science, its dominant knowledge system, or the West, not least because it was in the West that such revolutionary developments first occurred. Indeed, it was perhaps because the correlation between modernization/westernization, modern science, and improved living standards was so conspicuous and compelling that (the idea of) Western exceptionalism became commonplace. The West was and has been widely considered exceptional because its accomplishment of modernization, of which it was ahead of the rest of the world, led to its historic global ascendency.

Nonetheless, it is exactly such an evaluation of Western modernity, along with modern science as its dominant knowledge system—and which implicitly lies at the heart of Needham's questions—that I wish to problematize. I will offer a decolonial perspective by taking issue with the inadvertent Eurocentrism built into Needham's vision of modern science. The Eurocentrism implicated here manifests in the celebration of modern science as the apotheosis of knowledge and the concomitant regard of the West as the pioneer of that knowledge. In the process, I hope to demonstrate that what is called 'modern science' is but one version of science, and that its widespread acceptance as the world's predominant knowledge system—to the extent of being considered 'universal'—is but largely the result of power relations, specifically, coloniality.

In the next section, I begin the discussion by outlining Needham's views of modern science. This will be followed by Sect. 3 in which I will outline what a decolonial perspective entails. Taking cue from the work of decolonial scholars such as Quijano (2010) and Mignolo (2011), I invoke coloniality to refer to the malignant side of modernity which is coeval with it; hence the expression, 'modernity/coloniality'. From a decolonial perspective, I submit that the exaltation of modern science as 'knowledge par excellence' is a specific instance of knowledge coloniality.

Knowledge coloniality entails an inflation of one civilizational worldview to the extent of rendering it the gold standard by which to compare and evaluate other worldviews and the civilizations from which they derive. Decolonial scholars argue that knowledge coloniality serves to maintain coloniality—and hence, illegitimate political domination—insofar as it asserts the superiority of the colonizer's ways of knowing and being over those of others. This has been done in the modern era by the West with its self-conscious deployment and elevation of modern science, while deflating the value of other knowledges.

I proceed, then, to discuss the ambiguities around Needham on the subject of knowledge coloniality. Here, it seems that Needham was largely unaware of being ensnared by the latter, for while he stridently opposed all forms of cultural chauvinism—especially of the Western variety—and was deeply committed to the celebration of the world's diverse cultures, he appears still to be inadvertently mired in a Eurocentric trap. In other words, while Needham explicitly rejects 'modern science' as 'Western' and acknowledges it to be a multi-civilizational outcome, he remains of the view that it represents the pinnacle of knowledge.

Hence, while Needham rightly rejects the common variety of Western chauvinism, typically characterized by exclusively attributing to the West the accomplishment of modern science and modernity, he seems unsuspecting of the deeper and more subtle Western conviction that regards modern science as the universal fate of all civilizations. In other words, while it is Eurocentric to credit the West exclusively for the ostensible boon of modern scientific knowledge, is it not even more Eurocentric to inflate the value of such an accomplishment by believing that it represents the culmination of all human knowledge? Clearly, there are different expressions and degrees of Eurocentrism even if Needham was not aware of the full variety in circulation. Indeed, why should the knowledge of the world's many civilizations culminate in modern science? What—and who—accounts for such a teleology; and at what costs?

In conceiving of modern science as a superior knowledge system, it is clear that Needham was merely reflecting the general cultural mores and sentiments of his time. Nevertheless, as the following section explains, such taken-for-granted convictions—and their motivations—are exactly the stuff that animate decolonial research. Hence, although intellectually interesting, neither of Needham's questions would constitute the starting point for decolonial inquiry since the questions presuppose a certain epistemic teleology (i.e. the belief in modern science as knowledge par excellence) that decolonial scholarship would consider first necessary to understand and account for. In other words, from the perspective of decolonial thinking, Needham's questions assume and obfuscate what needs to be explored in the first place.

Be that as it may, decolonial scholarship would begin by asking: Who gets to have the last word on the nature of such a telos; why, and who benefits from it? As one interested in decoloniality, my preoccupation has been to examine and observe the different, multi-layered expressions of Western chauvinism and how they permit knowledge coloniality to take root. Therefore, instead of Needham's comparative and dialogical questions (respectively, again: why did modern science take root in the West and not elsewhere; and how did non-Western civilizations develop the knowledges and techniques that contributed to modern science?), the task of a decolonial re-examination would shift the focus and terms of the discussion. A decolonia approach to modern science would critically examine the relationship between coloniality and modern science: Why does modern science enjoy the privileged status it does in modern society, as reflected in Needham's questions? Who gains from it having such societal importance?

It should be clear that in this chapter on the need for extending Needham's vision of science from a decolonial perspective, my task is less about drawing attention to Needham's inadvertent Eurocentrism than about assessing the consequences of an unqualified and uncritical exaltation of modern scientific knowledge. What is gained—and lost—in the process of regarding modern science as the end-point of knowledge? This ultimately leads us to ask about the nature of modern science: What is modern science a science of, and what implications does its exaltation in society have for coloniality and its perpetuation?

I hope that by the end of this chapter, I would have made a strong case for decolonially extending Needham's vision of science.

2 The Background to Needham's Questions: Needham's Views of Science and Eurocentric History

Joseph Needham held a celebratory view of modern science, a view that appeared to be undergirded by the belief that it was a 'superior' as well as 'universal' form of knowledge. These beliefs underlie the two questions he proceeded to pose. The first question involved a comparison between Europe and non-Europe in the development of modern science: Why did modern science emerge in Europe and not elsewhere? The second question was dialogical and sought to understand the contributions of the non-West to the modern scientific enterprise; it may be articulated as follows: How did non-Western civilizations develop modern knowledge and techniques that contributed to the development of modern science?

It was therefore in the very nature of these two historical questions of Needham's that the non-West and the West would be brought together for dialogue and comparison, giving rise to the inter-cultural study of science and civilization. While such inter-cultural appreciation did seem to be Needham's intent, he was well-aware of the prevailing tendency in the West to see human history through an exclusively European perspective:

> The problems of the world will never be solved so long as they are considered only from a European point of view. It is necessary to see Europe from the outside, to see European history, and European failure no less than European achievement, through the eyes of that larger part of humanity, the peoples of Asia and indeed also of Africa. (Needham 1969, p. 11).

Needham felt that the Western tendency to impose their civilization upon others could be avoided, or at least mitigated, if Westerners could appreciate the commonalities other civilizations shared with theirs. Hence, his appeal: 'Their (Europeans) clear need today is to find in the non-European civilizations manifestations of those values to which they themselves feel so deeply committed, not to attempt, as formerly they did, to mould all other cultures into their own image' (Needham 1969, p. 116).

Needham was hopeful that such similarities could be found in certain 'universals', which he proceeded to identify as 'modern science' and 'modern technology'. As he wrote:

> Which parts of modern 'European' world civilization are universal, and which parts are locally and parochially European? Once the question is rightly stated the answer is obvious. The real universal factors are modern science and modern technology, together with the philosophies which made them possible. And it should be clearly understood that Europe did not give rise to 'European' or 'Western' science, but to universally valid world science. (Needham 1969, p. 12)

Needham further added, 'No doubt the greatest effect of modern science on human civilization has been to unify the whole world's surface. Among the tasks of integration which humanity is now facing in the progress towards higher forms of order and union, none can well be thought larger than the confluence of the EuroAmerican and Chinese civilisations' (Needham 1969, p. 96).

One cannot help but empathize with Needham here. His belief in the 'universality' of modern science and technology appears to stem from none other than his desire to see humanity progress 'towards higher forms of order and union'; in other words, an integration of humanity based upon a 'universally valid world science', as opposed, presumably, to Western cultural imperialism. Nonetheless, regardless of Needham's well-meaning humanist aspirations, it appears that his belief in the 'universal' nature of modern science and technology was based more on an ideal than actually-existing reality.

After all, are modern science and modern technology and the philosophies that spawn them really 'universal', as he claims? By the same token, is it true that Europe did not give rise to 'European' or 'Western' science but to 'universally valid world science'? What accounts for the fact that a cultural entity such as 'Europe' can produce 'universally valid'—rather than simply European or Western—knowledge? When human subjects, replete with their idiosyncratic whims and fancies, produce knowledge from their particular and parochial vantage points, what entitles them to claim universal validity? What turns a limited point-of-view into a supposedly universal one, a cultural way of seeing into a god's-eye-view of the world?

While it is true that modern science has unified the whole world's surface, what is this really a consequence of? It may be said that modern science has indeed become a 'universally valid world science' rather than 'European' or 'Western' science, but how has this come about? What are the mechanisms that have brought about such pervasive knowledge uniformity? Has this occurred by way of an integration of various cultural knowledges, or through an imposition of an ostensibly superior cultural knowledge system over others?

As Needham appears to have neglected such questions, which pertain to the sociology and politics of knowledge, I will now address them in the following section by introducing the general notions of coloniality/decoloniality and the Colonial Matrix of Power, before elaborating how they are implicated when discussing knowledge and, specifically, 'modern science'.

3 Modernity/Coloniality and the Colonial Matrix of Power

The opposition to Western colonialism and its associated ideological justifications has a considerable and veritable history which would include among them, the voluminous writings of Joseph Needham. Yet, while this is the case, the 'decolonial' scholarly tradition traces its origins to the work of Peruvian sociologist, Anibal Quijano, and other Latin American scholars reflecting upon the consequences of the historic New World discoveries on our contemporary global realities. Indeed, it is because decolonial scholars trace the history of modernity by connecting it to the New World conquests that they recognize modernity's coevalness with coloniality, hence, the compound formulation, 'modernity/coloniality', suggesting the two to be different sides of the same coin.

The latter modernity/coloniality interpretation of history is in obvious contrast to the Eurocentric view which regards modernity as the apotheosis of civilization. Such a triumphalist perspective sees only the wonders of modernity but is blind to its exorbitant costs, which are invariably borne by its victims at the periphery of the system: non-Western Others, women, and nature. Assessing matters from the perspective of those written out of History, therefore, decolonial scholars are aware that there cannot be modernity without coloniality. The boon of 'modernity' that first took root in the West and which is commonly feted in Eurocentric historical accounts cannot and could not have happened without simultaneous colonial exploitation elsewhere. Hence, again, the expression 'modernity/coloniality' to capture the simultaneity of the two phenomena.

On this score, it is worth mentioning that Quijano coined the term 'coloniality'—or the 'coloniality of power'—to explain the persistence of Western colonial domination despite the occurrence of formal decolonization across the world in the post-World War II era. Indeed, it seems reasonable to be asking why *de jure* decolonization and the emergence of ostensibly politically independent nation-states have not put an end to colonial relations of domination and exploitation.

It is here that Quijano's concepts of 'coloniality' and the 'colonial matrix of power' are relevant, offering an account that seems to distinguish the decolonial tradition from other strands of anti-colonial thinking. In the first place, whereas colonization is a historical phenomenon characterized by the explicit subjugation of weaker by stronger polities through overt military or political occupation, 'coloniality' refers to the logic of domination that animates such actions. Accordingly, whereas classical colonization has historically been predicated on direct and explicit military and political control, 'coloniality' has obviated the need for such forms of control, allowing colonial exploitation to continue without the direct rule of earlier times. It may correctly be said, therefore, that the occurrence of coloniality does not even need colonization even if the latter was once its most common expression. Clearly, this also explains why the state-centred decolonization movements in the aftermath of World War II have been inadequate, howsoever necessary they may have been to counteract colonialism. Indeed, the persistence of coloniality underscores the evident inadequacies of state-led processes of decolonization and explains why decolonial scholars consider 'decoloniality' to be the imperative of our times. Just as colonization necessitated decolonization, the persistence of coloniality today demands decoloniality as a fitting response for our times, involving an effort to de-link from the exploitative logics that have given us the modern world.

Having introduced 'coloniality' and 'decoloniality', we now extend our discussion to the related concept of the 'colonial matrix of power' (henceforth, CMP). As noted, because coloniality is essentially a *logos*—the logic undergirding all acts of colonial violence and expropriation—it is formless, capable of being expressed and applied in several important domains of life. Consistent with this, Quijano proposes that the maintenance of Western colonial power across the globe for over the past five centuries has been made possible because the West has controlled a matrix comprising several important elements that make up our social reality; hence, his formulation of the CMP.

The CMP contains four key institutional components which the West has had longstanding dominance; they include control over: (i) technology and the means of institutional violence, (ii) the global political and economic systems and their major institutions (e.g. the United Nations, International Monetary Fund, World Bank, World Economic

Forum) and terms on which they operate, including the control of money and finance, (iii) the nature of social forms and modes of sociation with nature and Others (e.g. the 'individual' person as the basic social unit; the human as 'master' of nature), and last, but most crucially, (iv) knowledge and human subjectivity.

Although the geopolitical parameters are today quickly shifting with the global re-emergence of China, the regulation of the CMP by the West for much of the past five centuries has safeguarded Western hegemony throughout that time. This has also ensured that the seed of coloniality was sown and perpetuated as the dominant cultural logic of the global system, hence, making the apparent paradox of 'colonization without colonies'—or 'colonization after de jure decolonization'—possible. Now that the logos of coloniality has become the dominant operating system, colonization that takes the form of military occupation or colonial governance has largely become redundant: the 'natives' can very well do it (to) themselves. While we have discussed classical colonization as an explicit and overt form of subjugation, of the space 'outside' and 'exterior' to us—the latter alludes to a form of colonization that is unseen and, accordingly, inconspicuous.

The colonization which we speak of is as such because it takes place at another level of our reality, the 'interior' domain. We refer specifically to mental, psychological, and knowledge colonization—the colonization of our 'inner space', our consciousness, our 'interiority'—which results in the control of human subjectivity. Such is an insidious form of colonization that is as powerful as it is hidden from view, for the colonization of our minds is the control of what 'we know' and perhaps even what 'we can know'. Indeed, because the consequences at stake are profound, the control of knowledge and human subjectivity is arguably the most pernicious form of coloniality in the CMP. It is here that our discussion of modern knowledge—modern science, in particular—is situated. It is to this topic of modern science and its relationship to coloniality—specifically, epistemic or knowledge coloniality—that we now turn.

4 Modern Science and Knowledge Coloniality

What does modern science have to do with the colonization of our consciousness? What are we to make of this seeming paradox, especially when 'science' (from the Latin, *scientia*, 'knowing') in its best interpretation, implicates the pursuit of truthful knowledge, whereas 'colonization' typically alludes to a diversion or detraction from it?

Decolonial thinking helps us to understand this apparent paradox by prompting us to examine the politics of knowledge, not least, the politics of science. Hence, instead of taking at face-value what proponents of science claim about it, our awareness of the machinations of the Colonial Matrix of Power impels us, at the least, to recognize the double-sided role of 'science' in modern society: its declared interest in the pursuit of 'truth' on the one hand and its somewhat more dubious 'political' interest of invoking 'truth' for the perpetuation of power on the other. The latter motivation, is, of course, never admitted.

The political and institutional element of modern science resides exactly in the fact that scientific claims about 'truth'—pretensions to 'truth'—are invariably political claims. 'Truth' is power because of what it represents: it is the very antithesis of 'falsehood', of 'erroneous' belief, of what is 'untrue'. Yet because 'truth' confers power, claims about 'truth' can be deployed simply for their political returns, regardless of their actual content. This raises the following questions: When do the institutional-political considerations of the project of modern science override the latter's epistemic concerns; when does the 'politics' of modern science hold sway over its 'epistemic' properties?

We noted above that the West has historically sustained and perpetuated coloniality by way of its control of the CMP. Indeed, the deployment of modern science for political outcomes verily exemplifies this. Moreover, I would submit that such a political invocation of science is not incidental but is, instead, a systemic function of modern science in its institutional role as gatekeeper of knowledge. That is not to imply that modern science is devoid of 'truth' content, or that it is lacking such aspirations; nor am I saying that it is wholly determined by political imperatives. Rather, I am simply highlighting the critical historical awareness brought upon by decolonial thinking, which is that institutional knowledge—as represented by Scientific Reason in the modern era—has been politicized to perpetuate empire.

After all, the control of knowledge (and by extension, knowledge coloniality) resides not so much in the production of 'truths' as such, but in the power to establish the *criteria* for what counts as 'truth'. The important difference, here, being between knowledge per se and the rules which govern knowledge. Indeed, knowledge coloniality ultimately entails the setting of rules, the creation of categories, and the institution of methods to govern what counts as 'legitimate', and then the imposition of such standards to regulate the flow and movement of knowledge. Consequently, when we speak of modern science as the predominant form of institutional knowledge in the modern era, what do we mean? What are its rules and its orientations? And who were its rule-makers?

It has become clear that Europe's early philosophers played a vital constructive role on this front, laying the foundations of modern science and establishing the bases by which all other (cultural) knowledges were to be compared and measured. Space considerations here limit our examination to just the examples of Descartes (1595–1650) and Kant (1724–1804), both of whom are widely considered to be the giants of modern philosophy.

Descartes is regarded as the father of modern rationalism which deeply influenced the development of modern science. He sought to put science on a sure footing and purge it of probability, uncertainty, metaphysics, and ultimately, hermeneutics. Descartes came to be convinced of the following: that true knowledge was necessary; mathematical knowledge alone was necessary; hence all knowledge had to be mathematical. Thus, he wrote: 'Those who seek the direct road to truth should not bother with any object of which they cannot have a certainty equal to the demonstration of arithmetic and geometry'. Accordingly, Descartes insisted that 'we should never allow ourselves to be persuaded excepting by the evidence of our Reason'. The second of his *Rules for the Direction of the Mind* states, 'We reject all such merely probable knowledge and make it a rule to trust only what is completely known and incapable of being doubted'.

Moreover, the methodological path on which Descartes embarked, Kant (1724–1804) would later seek to continue upon. But as Gilson (1937, p. 227) notes, 'By substituting empirical observation for abstract definitions as the first stage of philosophical knowledge, Kant was not shifting from mathematics to philosophy but from mathematics to physics. As Kant himself immediately concluded, 'The true method of metaphysics is fundamentally the same as that which Newton has introduced into natural science, and which has yielded such fruitful results'. With these simple words Kant marked out the line beyond which lies the waste land where no metaphysics can live. A new standard based on Newtonian science had been appointed the supreme judge of philosophy. But metaphysics is no more capable of physical than of mathematical demonstration... the verdict of its new judge was obliged to be a condemnation'.

Like Descartes, Kant felt that 'the time had come when men could no longer feel interested in any discipline for the sublimity of its ambitions, but only for the soundness of its demonstrations' (Gilson 1937, p. 225). Kant concludes, 'Our age is, in every sense of the word, the age of criticism, and everything must submit to it'.

The above excerpts briefly account for the foundations of modern science with reference to two of its most feted pioneers, Descartes and Kant. Both assume that the pronouncements of science are value-free, as if boomed from above, an absolute vantage point from which certitudes can be gleaned. The modern posture of being 'value-free' is thus equivalent to claiming a god's-eye-view of the world, albeit in the secularized lexicon of science. This god's-eye-view is what decolonial scholar, Santiago Castro-Gomez (2021), has referred to as 'the hubris of the zero-point', whereby knowledge is constructed from the perspective of an observer 'who observes without being observed', thereby eliciting a 'view from nowhere'.

The methodological orientation of being 'value-free'—of appearing to be disinterested despite being fully invested—was, indeed, what Descartes's and Kant's scientific method quite explicitly pursued. It led Descartes to the famous doctrine that 'our clear and distinct concepts are, as many 'simple natures', each endowed with a definite essence and wholly independent from the minds in which they dwell' (Gilson 1937,

p. 146). To be sure, then, Descartes and later, Kant, were pioneering the modus operandi of knowledge coloniality. By prescribing scientific methodologies which ostensibly purged values and uncertainty in human knowledge, they were alluding to the belief that the resulting knowledge was not idiosyncratic, subjective, or culture-bound, but universal.

It was exactly his pursuit of the certitudes of scientific knowledge and his aversion to uncertainty that motivated Descartes to regard mathematics to be first among sciences. Hence, his arrival at the following conclusion: 'that in our search for the direct road towards truth, we should busy ourselves with no object about which we cannot attain a certitude equal to that of the demonstrations of arithmetic and geometry' (quoted in Gilson 1937, p. 132). Descartes' scientific methodology, therefore, led to a kind of mathematization that dismissed all realities which were not demonstrable with mathematical precision. Descartes limits his interest to knowledge that is precise and certain beyond any possibility of doubt because his primary interest was colonial in nature, namely, that we should become the 'masters and possessors of nature', which Bacon, before him, had already encouraged.

Hence, writes Maritain: 'The mathematical knowledge of nature, for Descartes, is not what it is in reality, a certain interpretation of phenomena... This knowledge is, for him, the revelation of the very essence of things. These are analysed exhaustively by geometric extension and local movement. The whole of physics, that is, the whole of the philosophy of nature, is nothing but geometry... Thus, Cartesian evidence goes straight to mechanism. It mechanises nature' (cited in Schumacher 1990, p. 18).

Schumacher observes that Descartes had broken with tradition and made a clean sweep, undertaking to start afresh, which was a kind of arrogance that became the 'style'—the ego-logy—of European philosophy. As Maritain remarks, 'Every modern philosopher is a Cartesian in the sense that he looks upon himself as starting off in the absolute, and as having the mission of bringing men a new conception of the world'. In fact, Francis Bacon (1561–1626) had already made a similar plea in *The New Atlantis* (1626).

In contrast to Descartes, Kant not only considered Newton's method the only valid method but took it for granted that the real world was exactly in accordance with Newton's description of it, prompting Gilson

(1937, pp. 229–230) to write, '*The Critique of Pure Reason* is a masterly description of what the structure of the human mind should be, in order to account for the existence of a Newtonian conception of nature, and assuming that conception to be true to reality. Nothing can show more clearly the essential weakness of physicism as a philosophical method'.

5 Modern vs Traditional Science: Scientific Materialism vs Levels of Being

Despite their respective differences, Bacon, Descartes, and Kant had circumvented their scientific concerns to focus strictly on a reductive material, external, and quantifiable reality. Consequently, 'modern science', which has developed over the course of recent centuries, is a study of only the materialist world and operationalized by reductionist methods. Such reductive materialism fails to recognize the concept of different Levels of Being hierarchically stratifying the world, beginning with minerals, plants, animals, as recognized by EF Schumacher (1990). Corresponding with every progression to a 'higher' level of Being, there accrues a qualitative gain, and vice versa. Hence, on the level of minerals, there is no life, but as we move 'up' to the level of plants, there is the addition of the quality of life. Similarly, if we move from plants to the category of animals, we notice the latter having the additional qualities of movement and awareness.

It is worthwhile noting that even though the limited—materialist—emphasis of modern science has deprived it of deeper meaning, the latter has enjoyed pervasive global influence, for scientific materialism has yielded tremendous practical applications that have given rise to the radical transformations of the natural world. They have helped spawn the conveniences of modern-day existence, yet at the same time, their radical alteration of nature are now becoming deleterious and posing a danger to all planetary life.

From a decolonial perspective, modernity has its roots in and is sustained by coloniality, which as described by the CMP, has historically manifested by way of the West's control of the world's foremost political, economic, social, and knowledge institutions. Of its multiple expressions, we have submitted that knowledge coloniality is the most pernicious since it determines what one is 'permitted' to know. Hence, we have argued that apart from having an epistemic function centred on a 'truth' (empirical) concern, modern science also performs a second—arguably, more important—institutional function which is political. It duly performs this political—and colonial—function by playing its role as a gatekeeper of knowledge, disqualifying those knowledges that do not meet its ostensibly scientific criteria. In other words, as modernity's foremost knowledge institution, modern science is bestowed with the power to determine the standards of what counts as 'legitimate'. It is in this self-conferred institutional power to disqualify, de-legitimize, and subjugate other ways of knowing and being that modern science is implicated in knowledge coloniality.

We have seen how such arbitrary political determinations were made in the case of both Descartes and Kant, which disclose how coloniality operated in the realm of knowledge with the emergence of modern science in the West. It is this reductionist materialism that ignores knowledges, recognized in other cultures, of the distinctive structure of complex systems acknowledged at the biological, animal, and human levels by Schumacher among others. In contrast to the reductionist materialism of modern science, the traditional understanding was grounded in a qualitative, hierarchical distinction between the 'lower' and 'higher' aspects of reality, 'lower' and 'higher' Levels of Being.

For instance, Merchant (1983), who adheres to an ecofeminist position, notes that the scientific worldview supplanted the erstwhile organic cosmology that viewed the earth as 'living' and 'female'. She writes, 'Between the sixteenth and seventeenth centuries the image of an organic cosmos with a living female earth at its center gave way to a mechanistic world view in which nature was reconstructed as dead and passive, to be dominated and controlled by humans' (Merchant 1983, p. xvi). She adds: 'The female earth was central to the organic cosmology that was undermined by the Scientific Revolution and the rise of a market-oriented culture in early modern Europe' (Merchant 1983, p. xx).

It is instructive to observe coloniality being implicated on two counts here. First, it is implicated in the mechanistic and materialist view of reality that regards nature as a 'thing' to be brought under human control and dominion. Coloniality is built into such a view of reality in the way the latter 'objectifies' living nature, conceiving of the earth as 'dead' and 'passive' to render it controllable and exploitable for human benefit (Merchant 1983, p. xvi).

Coloniality is implicated a second time by the fact that such a materialist worldview and the variant of science it yields—scientific reductionism—has become predominant while disqualifying other cosmologies and other civilizational ways of knowing and being.

Hence, the materialism of modern science purged the vertical dimension from Western conceptions of reality. This began in the West, then by way of the coloniality of knowledge and consciousness, spread throughout the world. Unsurprisingly, this was a process which coincided with the universalization of modern science.

Consequently, while modern science is now ubiquitous, neither it nor its underlying philosophy of reductive materialism can be considered 'universal' in an immutable and historical sense. Rather, the foregoing discussion suggests that the emergence of modern science in the West occurred by virtue of socio-cultural and economic developments which were historically particular to the West. Indeed, Needham recognizes that the environment of Europe was conducive for the development of modern science, which marked a contrast with China:

> If modern science originated and developed wholly in the West, it was due very largely to the existence of favourable social and economic conditions there, conditions which did not exist in China. Conditions in China were, indeed, definitely inhibitory to the growth of modern science and its associated technologies. (Needham 1969, p. 96)

Secondly, we learn from the above discussion that modern science had emerged by way of a repudiation of the traditional ways of knowing and being, first in the West, then gradually spreading elsewhere via coloniality. To be sure, this development was tantamount to knowledge coloniality, involving the materialist ontology of modern science replacing the hierarchical distinction between lower and higher levels of reality (physical object, biological life, animal awareness, human consciousness). This resulted in a move towards the progressive reduction of the world to one level as the qualitative was displaced by the quantitative, wisdom by rationality, and the profound by the superficial, eventuating in the generalized loss of meaning encountered in modern existence. Science in the modern epoch thus ceased to be knowledge as an end-in-itself, as practical applications flourished to become the *raison d'etre* for science itself.

Be that as it may, it seems reasonable to argue that 'modern science' and 'modern technology' are not the 'universal' factors as posited by Needham. Certainly, these phenomena did not emerge out of a cultural vacuum, akin to a *creatio ex nihilo*, but were rather outcomes of a surreptitious cultural and historical transformation spawned by coloniality; knowledge coloniality, to be sure. And as noted, this first occurred in the West, with the negation of its own traditional sciences before it was perpetuated elsewhere. Therefore, if such a development were 'properly'—by which I mean, historically—accounted for, then it can be said that what is called 'modern science' and 'modern technology' were never 'universal' phenomena in the first place, but the result of a particular historical and cultural deviation from what was in fact common across the traditions of the world's civilizations.

Accordingly, what was in fact 'universal' was not modern science or technology as such, but the traditional civilizational wisdom which recognized the different Levels of Being stratifying and distinguishing between the various forms of non-life and life on our planet. It seems fair to say that 'modern science' and 'modern technology' only became 'universal' because of coloniality, which turned the unique cultural and historical experiences of the West into the fate of the rest of humankind.

The foregoing also suggests that modern science is but only one variant of science and, arguably, an 'inferior' one at that. It is confined only to the study of the lowest order of reality, the non-life of things, matter which can be seen, touched, and manipulated. Yet the consequence of such reductive scientific materialism was to dismiss as 'unreal' the realm that is definitive of our humanity, relegating it to the category of 'epiphenomena'. Since humanity—the quality of being human—adheres to an intangible symbolic order, dismissing the non-reductive aspects of existence as 'unreal' is therefore tantamount to ignoring large swathes—and certainly, the most significant and meaningful aspects—of human experience.

And while the commitment of modern science to an epistemology and ontology of materialism has undoubtedly yielded modern technologies that appear to bring 'nature' under human control, this is likely to be an illusion emerging from hubris. It is important to realize and fully account for the actual costs that the reductive scientific materialism of modern science has inflicted on planetary life. After all, the dangers of reducing life—human as well as nonhuman—to the status of 'objects' to render them more exploitable have become all too obvious with the drastic ecological and social crises of our times.

6 Postface: Decoloniality Extending Needham's Vision of Science

I hope that by adopting a decolonial perspective, I have drawn attention to the inadvertent blind-spots in Needham's famous questions and made a convincing case for the decolonial need to extend his vision of science. The discussion that runs through this chapter was initiated by Needham's statement on the 'universality' of 'modern science' and 'modern technology'. Are modern science and modern technology in fact universal, as Needham claims? Such is the line of questioning that prompted my efforts here to reveal the relationship of these modern institutions to coloniality.

The West has in the modern epoch sought to maintain its colonial hegemony over the modern world-system by regulating the Colonial Matrix of Power, of which the control of knowledge and human subjectivity is vital and, arguably, most important. I have referred to coloniality in the realm of knowledge as 'knowledge coloniality'.

Historically, the West has been implicated in knowledge coloniality by holding out and exalting modern science as the pre-eminent form of knowledge. In so doing, it has simultaneously deflated, de-legitimized, and disqualified the traditional, pre-modern knowledges of other civilizations, beginning with its own.

Notwithstanding its claims of 'objectivity', then, Western knowledge coloniality resides in the modern scientific pretension that its ways of seeing and knowing are transcendent of culture, as if emanating from a 'zero-point', which elevates it to the status of a reified, higher 'truth'. Indeed, it seems that it is only by uncritically accepting the validity of such a vantage point that claims about the 'universality' of modern science can be sustained.

In view of the foregoing, perhaps the answer to Needham's question—of why other civilizations did not develop modern science before the West—rests with the fact that their traditional wisdoms could not fathom the human devolution and degeneration that was to follow if they did. Perhaps the custodians of those civilizations knew better. Perhaps it is for that reason that Guenon writes of 'modern science' being 'constructed from the ruins of ancient sciences with the materials that had been rejected and left to the ignorant and the 'profane'' (Guenon 2009, p. 23).

What to do? The view from a decolonial perspective is clear enough: it is to epistemologically de-link from the CMP and to re-connect inter-culturally by resuscitating, reviving, and restoring all those traditional civilizational knowledges that have been subjugated and submerged by knowledge coloniality. This would involve restoring to science the cosmological integrity that characterized the traditional sciences, which would be a veritably fitting testimony to the inter-cultural and inter-civilizational study of science that Needham so avidly championed in his lifetime. Needham seemed to have already alluded to the appropriateness of such a move with the following observation about Chinese science:

'In general, we may say that Chinese scientific humanism, though all its achievements were made before there was any modern science in China, rested on two main bases: it never separated man from nature, and never thought of man apart from social man' (Needham 1969, p. 94).

References

Castro-Gomez, Santiago. 2021. *Zero-Point Hubris: Science, Race, and Enlightenment in Eighteenth-Century Latin America*. Lanham: Rowman & Littlefield.

Gilson, Etienne. 1937. *The Unity of Philosophical Experience*. New York: Charles Scribner's Sons.

Guenon, Rene. 2009. *The Essential Rene Guenon: Metaphysics, Tradition, and the Crisis of Modernity*. Bloomington: World Wisdom.

Merchant, Carolyn. 1983. *The Death of Nature*. New York: Harper and Row.

Mignolo, Walter. 2011. *The Darker Side of Western Modernity: Global Futures, Decolonial Options*. Durham: Duke University Press.

Needham, Joseph. 1969. *Within the Four Seas: The Dialogue Between East and West*. Routledge.

Quijano, Anibal. 2010. Coloniality and Modernity/Rationality. In *Globalization and the Decolonial Option*, ed. Walter Mignolo and Arturo Escobar. New York: Routledge.

Schumacher, E.F. 1990. *A Guide for the Perplexed*. New York: Abacus.

Part II

Cosmologies in Dialogue

5

The Circulation of Babylonian Astral Science

John Steele

1 Introduction

More than 5,000 cuneiform tablets recovered from sites in Iraq and its neighbours contain texts attesting to the practice of the astral sciences in ancient Mesopotamia. These texts show that during the first millennium BCE, Babylonian and Assyrian scholars made astronomical observations, developed and used methods to predict future astronomical phenomena, sought ways to describe the celestial world, and interpreted celestial phenomena through a variety of astrological methods. They also provide a glimpse—but not much more—into earlier Mesopotamian traditions of the astral sciences through the continued copying, use, and re-interpretation of earlier texts.[1]

[1] For a recent overview of Babylonian astronomy and astrology, see Steele (2018).

J. Steele (✉)
Department of Egyptology and Assyriology, Brown University, Providence, RI, USA
e-mail: john_steele@brown.edu

A. Bala et al. (eds.), *Multicivilizational Exchanges in the Making of Modern Science*,
https://doi.org/10.1007/978-981-97-3541-9_5

As one of the earliest and most detailed ancient sciences, Babylonian astronomy has sometimes been seen as the origin of science (see, e.g., the titles of van der Waerden's *Science Awakening: The Birth of Astronomy*[2] and Price's *Science since Babylon*[3]), and scholars have identified traces of Babylonian material in the astral sciences of a variety of different ancient cultures, including ancient Greek, Jewish, Egyptian, Indian, and even Chinese, although not all such claims have stood up to close scrutiny. Nevertheless, the importance of Babylonian astral science for discussions of the history of science and, in particular with regard to this volume, for discussions of scientific dialogues between cultures, is clear.

For our purposes, ancient Mesopotamia—the fertile area of land between and on the banks of the rivers Tigris and Euphrates and one of the areas where the first complex societies arose—may be divided into two regions. In the south lies Babylonia, stretching from somewhat north of Baghdad to the Persian Gulf. It was in this region that the first cities appear, where the cuneiform writing system developed, and where what we can think of as 'Mesopotamian society and culture' originated in the fourth and third millennia BCE. In the early second millennium BCE, kings from Babylon first unified Mesopotamia, forming the Old Babylonian Empire. By the first millennium BCE, Babylonia had mostly lost its political power: except for a period of about one hundred years between the late seventh and the late sixth century BCE, when it briefly rose again to dominate western Asia, Babylonia was for the most part subsumed into other empires: the Assyrian empire in the first part of the millennium, then coming under Achaemenid Persian, Greek, and finally Parthian rule following the Neo-Babylonian renaissance of the seventh and sixth century BC. Babylonia's northern neighbour, Assyria, centred in northern Iraq and southern Syria, was militarily and politically dominant in the late second and early first millennium, with an empire that stretched to the Mediterranean coast and, at times, reaching as far as Egypt. Babylonia and Assyria share a common language (Akkadian, a semitic language), albeit with different dialects, a writing system (cuneiform on clay tablets), and, in many respects, a culture. Indeed,

2 Van der Wearden (1974).

3 Price (1961).

in many ways, Assyria saw Babylonia as its cultural forebearer, and Babylonian scholarship remained extremely active throughout the first millennium BCE, despite the many political changes in the region.

My aim in this chapter is to explore the circulation of Babylonian astral knowledge across the ancient world, not simply to catalogue 'Babylonian elements' which show up in other traditions,[4] but rather to ask *why* certain parts of Babylonian astral science were embraced within another culture and others were not, and *how* they were integrated within the existing knowledge systems and practices of these cultures. Before doing so, however, it will be useful to make some brief remarks on the circulation of knowledge, with particular reference to what the editors of this volume have called 'Needham's Dialogical Question'.

2 Some Remarks on the Circulation of Scientific Knowledge

Needham's Dialogical Question—a question which, to my knowledge, was never directly posed or formalised by Needham himself—can be boiled down to asking did and, if so, how and in which ways did non-European science contribute to the development of modern science in Europe? Like the more well-known 'Needham Question' of why did modern science develop in Europe and why did it not develop in China (or elsewhere), the so-called Dialogical Question in this basic form can be criticised on various fronts.[5] For example, one could ask whether there is a single entity which we can call 'modern science' and whether we can assign its origin to a single place and time. And even if we can, should we define our history of science as looking for the roots of modern science? Doing so inevitably leads one to focus on those parts of earlier science which look (at least superficially) most like what scientists do today. As Rochberg has pointed out, we will tell a very different story of the

[4] For useful summaries of the evidence for Babylonian astral science in other cultures, see Pingree (1998), Brown (2018), and Rochberg (2020).

[5] For critiques and discussions of the 'Needham Question', see, for example, Sivin (1982), Hart (1999), Brennan and Lo (2016), Bala (2016), Shah (2016).

history of science when we construct that history forwards rather than backwards.[6]

At its most basic level, the 'Dialogical Question' is trivial: it is abundantly clear from many known cases that scientific knowledge has circulated widely throughout history. Let me give a couple of examples: (1) Greek papyri from the early first millennium AD contain planetary and lunar positions that were computed using Babylonian systems of mathematical astronomy. A few papyri and ostraca even contain written procedures for calculating using these Babylonian systems, which are more or less identical to Babylonian texts of the first millennium BC. These texts show that Babylonian mathematical astronomy was widely known in the Greek world and, crucially, that the Greek astronomers understood the Babylonian material sufficiently well to adapt it to work with local Greek calendars.[7] Many parameters of Babylonian lunar and planetary systems are also incorporated within texts such as Geminus's *Phenomena* and Ptolemy's *Almagest*. Indeed, one of these parameters, the mean length of the synodic month which lies at the heart of the Babylonian lunar System B is used by Hipparchus and Ptolemy, is embedded within the Jewish calendar and in a medieval European Book of Hours, and is even cited by Copernicus.[8]

(2) A number of technical developments made by medieval Islamic astronomers—in particular al-Ṭūsī and Ibn al-Shāṭir—were clearly known to Regiomontanus, Copernicus, and other astronomers of the European Renaissance. These developments became part of the toolbox of astronomy and can be found, for example, in Copernicus' lunar model.[9] In both of these cases, there can be no room for doubt that scientific knowledge passed from one culture to another. Indeed, at least in some cases, this was openly acknowledged by the participants at the time: the ancient Greek astronomer Geminus, for example, attributes various astronomical ideas and developments to the Babylonians[10]; Ptolemy states that some of the ancient observations he used were made in

[6] Rochberg (2015).

[7] Jones (1998, 2002).

[8] Kugler (1900), Aaboe (1955), Neugebauer (1989), Britton (2002).

[9] See, for example, Swerdlow and Neugebauer (1984), Saliba (2007).

[10] Steele (2020).

Babylon[11]; and Copernicus refers to the work of the Arabic astronomer al-Battānī and uses some of his observations. The more interesting way to read the 'Dialogical Question' is to think about *how* and *why* knowledge circulated, *who* was involved in this circulation, and *what* was done with the knowledge in the receiving culture.

Extensive work on the transmission of astronomy between cultures was undertaken by scholars during the twentieth century and many examples of the transmission of astronomical methods and parameters have been carefully documented. The identification of cases of transmission relies on detailed study of the evidence using careful methodology. In particular, it is insufficient to claim transmission simply based upon the prior development of an idea or method in one culture and an apparently similar one found in a later culture. For example, the identification of the equivalence of 19 solar years with 235 lunar months and its application in governing intercalation in a luni-solar calendar by the Babylonians in the fifth century BC and its use in Chinese calendrical systems from the first century BC onwards (and possibly earlier) does not imply transmission of Babylonian knowledge to China. We must allow for the possibility—indeed the likelihood in this case—of independent discovery. Nevertheless, many unreliable claims of transmission based only upon either or both the superficial similarity of methods or developments and/or scientifically trivial discoveries that could easily have been made independently appear in the literature, including in the work of Needham.[12]

In order to identify transmission it is necessary to show two things: a plausible historical circumstance in which the transmission could have taken place and the occurrence of identical and non-trivial parameters, methods, or other material in the two cultures which cannot be otherwise explained as independent discoveries. From a methodological perspective, the best markers of transmission are usually precise numerical parameters (ideally, ones that are slightly inaccurate) which are unlikely to have been identified independently. The Babylonian value

[11] Steele (2000a, pp. 93–100; 2004).

[12] For detailed critiques of several unfounded but widely repeated claims (including by Needham) concerning the supposed transmission of Babylonian astronomy and astrology to China, see Pankenier (2014) and Steele (2013, 2022).

for the length of the synodic month described above is a good example. It can be written in the sexagesimal system as 29;31,50,8,20 days,[13] which is precise in the last digit to $10/60^4$ or 1/1296000 days or (in modern terms) 1/150 of a second of time. It is highly unlikely that the exact same value would be obtained independently from two sets of observational data by two different (groups of) astronomers. Indeed, Ptolemy in *Almagest* IV.2 says that Hipparchus obtained this value from eclipse observations, but Ptolemy has to fudge the mathematics in order to obtain this value. Ptolemy says that Hipparchus found that 4267 synodic months equals 126,007 days plus one hour, divides the one by the other and finds 29;31,50,8,20 days for the mean synodic month. As Copernicus pointed out in *De revolutionibus* IV.4, however, the division actually yields 29;31,50,8,9,20 days. There can be little doubt, therefore, that this claimed Greek determination of the value is no more than a bodged after-the-fact confirmation of the Babylonian value. Aside from precise numerical parameters, other markers of transmission can include complete systems of calculation, reports of observations, star and constellation names, and certain technical developments such as systems of positional measurement.[14] It is worth noting that many claims for the transmission of knowledge made by ancient authors themselves are unreliable. In particular, it became a trope among Greek and Latin authors to attribute many things to the Egyptians and/or the Chaldeans (i.e. the Babylonians) when they are clearly Greek in origin.[15]

Thanks to the work of Neugebauer, Pingree, and others a clear methodology has been established for identifying the transmission of astronomical knowledge and many cases have been documented. As

---

[13] The sexagesimal number system, developed by the Babylonians and used by astronomers in the west up until today, is a base 60 system. I transcribe by using commas between sexagemical digits; a semicolon separates integers from fractions. Thus 29;31,50,8,20 $= 29 + 31 \times 60^{-1} + 50 \times 60^{-2} + 8 \times 60^{-3} + 20 \times 60^{-4}$.

[14] For further comments on the methodology for identify transmission, see Steele (2024).

[15] Sometimes, such ancient claims are merely mixed up. For example, in the *Tetrabiblos* Ptolemy describes three systems for the astrological concept known as the Terms in which each sign of the zodiac is divided into five unequal length parts each of which is associated with a planet. He attributes these schemes to the 'Egyptians', the 'Chaldeans', and 'an ancient manuscript'. However, his Egyptian system is very similar to the system known in texts from Babylonia whereas his Chaldean system is very different. See Jones and Steele (2011).

pointed out some years ago by Sabra, however, the study of transmission alone is insufficient to understand the movement of scientific knowledge as a historical process.[16] Instead, Sabra argues, we need to consider what he terms the 'appropriation' of knowledge by one culture from another and its subsequent 'naturalisation' within this new culture. Sabra's key insight is that it is the receiving culture which has the agency in the transmission of knowledge. To use his example, it was the medieval Islamic scholars who sought out, translated, and adapted Greek science, the Greeks did not impose this science on them (which should be obvious on simple chronological grounds, but the idea remains implicit in many discussions), nor did the knowledge itself somehow push its way forward.[17] Once we understand this fact, we see that it is the receiving culture—or at least individuals or groups within it—that chooses which knowledge, which texts, ideas, and methods, to appropriate, selecting from a wider menu of available material. Receiving cultures do not, on the whole, scramble to preserve whatever crumbs they can find from the source culture but instead pick and choose what they will, discarding other material for a variety of reasons (however much *we* may wish that they had preserved it). Following this process of appropriation, Sabra points out that knowledge then becomes 'naturalized'—adapted to work within existing systems of knowledge and worldviews, incorporated within these systems, and often ceasing to exist as 'foreign' knowledge instead simply becoming part of the general body of scientific knowledge.

Sabra's groundbreaking approach to understanding the circulation of scientific knowledge outlined in his 1987 paper is, I think, not as well known outside of specialists of the early exact sciences as it should be (probably in part because he frames his discussion around the case of Greek and medieval Islamic science), but it has important consequences for how we think of the so-called dialogical question at the core of this volume.[18]

---

[16] Sabra (1987).

[17] Sabra (1987, p. 225) writes: 'Greek science was not thrust upon Muslim society any more than it was later upon Renaissance Europe'.

[18] See also the important discussions by Pingree (1992), Ragep (1996), and Brentjes et al. (2014).

The study of the circulation of knowledge, therefore, is the study of the transmission, selection, translation, adaptation, incorporation, and normalisation of knowledge. As we will see in the three case studies I present here, the receiving culture usually has access to a wide array of material from the source culture but selects what material to appropriate and then adapts that material to fit within existing systems of knowledge.

3 Overview of Babylonian Astral Science

The majority of preserved Babylonian astronomical and astrological texts are the output of ongoing astronomical practice: records of astronomical observations, collections of predicted astronomical phenomena, tables of calculated astronomical data, horoscopes, etc. Only a relatively small number of tablets contain what I term 'reference texts', i.e. texts which provide the necessary background to this practice: texts presenting procedures for calculating astronomical phenomena, lists of stars, collections of astrological omens, instructions for interpreting astrological data, etc. It should be noted that neither group of texts contain any reflections upon things like how observations are to be made, how methods of computation were derived, or the underlying 'theories' of the astronomy, nor do they deal with the relative merits of different methods, attribute methods to individuals, or the relationship between astronomy, astrology, and other disciplines. Questions about such matters can only be answered by inferences from the practice itself.

Babylonian astral science can be divided into several main areas:

(a) *Schematic astronomy*. Schematic astronomy aims to present a schematic and self-consistent description of the celestial world. It is founded on a simplified 360-day calendar made up of twelve 30-day months. The content of schematic astronomical texts includes ordered lists of stars which tie things like their date of first appearance to dates in the schematic calendar and simple numerical schemes for the duration of visibility of the moon after sunset or before sunrise on each day of the month, the variation of the length of day and night over the year, and the variation in the length of the shadow case by a gnomon. At the heart of the schematic astronomical tradition lies the text MUL.APIN, which was written sometime before the seventh century BC and was widely copied, read, and used throughout Mesopotamia until the end of cuneiform astronomy.[19]

(b) *Observational astronomy*. Regular and systematic observations of selected astronomical phenomena were made throughout the Late Babylonian period. These observations include the dates of passages of the moon past select reference stars (known as 'Normal Stars' in modern scholarship) and planets and the distance between the two, measurements of the time interval between sun and moon crossing the horizon at six specific moments each month (known as the 'Lunar Six'), the dates and positions of the synodic phenomena of the planets (first visibility, first station, acronychal rising, second station, last visibility), the dates and passages of the planets past the Normal Stars with measurements between the two, and lunar and solar eclipses. Irregular astronomical phenomena were only occasionally observed and reported. Detailed records of observations were recorded in texts known today as Astronomical Diaries, which were themselves used to compile Compilation Texts and Goal-Year Texts.[20]

(c) I *Goal-Year astronomy*. This term refers to the practice of predicting future astronomical phenomena by applying known-periodicities to previous occurrences of the same phenomena. For the planets, it is simply a matter of identifying a periodicity after which the phenomena will recur on the same day and at the same position. For example, after 59 years, both Saturn's synodic phenomena and the passages of the planet by the Normal Stars will occur on more or less the same date in the Babylonian calendar. Thus, one simply needs to look back to the Diaries 59 years earlier in order to know

[19] For an edition, translation, and study of MUL.APIN, see Hunger and Steele (2019). For the continued use of MUL.APIN in the late period, see Steele (2021).

[20] For a summary of these types of text and their relationship to one-another, see Hunger (1999).

when the phenomena will occur.[21] The prediction of the Lunar Six and of lunar and solar eclipses is slightly more complicated but in both cases it relies on observations made 223 months earlier.[22] The Goal-Year Texts contain collected observational data to be used in making predictions for a given year which are themselves recorded in Almanacs and Normal Star Almanacs.

(d) *Mathematical astronomy*. By the fourth century BC the Babylonian scholars had developed methods of astronomical prediction which relied upon arithmetical calculation without direct empirical input. Two basic methods were developed. In one, known as System A, the synodic arcs between successive phenomena of the same kind were determined by piecewise constant functions dependent upon the initial position in the zodiac.[23] In the other, known as System B, the synodic arcs were determined by linear zigzag functions with constant difference. The results of mathematical astronomy are found in texts known as Synodic Tables; procedure texts contain instructions for computing the Synodic Tables. Because of the mathematical complexity and astronomical knowledge embedded within Systems A and B, modern scholars have both praised them and often described them as lunar and planetary theories. However, Babylonian mathematical astronomy does not present a 'theory' of lunar or planetary motion. Rather, it is a system of computation.

(e) *Omen based astrology*. The traditional form of astrology in Babylonia was the omen. An Omen is a two-part statement linking an 'observable' phenomenon with a predicted outcome ('if X: Y').[24] Many texts containing celestial omens are known, including the long (ca. 70 tablet) omens series *Enūma Anu Enlil*, which contains omens

---

[21] See Steele (2011). In practice, some corrections need to be made to take into account intercalation and inaccuracies in the identified periods. For evidence of how the Babylonians made these corrections, see Gray and Steele (2008, 2009).

[22] For the lunar six methods, see Brack-Bernsen (1999, 2011). For the eclipse methods, see Steele (2000b) and Brack-Bernsen and Steele (2005).

[23] See Ossendrijver (2012) for a detailed discussion of these methods.

[24] I put 'observable' in quotation marks because some omens can never actually occur, though one could imagine them as possible phenomena. The conceptual and semantic relationship between the two parts of the omen is a complex issue, but does not concern us here. For a discussion of these issues, see Rochberg (2016).

drawn from general lunar phenomena, lunar eclipses, general solar phenomena, solar eclipses, metrological events, comets, meteors, and stellar phenomena, and planetary phenomena. In addition to the omen texts, a large number of commentaries are preserved which provide explanations of particular terminology found in the omens and reinterpretations of some of the omens.

(f) *Late astrology*. During the Late Babylonian period, many new forms of non-omen astrology were developed. These new forms include texts giving astrological associations between, for example, the signs of the zodiac and cities, parts of the human body, and medical ailments, various forms of medical astrology, and horoscopic astrology.

It is important to note that during the Late Babylonian period these various different types of astronomy coexisted and were practised by the same individuals. In particular, the development of mathematical astronomy did not replace goal-year methods of prediction.

4 Case Studies

The Greco-Roman World

Considerable and convincing evidence exists for the circulation of Babylonian astronomical knowledge to the Greco-Roman world. This circulated knowledge ranges from basic concepts, such as the system of twelve equal-length zodiacal signs, the names and identities of several constellations, and the degree as a unit of measurement to systems of mathematical astronomy, concepts from schematic astronomy, accounts of specific dated astronomical observations, and certain astronomical concepts and techniques. This circulation mostly took place during the Hellenistic period following Alexander's conquest of Babylonia, a time period in which there was ample opportunity for Babylonian scholars to travel west and for Greek speakers to visit Babylonia. Indeed, a considerable Greek community lived in the major Babylonian cities during this period.

The wide range of Babylonian astronomical knowledge which appears in Greek sources indicates that Greek scholars had some form of access to most aspects of Babylonian astronomy. This access must have involved both personal contact between (some) Greek and Babylonian scholars as well as through the circulation and translation of texts.[25] But given the amount of Babylonian material available, even if only in poor summary, the interesting questions are what did different (groups of) scholars take from this material, why did they choose this material rather than other material, and how did they use it?

Let me first consider mathematical astronomy. We can trace three separate ways in which Babylonian mathematical astronomy circulated in the Greco-Roman world: the extraction and use of parameters from the Babylonian systems, the extraction and use of individual astronomical functions from these systems, and the wholescale adoption of the systems for computing astronomical data. As is well known, Hipparchus (second cent. BC) used many of the core parameters from Babylonian lunar System B in his own lunar theory.[26] Hipparchus' lunar theory is, of course, geometrical, using epicycles and eccentric circles to model the moon's motion at any arbitrary moment rather than the Babylonian approach of using arithmetical functions to compute the moon's position at syzygy, with interpolation methods to determine positions not at syzygy. Thus, all Hipparchus takes from the Babylonian system are the parameters embedded within the system, not the system itself. Unfortunately, Hipparchus' own presentation of his lunar theory is lost and we must rely upon the detailed account of it given by Ptolemy (second cent. AD). Ptolemy gives no details of how Hipparchus obtained these parameters—whether Hipparchus himself extracted them from the

[25] The technical terminology used in Babylonian astronomical texts as well as the large amount of tacit information that is needed to understand their contents requires that a reader be guided through the Babylonian texts, at least at first. Late Greek readers may have encountered Babylonian astronomy and astrology through translations and summaries by earlier Greek scholars. See Steele (2004).

[26] Kugler (1900), Aaboe (1955).

Babylonian lunar system or whether they were already circulating in Greek at Hipparchus' time. Ptolemy does not indicate that the parameters are Babylonian. Indeed, he implies that they were derived by Hipparchus from observation, which is clearly not the case, and given that Hipparchus' work was still circulating when Ptolemy was writing, he is unlikely to have done this if Hipparchus had been explicit about the parameters' Babylonian origin. Thus, following and perhaps already at the time of Hipparchus these parameters had been fully naturalised into Greek astronomy: they were no longer either seen as Babylonian or tied to the Babylonian lunar System B.

Whereas Hipparchus only used parameters from lunar System B, Geminus (first cent. BC) in his *Phenomena*, an introduction to astronomy, described in detail the arithmetical function for daily lunar velocity (col. F*) of Babylonian lunar System B.[27] Unlike Hipparchus, Geminus says that the mean value of this function 'had been found by the Chaldeans' (i.e. the Babylonians), preserving therefore both the Babylonian function itself and its Babylonian origin. Nevertheless, the function is presented on its own and out of context, whereas in Babylonia it is always integrated within the full computational system.

Greek papyri from Roman Egypt dating from the second century AD onwards present a different story. Many papyri are known which contain astronomical data that was computed using systems of Babylonian mathematical astronomy.[28] These systems are used in the full form, identically to how they were used in Babylonia, with the exception that they have been adapted to work with the Greek calendar. In addition to the papyri containing computed data, a few papyri contain procedure texts, written in Greek, which explain how to compute. Again, these procedure texts fully mirror their Babylonian counterpoints. Neither the procedure texts nor the texts containing computed data state that the systems of computation are Babylonian. Given their archaeological context and links with other papyri, in particular with papyri containing horoscopes, it seems very likely that these texts were written by practising astrologers in provincial towns. Thus, we have at least three distinct

27 Evans and Berggren (2006), Steele (2020).

28 Jones (1998, 2002).

contexts in which we can detect Babylonian mathematical astronomy: first, high-level astronomers such as Hipparchus and Ptolemy using useful parameters from Babylonian mathematical astronomy within their own geometrical models, stripping these parameters from their computational context within the original Babylonian systems, and presenting them simply as pieces of data with no acknowledgement of their origin; second, Geminus, the author of an introductory work in astronomy, presenting a Babylonian function from one of these systems of mathematical astronomy, correctly identifying its origin but implying that it is still useful within the context of Greek astronomy; and third, practising astrologers who used the Babylonian systems—presumably because they were considered either simpler to calculate with or more accurate than the available Greek astronomical theories—without stating and probably without even knowing that these systems were Babylonian.

Looking outside of mathematical astronomy we see further contexts in which authors know about or choose to refer to other types of Babylonian astronomy. The general historian Diodorus, for example, presents an at times garbled account of 'Chaldean' astronomy which draws on the traditions of schematic astronomy and omen-based astrology, though sometimes mixing in elements of Egyptian astronomy.[29] But he says nothing about mathematical astronomy, which would almost certainly not have been understandable to his readers. Diodorus draws on those aspects of Babylonian astral science which were appropriate to his aim in writing his book.

India

Many claims for the circulation of different aspects of Babylonian astral science to India have been made, including celestial omens, aspects of schematic astronomy, and parts of mathematical astronomy.[30] In the following, I want to consider a single example in order to examine the ways in which the Babylonian material was adapted so that it worked

[29] Jones and Steele (2018).

[30] See, for example, Pingree (1973, 1987, 1998, 2007). For criticisms of some of these claims, see, for example, Falk (2000) and Brown (2018).

within existing frameworks of Indian astronomy. This material is found in the chapter 17 of the *Pañcasiddhāntikā* of Varāhamihira.[31]

The *Pañcasiddhāntikā* was written in the sixth century AD. Varāhamihira states that the work is based upon five *siddhāntas*—the Pauliśa, the Romaka, the Vāsiṣṭha, the Sūrya, and the Paitāmaha—although Neugebauer and Pingree point out that the work also draws upon other works.[32] It contains a number of procedures for the calculation of lunar, solar, and planetary positions and phenomena drawn from these *siddhāntas*; sometimes alternative procedures are given for a calculation using the methods of the different *siddhāntas*. Chapter 17 concerns the calculation of the motion and synodic phases of the five planets. As discussed by Neugebauer and Pingree and others, there are clear relations between some of this material and Babylonian systems of mathematical astronomy.

Let us consider the part of this chapter concerning Jupiter as an example. This section reads:

> For Jupiter subtract from the *ahargaṇa* 34 (days) and as many *nāḍis* and divide (the remainder) by 399; the quotient is (the number of) its risings. The (remaining) days are put down. Add to (these) days 1/9th (of a day for every) rising. Multiply the (number of) risings by 6^2 (= 36) and divide (the product) by 391; (the remainder) is called the *pada*. Add 18 to the remainder. Put down the mean and the true segments, in order, in two places. (Decide) from the difference between them. If the true is less, one should add it to the days; otherwise, if the mean segment (is less), subtract it. Until the *pada* (equals) 180, (Jupiter) is in the negative segment 1456; until (it increases by) 195, Jupiter is in the positive (segment) 1265 (or 1456 – 65 = 1391); for 16 Jupiter is in the third, negative segment 1486. Multiply by 5 and divide by 8 (or 83); it rises to the east in so many minutes (of arc). In the first segment of Jupiter they are 9½° of Virgo (= 159;30°) in the second segment half a circle (= 180°); (and in the third) 20½°. In 60 days (Jupiter) traverses 12°, in 40 (days) 4 (degrees), and in 24 (days) 2 (degrees); it moves retrograde 6° in 56 (days) and 6 (degrees) in 60 (days); in direct motion (it goes) 12 (degrees) in 80 (days);

[31] For an edition, translation, and commentary of the *Pañcasiddhāntikā*, see Neugebauer and Pingree (1970–1971).

[32] Neugebauer and Pingree (1970–1971, part I, p. 9).

9 (degrees) in 50 days; then it sets; staying (set it travels) 7 (degrees) in one month (i.e., 30 days); its accurate rising is on the last day of the month (i.e., on the 29th). Jupiter[33]

The aim of the procedure is to calculate the position of Jupiter on a given date. The procedure begins by calculating the number of complete synodic cycles which have elapsed since the calculational epoch. This is done by manipulating the *ahargaṇa* (the number of accumulated days since epoch): first a correction is made to account for the fact the first appearance of Jupiter nearest to the epoch, which will be the basis of the further calculation, took place several days after the epoch, and then by dividing by 399, which is taken to be the number of days in Jupiter's synodic cycle; the remainder will then be the number of days after first appearance on our calculation date. Next the procedure invokes the period relation that Jupiter will circle the zodiac exactly 36 times over the course of 391 synodic cycles. This period relation is precisely that which underlies all of the Jupiter systems in Babylonian mathematical astronomy. The next part of the procedure introduces further corrections in the calculation which are hard to understand. What is notable, however, is that these corrections are determined by the position of Jupiter in the zodiac, with the zodiac divided into three zones. The division of the zodiac into two or more zones is typical of System A schemes in Babylonian mathematical astronomy. Several Babylonian System A schemes are known: currently known schemes use two or four zones; a related system attested in a Greek papyrus uses six zones.[34] Finally, the procedure sets out a scheme for the number of days and degrees of longitude between subdivisions of the synodic arc. Jupiter's motion is modelled as a piecewise constant function tied to the synodic phenomena of first appearance, first station, second station, and last appearance, but with intermediary jumps in speed between some of these points. Similar (though not identical) piecewise constant schemes are known in Babylonian mathematical astronomy.

---

[33] *Pañcasiddhāntikā* 17 6–13; translation by Neugebauer and Pingree (1970–1971, part I, pp. 159–161).

[34] Britton and Jones (2000).

The appearance of the period relation 391 synodic cycles = 36 zodiacal revolutions, the splitting of the zodiac into a small number of zones, and the piecewise constant function for the planet's motion within its synodic cycle led Pingree to claim that the Indian system is 'a direct descendent of the Babylonian System A … though with relatively minor deviations'.[35] Brown is more circumspect, arguing that the Babylonian origin was not direct and that 'it is just as likely, however, that these constant-velocity-zone models … derive from Egyptian adaptations of Babylonian-style arithmetic'.[36]

Neither Brown nor Pingree, however, highlight the fact that irrespective of whether the basis of the scheme came directly from Babylonian material, was 'filtered through Greek intermediaries',[37] or derived from adaptations made in Egypt, it was subsequently adapted to work within the tradition and methods of Indian *siddhāntas*. In particular, the original Babylonian systems (and their counterparts known from Greco-Roman Egypt) do not use the concept of an epoch or start date where the planet's position is assumed to be known. Instead, all calculations are performed locally from the last occurrence of a phenomenon. Consider the following passage from a Babylonian procedure text for Jupiter:

> (First) appearance to (first) appearance. From 25 Gemini until 30 Scorpio you add 30. From 30 Scorpio until [25] Gemini you add 36.[38]

The procedure states that in order to calculate the position of Jupiter at its first appearance, we take its position at its previous first appearance and depending upon which of two zones of the zodiac it is located within we add either 30° or 36° to that position and to get the planet's position at its upcoming first appearance. This calculation is purely local: we are not moving forward many cycles from a date in the far past when it is assumed that Jupiter's position was known, as in the Indian system with the *ahargaṇa* count, but instead using *only* the previous occurrence of the phenomena. This previous occurrence is itself calculated from the

35 Pingree (2007, p. 47).

36 Brown (2018, p. 709).

37 Neugebauer and Pingree (1970–1971, p. 10).

38 BM 34,081 + Obv. I 8′–9′; translation adapted from Ossendrijver (2012, p. 249).

occurrence before that, and so on, but there is no starting point—no epoch—for the calculation. Transforming the Babylonian system into its Indian counterpart, therefore, required both an initial value for the planet's position at the assumed epoch—something that cannot have come from the Babylonians since it was not part of their computational method but which could have come from other Indian computational systems in the *siddhānta* tradition—and a sufficient understanding of the internal working of the Babylonian systems to make the adaptation.

The Book of Enoch and the Dead Sea Scrolls

Several chapters of the Book of Enoch contain astronomical and astrological material[39]; related material is also found in a number of scrolls found in the caves at Qumran.[40] Many aspects of this material have been traced back to a Babylonian origin, including schemes for the variation in the length of daylight, the duration of visibility of the moon on different nights of the month, and some astrological material.[41] I will briefly examine two cases which highlight both the process of the selection of Babylonian material and the adaptation of this material into existing calendrical and cultural frameworks by early Jewish authors.

1 Enoch 72 presents a prose account of a mathematical scheme for the change in the length of day and night and the 'gate' through which the sun rises over the course of the year. For example, verses 6 to 11 read:

> This is how it emerges in the first month through the large gate; it emerges through that fourth one of those six gates on the side where the sun rises. In that fourth gate from which the sun emerges in the first month there are thirty open windows from which heat comes out when they are open at their times. When the sun rises in the sky it emerges through the fourth gate for thirty mornings, and it accurately goes down in the fourth gate on the west side of the sky. During those days day

---

[39] For these chapters, see Nickelsburg and VanderKam (2012) and Badalanova Geller (2010).

[40] For these scrolls, see Talmon et al. (2001) and Drawnel (2011).

[41] For detailed discussions with references to earlier studies, see Ben-Dov (2008), Drawnel (2011), Nickelsburg and VanderKam (2012), and Jacobus (2015). See also my comments in Steele (2024).

> becomes longer than day and night becomes shorter than night for thirty days. On that day the daytime is two units more than the night – one-ninth, with the daytime being exactly ten parts, and the night exactly eight parts. The sun emerges through the fourth gate and sets through the fourth (gate).[42]

The scheme states that at the beginning of the first month of the year, the sun rises in the fourth gate and that daylight lasts for ten parts and night for eight 'parts' (implying that a full day plus night contains eighteen parts). The text continues with similar passages for each month of the year (see the summary in Table 1). According to this scheme, the length of day and night change linearly by one unit per month between maximum and minimums of 12 and 6, meaning that night is twice as long as day at the winter solstice and vice versa. This daylight scheme clearly echoes earlier Babylonian schemes. MUL.APIN II ii 43–iii 12, for example, presents a numerical scheme for the length of night and the daily change in the duration of visibility of the moon which can be directly compared with the scheme in 1 Enoch 72. Here are the first few lines of this scheme:

Table 1 Scheme for the gate in which the sun rises and the length of day and night given in 1 Enoch 72

Month	Days in month	Gate	Length of day (at the beginning of day 1)	Length of night (at the beginning of day 1)
1	30	4	9	9
2	30	5	10	8
3	31	6	11	7
4	30	6	12	6
5	30	5	11	7
6	31	4	10	8
7	30	3	9	9
8	30	2	8	10
9	31	1	7	11
10	30	1	6	12
11	30	2	7	11
12	31	3	8	10

42 1 Enoch 72:6–11; translation by Nickelsburg and VanderKam (2012, p. 416).

> On the 1st of Month I a nighttime watch is 3 *minas* 10 *shekels*; 12 UŠ 40 NINDA setting of the moon.
> On the 15th of Month I a nighttime watch is 3 *minas*; 12 UŠ rising of the moon.
> On the 1st of Month II a nighttime watch is 2 5/6 *minas*; 11 UŠ 20 NINDA setting of the moon.
> On the 15th of Month II a nighttime watch is 2 2/3 *minas*; 10 UŠ 40 NINDA rising of the moon.[43]

This passage gives the length of night for the 1st and 15th day of each month of the year using the units *minas* and *shekels*, where there are 60 *shekels* in a *mina* and 6 *minas* in a complete day plus night. The units *minas* and *shekels* are in fact units of weight, the implication being that they correspond to the weight of water in a water clock such that weight and time are directly proportional.[44] Like the scheme in 1 Enoch, the length of night follows a linear zigzag function with maximum at winter solstice which is twice the length of the minimum length of night at the summer solstice. There are, however, differences in the two schemes (see Table 2). First, the two schemes use different units: *mina* which correspond to one-sixth of a day plus night in MUL.APIN and 'parts' which are one-eighteenth of a day plus night in Enoch. Second, the MUL.APIN scheme places the equinoxes and solstices in the Middle of months I, IV, VIII, and X, whereas in Enoch they are placed on the first day of the month. But most importantly, the Babylonian scheme operates with the 360-day schematic calendar in which all of the months contain 30 days whereas the scheme in Enoch uses the 364-day calendar in which every third month contains 31 days. Despite these differences, however, it is clear that the two schemes are related: they both assume a linear monthly change in the length of day and night (tacitly ignoring the difference between 31—and 30-day months in the Enoch scheme), and they both assume that the maximum length of night is twice the minimum length of night. However, this ratio for the extreme length of night is far too big

[43] MUL.APIN II ii 43–44; translation by Hunger and Steele (2019, p. 155).

[44] For this relationship and a discussion of the evidence for Babylonian waterclocks, see Brown et al. (1999–2000).

for the latitude of Babylonia or anywhere in the ancient Jewish world. The use of this inaccurate value along with the zigzag function strongly points to the Babylonian origin of the Enoch scheme, which was been adapted to work within the framework of the Jewish 364-day calendar tradition.[45]

Texts from Qumran provide another example of the adaptation of a Babylonian scheme to a local context. The so-called Aramaic Astronomical Book (4Q208–4Q211) presents a scheme for the duration of visibility of the moon throughout the month. The moon's visibility is represented by a linear zigzag function in units which are fractions of sevenths of the length of the night. The zigzag function has a period of 28 days and is constructed on the assumption that the moon is visible all night on the 14th day of the month. This can be compared with Babylonian schemes for the duration of visibility of the moon found in *Enuma Anu Enlil* 14 and MUL.APIN. In these schemes, the duration of visibility is also represented by a linear zigzag function but with a period of 30 days. The scheme uses the time units UŠ and NINDA, where there are 360 UŠ in a day plus night and 60 NINDA in an UŠ. Once more,

Table 2 Scheme for the length of night in 1 Enoch 72 and MUL.APIN

Month	Days in month (Enoch)	Length of night Enoch (day 1)	Days in month (MUL.APIN)	Length of night MUL.APIN (day 1)	Length of night MUL.APIN (day 15)
1	30	9	30	3 1/6 *mina*	3 *mina*
2	30	8	30	2 5/6 *mina*	2 2/3 *mina*
3	31	7	30	2 1/2 *mina*	2 1/3 *mina*
4	30	6	30	2 1/6 *mina*	2 *mina*
5	30	7	30	2 1/6 *mina*	2 1/3 *mina*
6	31	8	30	2 1/2 *mina*	2 2/3 *mina*
7	30	9	30	2 5/6 *mina*	3 *mina*
8	30	10	30	3 1/6 *mina*	3 1/3 *mina*
9	31	11	30	3 1/2 *mina*	3 2/3 *mina*
10	30	12	30	3 5/6 *mina*	4 *mina*
11	30	11	30	3 5/6 *mina*	3 2/3 *mina*
12	31	10	30	3 1/2 *mina*	3 1/3 *mina*

[45] For the Jewish (*not* Babylonian) origin of the 364-day calendar, see Steele (2024) which also contains references to earlier discussions of this topic.

the zigzag function is constructed by assuming that the moon is visible all night in the middle of the month, which is here taken to be on the 15th day. Thus, we seem to have two related but not identical schemes. It is, of course, possible that the Qumran scheme was developed independently of the Babylonian one. But, given the clear evidence for knowledge of other aspects of Babylonian astronomy in the early Jewish world (and specifically amongst the community at Qumran), it seems more likely that we should see the Qumran scheme as an adaptation of the Babylonian scheme to accord with local needs and traditions. The 28-day period for lunar visibility assumed in the Qumran scheme is probably tied to the Jewish seven-day week, which itself meshes with the 364-day calendar attested in other scrolls from Qumran as well as in the Book of Enoch. A 28-day period is no more artificial (or, in modern terms, no more inaccurate) than the 30-day period of the Babylonian scheme; both have their origin in simple calendrical schemes which are tied to their local cultural context.

5 Final Remarks

The three brief case studies presented above illustrate two of the key features of the circulation of (scientific) knowledge: the selection of what material to appropriate and the adaptation of that material to fit and be useful within existing knowledge systems. Thus, technical Greek astronomers took Babylonian mathematical astronomy from which they extracted parameters that were useful within the geometrical astronomical theories that they were developing; Greek astrologers took complete systems of Babylonian mathematical astronomy, adapted them to work with the local calendar and used them to generate the astronomical data that they needed to produce horoscopes; Greek historians and encyclopedists, by contrast, drew mainly on Babylonian schematic astronomy. Indian astronomers took Babylonian mathematical astronomy and adapted it to work with their system of an epoch. Early Jewish scholars appropriated parts of Babylonian schematic astronomy, adapting it to work with the Jewish 364-day calendar, as well as aspects of Babylonian astrology.

It seems safe to suppose that a broad range of Babylonian astronomical knowledge was available (either directly or indirectly) to all of these scholars, so we must see each case of the appropriation of one type of astronomy and not of others as a choice by the receiving culture. Jewish scholars drew on the schematic astronomical tradition because it provided a unified, simple, and self-consistent mathematical description of the celestial realm which was easily adaptable to the Jewish 364-day calendar and the seven-day week. By contrast, Babylonian mathematical astronomy, although much more scientifically complex and 'accurate' from a modern (and even an ancient) perspective, does not provide a unified, self-consistent descriptive model, nor is it easily adaptable to the 364-day calendar. In short, Babylonian mathematical astronomy was neither useful nor fit in with existing Jewish astronomical and calendrical traditions. At the other extreme, Ptolemy, Hipparchus, and Greek technical astronomers needed accurate and precise values for various astronomical parameters (mean motions, period relations, etc.). The parameters underlying Babylonian schematic astronomy are clearly not precise or accurate—they were not intended to be—which made them of no use to these astronomers, whereas the parameters embedded within Babylonian mathematical astronomy are precise and in general quite accurate. Thus, it was to mathematical astronomy that Greek technical astronomers naturally turned.

The evidence from the ancient and medieval period, including the three case studies presented here to which many more examples could be added, shows that the circulation of astronomical knowledge—and, by analogy, the circulation of scientific knowledge, or even of knowledge more broadly—is the norm across almost all cultures, both throughout history and today. The interesting question for historians of science is not whether knowledge circulated but rather *how* did the process of circulation happen, *why* was particular knowledge appropriated by particular groups or individuals, and *in what ways* was the knowledge subsequently adapted and naturalised. Accepting the basic fact of the circulation of knowledge should not be controversial but, sadly, it is sometimes denied by some scholars for ideological reasons. Often this is because these scholars see the circulation of knowledge as somehow denying the importance or originality of one culture or another.[46] In response, I would like to end by quoting a short remark by David Pingree, who's work on the circulation of knowledge has often been misread and attacked by such scholars. Pingree is referring to the case of India, but his words are equally applicable to the dialogical nature of science more generally:

> The occurrence of Mesopotamian, Greco-Babylonian, and Greek material in Indian astronomy does not in any way detract from Indian originality; it merely shows that, historically, India has been and still is an active player in the international development of the astral (and other) sciences. Indian material is found in Iran, Central Asia, the Far East, the Islamic countries, Byzantium, and Western Europe. Like every other culture in Eurasia (and now in the world) it has learned from others and taught others.[47]

References

Aaboe, A. 1955. On the Babylonian Origin of Some Hipparchian Parameters. *Centaurus* 4: 122–125.

Badalanova Geller, F. 2010. *2 (Slavonic Apocalypse of) Enoch: Text and Context*. Max-Planck-Institut für Wissenschaftsgechicht Preprint 410. Berlin: Max-Planck-Institut für Wissenschaftsgechicht.

Bala, A. 2016. The Needham Question and Southeast Asia: Comparative and Connective Perspectives. In *The Bright Dark Ages: Comparative and Connective Perspectives*, ed. Bala Arun and Duara Prasenjit, 199–217. Leiden: Brill.

Ben-Dov, J. 2008. *Head of All Years: Astronomy and Calendars at Qumran in their Ancient Context*. Leiden: Brill.

[46] As emphasised by Pingree (1992), these ideological objections come both from scholars who want to see modern science as purely as western (i.e., Greek and European) and scholars who want to see non-western science as anticipating western science—both of which are forms of what Pingree terms 'Hellenophilia'.

[47] Pingree (2007, p. 48).

Brack-Bernsen, L. 1999. Goal-Year Tablets: Lunar Data and Predictions. In *Ancient Astronomy and Celestial Divination*, ed. N.M. Swerdlow, 149–177. Cambridge, MA: The MIT Press.

Brack-Bernsen, L. 2011. Prediction of Days and Pattern of the Babylonian Lunar Six. *Archiv Für Orientforschung* 52: 156–179.

Brack-Bernsen, L., and J.M. Steele. 2005. Eclipse Prediction and the Length of the Saros in Babylonian Astronomy. *Centaurus* 47: 181–206.

Brennan, A., and N.Y.S. Lo. 2016. The Descent of Theory. In *The Bright Dark Ages: Comparative and Connective Perspectives*, ed. Bala Arun and Duara Prasenjit, 21–39. Leiden: Brill.

Brentjes, S., A. Fidora, and M.M. Tischles. 2014. Towards a New Approach to Medieval Cross-Cultural Exchanges. *Journal of Transcultural Medieval Studies* 1: 9–50.

Britton, J.P. 2002. Treatments of Annual Phenomena in Cuneiform Sources. In *Under One Sky: Astronomy and Mathematics in the Ancient Near East*, ed. J.M. Steele and A. Imhausen, 21–78. Münster: Ugarit-Verlag.

Britton, J.P., and A. Jones. 2000. A New Babylonian Planetary Model in a Greek Source. *Archive for History of Exact Sciences* 54: 349–373.

Brown, D. 2018. *The Interactions of Ancient Astral Science*. Bremen: Hempen Verlag.

Brown, D., J. Fermor, and C. Walker. 1999–2000. The Water Clock in Mesopotamia. *Archiv für Orientforschung* 47: 130–148.

Drawnel, H. 2011. *The Aramaic Astronomical Book (4Q208–4Q211) from Qumran: Text, Translation, and Commentary*. Oxford: Oxford University Press.

Evans, J., and J.L. Berggren. 2006. *Geminos's Introduction to the Phenomena*. Princeton: Princeton University Press.

Falk, H. 2000. Measuring Time in Mesopotamia and Ancient India. *Zeitschrift der Deutschen Morgenländischen Gesellschaft* 150: 107–132.

Gray, J.M.K., and J.M. Steele. 2008. Studies on Babylonian Goal-Year Astronomy I: A Comparison Between Planetary Data in Goal-Year Texts, Almanacs and Normal Star Almanacs. *Archive for History of Exact Sciences* 62: 553–600.

Gray, J.M.K., and J.M. Steele. 2009. Studies on Babylonian Goal-Year Astronomy II: The Babylonian Calendar and Goal-Year Methods of Prediction. *Archive for History of Exact Sciences* 63: 611–633.

Hart, R. 1999. Beyond Science and Civilization: A Post-Needham Critique. *East Asian Science, Technology, and Medicine* 16: 88–114.

Hunger, H. 1999. Non-Mathematical Astronomical Texts and Their Relationships. In *Ancient Astronomy and Celestial Divination*, ed. N.M. Swerdlow, 77–96. Cambridge, MA: The MIT Press.

Hunger, H., and J.M. Steele. 2019. *The Babylonian Astronomical Compendium MUL.APIN*. Abingdon: Routledge.

Jacobus, H.R. 2015. *Zodiac Calendars in the Dead Sea Scrolls and Their Reception: Ancient Astronomy and Astrology in Early Judaisim*. Leiden: Brill.

Jones, A. 1998. Studies in the Astronomy of the Roman Period III Planetary Epoch Tables. *Centaurus* 40: 1–41.

Jones, A. 2002. Babylonian Lunar Theory in Roman Egypt: Two New Texts. In *Under One Sky: Astronomy and Mathematics in the Ancient Near East*, ed. J.M. Steele and A. Imhausen, 167–174. Münster: Ugarit-Verlag.

Jones, A., and J.M. Steele. 2011. A New Discovery of a Component of Greek Astrology in Babylonian Tablets: The 'Terms'. *ISAW Papers*, 1.

Jones, A., and J.M. Steele. 2018. Diodorus on the Chaldeans. In *The Scaffolding of Our Thoughts: Essays on Assyriology and the History of Science in Honor of Francesca Rochberg*, ed. C.J. Crisostomo, E.A. Escobar, T. Tanaka, and N. Veldhuis, 333–352. Leiden: Brill.

Kugler, F.X. 1900. *Die Babylonische Mondrechnung*. Freiburg: Herser'sche Verlagshandlung.

Neugebauer, O. 1989. From Assyriology to Renaissance Art. *Proceedings of the American Philosophical Society* 133: 391–403.

Neugebauer, O. 1970–1971. *The Pañcasiddhāntikā of Varāhamihira*. Copenhagen: Munsgaard.

Nickelsburg, G.W.E., and J.C. VanderKam. 2012. *1 Enoch 2: A Commentary on the Book of 1 Enoch Chapters 37–82*. Minneapolis MN: Fortress Press.

Ossendrijver, M. 2012. *Babylonian Mathematical Astronomy: Procedure Texts*. New York: Springer.

Pankenier, D.W. 2014. Did Babylonian Astrology Influence Early Chinese Astral Prognostication *Xing Zhan Shu* 星占術. *Early China* 37: 1–13.

Pingree, D. 1973. The Mesopotamian Origin of Early Indian Mathematical Astronomy. *Journal for the History of Astronomy* 4: 1–12.

Pingree, D. 1987. Venus Omens in India and Babylon. In *Language, Literature, and History: Philological and Historical Studies Presented to Erica Reiner*, ed. F. Rochberg-Halton, 293–315. New Haven: American Oriental Society.

Pingree, D. 1992. Hellenophilia versus the History of Science. *Isis* 83: 554–563.

Pingree, D. 1998. Legacies in Astronomy and Celestial Omens. In *The Legacy of Mesopotamia*, ed. S. Dalley, 125–138. Oxford: Oxford University Press.

Pingree, D. 2007. Mesopotamian and Greek Astronomy in India. In *Expanding and Merging Horizons: Contributions to South Asian and Cross-Cultural Studies in Comemoration of Wilhelm Halbfass*, ed. K. Preisendanz, 41–50. Vienna: Österreichischen Akademie der Wissenschaften.

Price, D. de Solla. 1961. *Science Since Babylon*. New Haven: Yale University Press.

Ragep, F.J. 1996. Introducton. In *Tradition, Transmission, Transformation: Proceedings of Two Conferences on Pre-Modern Science Held at the University of Oklahoma*, ed. F.J. Ragep, S.P. Ragep, and S. Livesey, xv–xxxiv. Leiden: Brill.

Rochberg, F. 2015. Conceiving the History of Science Forward. In *The Frontiers of Ancient Science: Essays in Honor of Heinrich von Staden*, ed. B. Holmes and K.-D. Fischer, 515–531. Berlin: de Gruyter.

Rochberg, F. 2016. *Before Nature: Cuneiform Knowledge and the History of Science*. Chicago: University of Chicago Press.

Rochberg, F. 2020. The Babylonian Contribution to Greco-Roman Astronomy. In *Hellenistic Astronomy: The Science and its Contexts*, ed. A.C. Bowen and F. Rochberg, 147–159. Leiden: Brill.

Sabra, A.I. 1987. The Appropriation and Subsequent Naturalization of Greek Science in Medieval Islam: A Preliminary Statement. *History of Science* 25: 223–342.

Saliba, G. 2007. *Islamic Science and the Making of the European Renaissance*. Cambridge, MA: The MIT Press.

Shah, M.H. 2016. Rethinking the Needham Question: Why Should Islamic Civilization Give Rise to the Scientific Revolution. In *The Bright Dark Ages: Comparative and Connective Perspectives*, ed. A. Bala and P. Duara, 217–235. Leiden: Brill.

Sivin, N. 1982. Why the Scientific Revolution did not Take Place in China—or didn't it? *Chinese Science* 5: 45–66.

Steele, J.M. 2000a. *Observations and Predictions of Eclipse Times by Early Astronomers*. Dordrecht: Kluwer.

Steele, J.M. 2000b. Eclipse Prediction in Mesopotamia. *Archive for History of Exact Sciences* 54: 421–454.

Steele, J.M. 2004. Applied Historical Astronomy: An Historical Perspective. *Journal for the History of Astronomy* 35: 337–355.

Steele, J.M. 2011. Goal-Year Periods and Their Use in Predicting Planetary Phenomena. In *The Empirical Dimension of Ancient Near Eastern Studies—Die Empirische Dimension Altorientalischer Forschungen*, ed. G. Selz and K. Wagensonner, 101–110. Wien: LIT Verlag.

Steele, J.M. 2013. A Comparison of Astronomical Terminology, Methods and Concepts in China and Mesopotamia, With Some Comments on Claims for the Transmission of Mesopotamian Astronomy to China. *Journal of Astronomical History and Heritage* 16: 250–260.

Steele, J.M. 2018. Babylonian and Assyrian Astral Science. In *The Cambridge History of Science, Volume 1: Antiquity*, ed. A. Jones and L. Taub, 73–98. Cambridge: Cambridge University Press.

Steele, J.M. 2020. Geminos and Babylonian Astronomy. In *Instruments—Observations—Theories: Studies in the History of Early Astronomy in Honor of James Evans*, ed. C. Carman and A. Jones, 205–212. New York: ISAW.

Steele, J.M. 2021. The Continued Relevance of MUL.APIN in Late Babylonian Astronomy. *Journal of Ancient near Eastern History* 8: 259–277.

Steele, J.M. 2022. The Influence of Assyriology on the Study of Chinese Astronomy in the Late Nineteenth and Early Twentieth Centuries. In *Overlapping Cosmologies in Asia*, ed. B.M. Mak and E. Huntington, 17–44. Leiden: Brill.

Steele, J.M. 2024. Babylonian Astronomy and Enoch: Some Comments from the Perspective of the History of Babylonian Astronomy. In *The Aramaic Books of Enoch and Related Literature from Qumran*, ed. H. Drawnel, 108–137. Leiden: Brill.

Swerdlow, N.M., and O. Neugebauer. 1984. *Mathematical Astronomy in Copernicus's De Revolutionibus* Berlin: Springer.

Talmon, S., J. Ben-Dov, and U. Glessmer. 2001. *Qumran Cave 4 XVI. Calendrical Texts*. Oxford: Clarendon.

van der Waerden, B.L. 1974. *Science Awakening II: The Birth of Astronomy*. Leiden: Noordhof.

6

Scientific Exchanges with Qing China and the Formation of a Local Science in Eighteenth-Century Korea

Jongtae Lim

1 Introduction

The National Palace Museum of Korea has a horizontal sundial, which I believe was carved into a stone in 1789 by the Royal Astronomy Bureau of the Chosŏn Dynasty (Fig. 1).[1] This artifact, which represents the scientific vitality of the Bureau at that time, was not based on a design created by Korean astronomers. Rather, it was a remake of a sundial imported from China. In fact, the Chinese original is also kept in the same museum; an inscription tells that it was made in 1636 under the supervision of Li Tianjing 李天經, Adam Schal von Bell, and Giacomo Rho, leaders of the Chongzhen calendrical reform of the late Ming period in China.

J. Lim (✉)
Department of Science Studies, College of Natural Sciences, Seoul National University, Seoul, South Korea
e-mail: jtlimbabo@snu.ac.kr

[1] Researchers since Yi (1985) have determined the date of this plane sundial in the first half of the eighteenth century. However, a recent research, supported by sufficient evidence, suggests that it was made in 1789. For more details, see Kim Seulki (2023).

A. Bala et al. (eds.), *Multicivilizational Exchanges in the Making of Modern Science*,
https://doi.org/10.1007/978-981-97-3541-9_6

Fig. 1 Horizontal sundial according to the new Western method 新法地平日晷 (1789, Royal Bureau of Astronomy, Chosŏn Dynasty). 38.3 × 59.0 × 16.0 cm. National Palace Museum of Korea. https://www.gogung.go.kr/gogung/main/main.do

We don't know precisely when this monument was brought to Korea,[2] but in the mid-eighteenth century it suddenly caught the attention of the court, and finally witnessed its own reproduction. Yet the Korean remake differed from the original in one important respect—the astronomers adjusted the sundial's projection to the latitude of Seoul, the value of which was inscribed in a side of the sundial: 'Elevation of the North Pole in Seoul is 37 degrees and 39 minutes high.' This modification represented one of the main concerns of Korean astronomers in the late eighteenth century: adjusting the official calendrical system of Qing China to Korea's locality.

There is an irony here that leads us to the main question of this chapter. The value of the latitude of Seoul, the symbol of the localizing spirit, was not measured by Korean astronomers. It was measured by a Qing astronomer named He Guozhu 何國柱, who was sent to Korea in 1713 as a member of the imperial embassy to gather geographical information about Korea for use in the Qing project of mapping the entire

[2] Yi (1985) suggests that it might have been brought by the Korean Crown Prince Sohyŏn, when he returned from the Qing, where he had been a war captive since 1637, to Korea in 1645. Yi surmises that the sundial was among the gifts bestowed to the Prince by the Jesuit Adam Schall von Bell. However, there is not sufficient evidence to substantiate this conjecture.

empire and the areas under its political influence. This is now known as the *Kangxi Atlas* project. How, then, had the Qing astronomer's measurement, spurred by the geopolitical ambition of the Qing dynasty, come to symbolize the localizing spirit of the astronomers of its tributary state?

Showing at once the distinctive identity of Korea's official astronomy and its historical and scientific link with the Qing imperial science, Seoul's latitude inscribed on the sundial represents a central feature of the cultural dynamics governing the scientific exchanges between China and Korea in the eighteenth century. This chapter elaborates on this cultural dynamics, as seen from the Korean side, with focus on its earlier phase, which began around the turn of the eighteenth century and culminated in He Guozhu's mission to Korea in 1713. The stabilization of the Qing imperial order at that time allowed increased opportunities for astronomers, scholar-officials, and Western priests of China and Korea to exchange knowledge and techniques through government projects in geography and astronomy. And, as a consequence, the geographical information of Korea flowed to the Qing side, while the Sino-European astronomy and mathematics of the Qing were transferred to Korea.

This chapter highlights two main features of these scientific exchanges in the early eighteenth century. First, this case shows a highly political nature of the border crossing of knowledge and techniques; they could move between the Qing and Chosŏn border only under certain political constrains, which, I argue, at once promoted and limited the scientific exchanges between the two states. Second, this case illustrates how the moved entities acquired a new cultural meaning in the host community, while at the same time remaining connected to, and thus commensurable with, the sciences of the original community. Although Korean actors sought to make these scientific exchanges with the Qing a unique opportunity to construct their own scientific tradition, their identity politics never went so far as to build a radically different, self-sufficient national science.

2 Political Context at the Turn of Eighteenth Century

The turn of the eighteenth century saw a dramatic rise of scientific exchanges between Qing and Chosŏn. In a period when scientific undertakings were more openly political than they are now in both their pronounced aims and the knowledge they produced, the Qing-Chosŏn scientific exchange was deeply structured by the political relations between the two states, each having its own political interests to protect and promote. As previously mentioned, scientific interactions between the two states in this period were spurred by the amelioration of mutual suspicion and antagonism that had characterized their relations in the seventeenth century (Lim 2016).

The earlier antagonistic relation between the two states began from the Qing's conquest of Korea in 1637, by which the Qing replaced the Ming dynasty as the new suzerain state to which Korea had to pay tribute. The court and ruling elites of Korea did not serve the new master wholeheartedly, however, retaining their secret loyalty to the cause of Ming restoration. After the Qing successfully consolidated its rule over the former Ming territory and the last hopes for a Ming restoration died during the second half of the seventeenth century, the Ming loyalism of the Chosŏn elites became internalized (Hŏ 2019). They considered their own dynasty as the legitimate heir of the Ming, and by extension, to the entire Middle Kingdom lineage handed down from Chinese antiquity.

However, it was both the consolidation of Qing rule and the internalization of Korean Ming loyalism that signaled the stabilization of their relationship. With all threat of a potential Ming restoration obsolete, the Qing withdrew its formerly suspicious eye from Korea, lifting harsh regulations and granting various political and economic benefits. The Koreans on their side accepted the reality of Qing's rule and resumed their former business of bringing Chinese culture to Korea, while quietly maintaining their internalized hostility toward the Qing (No 2005). The exchanges between the two countries, which had previously been limited to the tribute trades of silk, ginseng, and silver, gradually broadened to what might be called the cultural arena, including various fields of science, notably geography, astronomy, and mathematics (Lim 2016), about which I will now proceed to examine one by one.

3 Geographical Exchange and Its Limits

As scientific fields of obvious political utility, geography and cartography show how the amelioration of Qing-Chosŏn relations at the turn of the eighteenth century provided the two states with motivations and opportunities for geographical surveying, particularly of their borderlands. It also showed how the inherent tension between the two states, each with its own territorial interests, limited the flow of geographical information between the two at a superficial level.

The strict political hierarchy between the two states gave the Qing the upper hand in their geographical contacts, while the Koreans assumed a defensive posture toward what they viewed as imperial encroachment into the borderland. The Qing's geographical interest in the Qing-Chosŏn borderland heightened during the last decades of the seventeenth century. Previous research identifies several overlapping factors behind this increased focus. Culturally, the border area, which is now called Manchuria, particularly the area around Mt Changbai 長白山 (Mt Paektu 白頭山 for Koreans) was to the Qing imperial house the sacred place of their tribal origin. Yet even until the turn of the eighteenth century, the Qing lacked sufficient knowledge of the area, due to both its rugged terrain and its politically sensitive location bordering Korea. Although the two states had a consensus that the two rivers, the Yalu and the Tumen, which originated in Mt Changbai, determined their natural boundaries, the upper reaches of those rivers were not clearly known. Economically, this area produced a diversity of natural resources, particularly the precious medicinal plant called *ginseng*, which had economically supported the rise of the Jurchen-Manchu state. For these reasons, the Qing put this area under close state control, banning trespassing and unauthorized collecting of *ginseng*. Many Korean smugglers, however, crossed the border in search of *ginseng* and other valuable resources, causing troubles with the Manchus residing there, which often led to serious diplomatic problems between the two courts (Li 2008, pp. 59–68; Kim 2007).

As the Qing successfully consolidated its rule over the Chinese mainland and turned its territorial ambition into the inner Asian region in the late seventeenth century, Qing geopolitical interests in Manchuria assumed a more scientific form. Geographical and cartographical surveys of this region were attempted, usually as part of an empire-wide survey project. In 1679, for example, the Kangxi emperor ordered geographical survey of Mt. Changbai and the Qing-Chosŏn border, the results of which were incorporated into the *Comprehensive Gazetteer of Shengjing* (盛京通志) compiled in 1684. In the first decade of the eighteenth century, the same emperor launched a more ambitious project of mapping the empire, this time systematically employing the European survey methods introduced by Jesuit fathers and astronomical officials in the Bureau of Astronomy. During this cartographic project, which would produce the famous *Kangxi Atlas*, two major expeditions to the Qing-Chosŏn border were made. First, in 1709, the Qing survey team, including three Jesuit missionaries, Pierre Jartoux, Jean-Baptiste Régis, and Xavier Ehrenbert Fridelli, were sent to the border areas Fengcheng and Hunchun. This trip occasioned Jartoux's famous report of what he called 'a Tartarian plant called *Gin-seng*' to European scientific communities, which was published in the *Philosophical Transaction* of the Royal Society of London in 1713 (Jartoux 1713; Cams 2017, pp. 111–113). In 1711 and 1712, the Kangxi emperor again sent an official, named Mu-ke-deng (General Commander of Daxing Ula), whose mission was to find and officially document the origin of the two border rivers around the top of Mt. Changbai, and thus to fix the border between the two states (Cams 2017, pp. 118–122; Li 2008, pp. 140–148).

The Chosŏn court became very apprehensive about the Qing's repeated attempts to survey the border area. From the 1680s, the Qing requested the Chosŏn court to allow their officials access to Korean territory in order to find paths approaching Mt. Changbai. Although the Qing's request was purely geographical in motivation, the Koreans took it as a sign of a Qing hidden military plot. Many believed the Qing were searching for potential military routes from Beijing to Ningguta, the place of their tribal origin, to be taken when they would eventually flee from the Mongols—an imaginary scenario entertained by Koreans, which partly reflected the ongoing conflicts between the Qing and the Zhungar Khanate. After a series of successful Korean evasions, the Qing was finally able to access the mountain from Korean territory in 1712. The Qing and Korean officials concluded this expedition by jointly building the famous stele to mark the border at what they thought was the watershed of the two rivers (Li 2008, pp. 148–170; Pae 1998, pp. 64–93).

The Qing sent yet another mission to Korea the following year, this time directly to the Chosŏn court in Seoul. As mentioned earlier, an astronomer named He Guozhu, the Calendar Manager of Five Bureaus (*Wuguan sili* 五官司曆), accompanied this embassy headed by Manchu officials Aqitu (Imperial Bodyguard of the First Rank) and Mu-ke-deng. This mission was obviously motivated by a geographical aim and served as an extension of Mu-ke-deng's former expeditions to Mt. Changbai in previous years. This intention was unmistakable to the Korean side, particularly given the inclusion of He Guozhu, a young astronomical official in the Imperial Bureau of Astronomy, about which the Chosŏn's official history commented as 'unprecedented' (*Sukchong sillok* 肅宗實錄 the 16th day, 5th month of the 39th reign year [hereafter reference to *Sukchong sillok* is in the form of *SS* date/month/reign year]—see Chang 2023, pp. 106–109). He Guozhu conducted a series of astronomical measurements on his trip, including the famous observation of Seoul's latitude, and during a stay of around ten days in Seoul, Mu-ke-deng repeatedly asked the Chosŏn court, at the behest of the Kangxi emperor, for access to their geographical archives. The items they requested included a hydrographic map portraying the Mt. Changbai area, a national map of Korea, and the list of distances of towns from the capital (*SS* 27/Intercalary 05/39; 02/06/39; 04/06/39). These requests, which were certainly embarrassing to the Korean side, were made with a clear objective: to produce a map of Korea to be included in the *Kangxi Atlas*. The Chosŏn court did everything they could to evade the requests. While refusing to give most of the requested information under the pretext that Chosŏn had no such sophisticated geographical archive, the Chosŏn court could not but provide a national map of Korea. They responded to this dilemma by offering an outdated and inferior map, implying it was the best map available while keeping their most updated official map unseen and unmentioned (*SS* 02/06/39).

The Chosŏn map included in the *Kangxi Atlas* was certainly less accurate than the best one that the contemporary Chosŏn government could provide (Fig. 2). This episode thus illustrates how the political relations between the two states simultaneously allowed and limited the exchanges of knowledge between them. The geographical contacts between the Qing and Chosŏn at the turn of the eighteenth century greatly increased the knowledge of the border area on both sides. However, the political tensions and mutual suspicions that plagued the representatives of the two courts limited the flow of Korean knowledge to the Qing side only to a superficial level.

The same can be said about the transactions in the opposite direction. The political tension caused by the Qing geographical survey certainly aroused the interest of the Korean elites in the northern frontier. However, Chosŏn's inferior position in relation to the Qing prevented it from making a symmetrical attempt to survey into the Qing's hinterland.

Fig. 2 *Kangxi huangyu quanlantu* 康熙皇輿全覽圖. Leiden University Libraries Digital Collection. http://hdl.handle.net/1887.1/item:3285048

Fig. 3 *Yogye kwanbang chido* 遼薊關防地圖 (Borderland Map of Liaodong and Beijing). Kyujanggak Institute for Korean Studies, Seoul National University

The lack of information about this strategically important area forced the concerned Chosŏn officials to rely on old sources. In 1706, a Korean high official, Yi Imyŏng 李頤命, supervised the production of a map of China, portraying the area from Liaodong to Beijing, titled *Yogye kwanbang chido* 遼薊關防地圖 (Borderland Map of Liaodong and Beijing) (Fig. 3). This map combined information from a heterogeneous set of sources, including the Koreans' knowledge of their frontier area as well as the tributary route to Beijing, but a considerable part of it derived from the works of military geography compiled in the late Ming period, which Yi had acquired during his mission to Beijing the previous year (Pae 1998). Although being filled with outdated information from the Ming period, the *Yogye kwanbang chido* nevertheless shows a high level of understanding by Korean officials of the border area, especially regarding the Korean hinterland, information that Mu-ke-deng and other Qing officials desperately sought after at that time. Interestingly, Yi Imyŏng was one of the high officials who did everything he could to evade Mu-ke-deng's embarrassing request of Korea's geographical sources in 1713. After all, the knowledge of the border area remained incomplete in both the Korean and the Qing sides.

4 Learning the Qing Astronomy

He Guozhu's mission to Seoul in 1713 was largely geopolitical in its motivation, but it inadvertently sparked interest in calendrical astronomy among the high officials at the Chosŏn court. Upon learning that the imperial embassy had included an astronomer-official among their personnel—an unusual addition—the Chosŏn court paid a close attention to the person named He Guozhu and his assignments in the mission. Despite their embarrassment over the geographical request of the imperial mission, the Chosŏn court found at least one positive point in the inclusion of the astronomer in the embassy. He Guozhu's presence presented Chosŏn officials with a great opportunity to learn the Qing's calendrical system, without having to dispatch its own astronomers to Beijing, a costly scheme that the Chosŏn government had previously adopted (Lim 2013).

Specifically, what, then, did the court expect its astronomers to learn from the Qing astronomer? The focus of attention was directed at what the reception official described as 'a piece of semi-circle shaped instrument (半圓儀器一件)' that He Guozhu had brought for astronomical measurements, and it was perhaps with this instrument that he measured the abovementioned value of Seoul's latitude.[3] Indeed, He Guozhu's mission left a great impact on the observational aspects of Korea's calendrical astronomy. It was this impact of the imperial mission, which set in motion the process of localizing Qing-Jesuit astronomy in Korea, a process that would continue for a century thereafter. Yet, the cultural dynamics of the scientific interaction in the field of astronomy could not be described simply as that of action and response.

It should first be noted that the Chosŏn court's interest in the Qing calendrical astronomy was not a complete novelty at the time, but rather predated the interactions associated with this particular mission. The Qing's calendrical system, or the Jesuit-created *Shixian li* 時憲曆 astronomy, was one of the few Qing cultural items that the Chosŏn government had paid relatively close attention to and made efforts to learn since its promulgation in the mid-seventeenth century. Initially, the Korean court's attitude toward the Qing astronomy was not one of wholehearted acceptance. The kings and high officials regarded the project of learning the *Shixian li* only as a political imperative dictated by the hierarchical relationship between the Qing and Chosŏn, which had been established in 1637, as a result of the Qing conquest of Korea.

[3] However, an official record from the late eighteenth century identified He Guozhu's instrument differently as a 'large quadrant' (Sŏ Hosu 1796/2004). We don't have clear evidence to identify the instrument(s) He Guozhu brought to Korea.

Given the political function of calendrical astronomy, which symbolized the cosmic power of China's Son of Heaven, Chosŏn, now being a tributary state of the Qing, had no choice but to adopt the Qing's calendrical system in order to synchronize its almanac with the almanac issued by the Qing emperors. After several years of efforts by Korean astronomers who were sent to Beijing to learn the *Shixian li* method, Chosŏn was able to promulgate its own *Shixian li* almanac in 1653 (Lim 2012).

However, the early court support for the Qing system never exceeded the bare minimum of political expediency. After its Astronomy Bureau was able to produce an almanac in which months and dates were arranged identically with the Qing almanac, and thus the minimum political requirement was barely satisfied, the Chosŏn court did not support the Astronomy Bureau's proposal for furthering the astronomy project into more advanced areas, such as calculating eclipses and planetary motions, in order to produce 'Ephemeris for the Seven Planets (*ch'ilchŏng ryŏk* 七政曆),' according to the *Shixian li* method. The deadlock lasted for half a century thereafter. Only in the first decade of the eighteenth century, several years before He Guozhu's mission in 1713, did the Chosŏn court resume its support for the Astronomy Bureau's project of learning Qing astronomy (Kim Seulki 2017; Lim 2012).

Apparently triggered by an embarrassing discrepancy between the almanacs of Chosŏn and the Qing for the year of Kangxi's 44th reign (approximately 1705 in the Gregorian calendar), the renewed interest of the Chosŏn court in Qing astronomy was fundamentally conditioned by significant changes in Qing-Chosŏn political relations. Weakening of the hostile political factors, having thus far blocked the court's support of Qing astronomy, provided the astronomers and their patrons at the court with new opportunities to resume the project of learning the higher level of Qing astronomy that had remained unstudied for half a century. The court sent an astronomer named Hŏ Wŏn 許遠 to Beijing twice, in 1705 and 1708. He contacted a Qing astronomer named He Junxi 何君錫, then an aged official of the Imperial Astronomy Bureau famous for his former allegiance to the old system in the calendar controversy in early Kangxi period (Chang 2023; Lim 2013). From He Junxi, the Korean astronomer was able to learn how to calculate planetary motions according to the new method. From the year 1708, the Chosŏn's Bureau of Astronomy was able to issue its own 'Ephemeris for the Seven Planets,' according to the new *Shixian li* method. In 1711, based upon his knowledge learned from the Chinese astronomer, Hŏ Wŏn compiled a manual of calculating ephemeris for his fellow astronomers in the Astronomy Bureau (Kim Seulki 2017; Lim Jongtae 2013).

The Chosŏn astronomers' successful acquisition of the advanced level of Qing-Western astronomy signaled the outset of Koreans' effort to localize the imperial astronomy. It was a two-tiered process, in which, on the one hand, Qing astronomy was technically accommodated to Chosŏn's locality, while, on the other hand, a new cultural identity was invented, by which to transform the Manchu-Western astronomy into Chosŏn's own cultural asset.

However, the cultural justification of the Chosŏn elites for accepting and appropriating the Qing astronomy did not take the form of imbuing it with a new national identity, as is often the case in modern nationalist projects (Harrison and Johnson 2009). Instead, they relied on the Ming loyalist ideology that was prevalent in Korea at the time, especially after the sixtieth anniversary of the fall of the Ming in Beijing in 1644. Since the Qing had promulgated the *Shixian li* system on the basis of the calendrical treatises compiled in the late Ming period under the auspices of the Ming's Chongzhen emperor, Korean elites decided to view *Shixian li* astronomy as a noble legacy of Ming China that had been handed down to Chosŏn, making Korea the sole remaining protector of Middle Kingdom culture. This was the logic behind the reproduction of Adam Schall's star map and Matteo Ricci's world map by the Chosŏn Bureau of Astronomy in 1708, the year when the Bureau promulgated its first ephemeris using the *Shixian li* method. In the preface to the reproduction of these two Jesuit maps, Ch'oe Sŏkchŏng 崔錫鼎, then the Chief State Councilor and himself a great patron of astronomy as well as a competent mathematician, equated these two late Ming Jesuit works with the 'world-ordering of Wu the Great and the calendar of the Zhou dynasty' (*wubong churyŏk* 禹封周曆) (Lim 2012, pp. 221–223).

With this new cultural justification for adopting Qing astronomy, the Bureau of Astronomy, particularly its leading astronomer Hŏ Wŏn, seemed to try to move the project of learning Qing astronomy into the next stage: constructing a time keeping system according to the *Shixian li* method. Since time keeping depended on astronomical phenomena as observed in a given locality, such as the meridian stars seen at the 24 solar terms, it required the systematic observation of stars to implement the local time keeping system. The Astronomy Bureau's increased interest in observation can be glimpsed in the fact that Hŏ brought from Beijing in his 1708 mission the treatise on astronomical instruments, compiled by Ferdinand Verbiest, titled the *Lingdai yixiang zhi* 靈臺儀象志 (Treatise on Astronomical Instruments Installed at the Imperial Observatory, 1674), and the project of its republication was already underway when He Guozhu came to Seoul in 1713 with his 'semi-circle shaped' instrument. It was a very timely overlap, since it provided Hŏ Wŏn and his patrons with a good opportunity to remind the king and his court of the importance of instrumentations in perfecting the official calendrical system (Kim, Seulki 2017, pp. 460–461).

The court thus decided that Hŏ Wŏn be sent to the envoy's residence and learn from him about astronomical instruments and meridian star observation. A skillful artisan-officer named Ch'oe Ch'ŏnyak 崔天若 would accompany the astronomer, showing that learning and then reproducing the envoy's instrument were the main objectives of the Chosŏn court (*Sŭngjŏngwŏn ilgi* 承政院日記 [The Daily Records of Royal Secretariat of Chosŏn Dynasty, hereafter SJWIG], 15/Yun 05/Sukchong 39 [1713]).

The meeting between He Guozhu and Hŏ Wŏn, as described in Chosŏn's official chronicles, showed how the Korean side took a modest attitude toward whom they considered the imperial authority in calendrical astronomy. At first, the king even ordered Hŏ Wŏn to hide his identity as a government astronomer and to behave as a layman who happened to have an interest in astronomy. The king worried that the Qing might make a political issue after knowing of Chosŏn having its own calendrical system, a possible encroachment on the imperial prerogative in calendrical science (*SJWIG* 15/Yun 05/39).

It was needless to worry, however. He Guozhu was the son of the Chinese astronomer, He Junxi, who had helped Hŏ Wŏn during his previous visit to Beijing. Perhaps because of this connection, the meeting between the two parties was very fruitful. Hŏ Wŏn even accompanied the imperial astronomer on his return journey to the border town, 'learning exhaustively the difficult part of the meridian star method and the usage of the astronomical instrument' (*Pibyŏnsa tŭngnok* 備邊司謄錄 [Records of the Border Defense Council], 08/01/Sukchong 39). The court immediately assigned the artisan Ch'oe Ch'ŏnyak to make a copy of the instrument, and the next year saw the completion of the printing of Verbiest's *Yixiang zhi*.

Shortly thereafter, the promulgation of Chosŏn's local time keeping system was also in order. However, for reasons unknown to us, the project never seemed to be fully realized at that time. What we do know is that the Bureaus of Astronomy would soon be involved in yet another phase of learning the Qing astronomy, as the Qing promulgated the new calendrical system based on the *Thorough Investigation of Calendrical Astronomy* (*Lixiang kaocheng* 曆象考成) in the next decade. Notably, He Guozhu's value of the latitude of Seoul was incorporated into this new Qing system along with those of other places under the Qing Empire. The project of localizing Qing astronomy would resume in earnest in the second half of the eighteenth century, when Korean astronomers fully grasped the astronomy of *Lixiang kaocheng*, in which the Chosŏn Kingdom was, at least astronomically, a part of the Qing Empire.

5 Emergence of the 'Eastern Computation'

If the official description of Hŏ Wŏn's astronomical session with He Guozhu presented it as a one-way transfer of advanced knowledge from the latter to the former, we have yet another record of a scientific exchange occasioned by He Gouzhu's mission that provides a very different, much more reciprocal reflection of the inter-relational dynamics of the encounter. Around 1724, more than ten years after the imperial embassy, a Korean mathematical official of the Ministry of Revenue, Hong Chŏngha 洪正夏, recorded a long dialogue between the imperial astronomer and two Korean mathematicians, including himself, in his mathematical work, entitled the *Kuil chip* 九一集 (Treatise on Nine and One).

At the beginning, Hong gave a very brief introduction to the meeting to show that the meeting had really happened. It reads: 'On the 29th day of the intercalary 5th month of the year Gyesa (1713), I, together with Mr. Yu Susŏk, entered the envoy's residence, where we discussed computational methods (*san* 算) with the Calendar Manager of Five Bureaus, He Guozhu' (Hong Chŏngha, date unknown, Book 3, 73a). Hong then directly introduced a series of mathematical problems and solutions that were exchanged between the two parties.

Hong structured his portrayal of the meeting in a highly symmetrical way. The exchange, as described in the *Kuil chip*, was a significant encounter of the two mathematical traditions (Jami 2011, pp. 277–279). The political tension between the two states also featured in Hong's account. After He Guozhu raised several problems of simple divisions and proportions, to which Korean mathematicians provided correct answers, the Chief Imperial Envoy, Aqitu, asked the two Koreans to test He Guozhu's mathematical ability. Aqitu praised He Guozhu by boasting that 'Our Calendar Manager ranks the fourth in the world [all under heaven] in computational skill; his tummy is full of computational methods (*suanfa* 算法).' Apparently provoked by this challenge, Hong raised a difficult question dealing with a cube inscribed in a sphere, for which the imperial astronomer reportedly failed to provide the correct answer and solution. As if to confirm his victory, Hong added his solution to the problem. His solution used the celestial origin (*tianyuan* 天元) method, a technique for solving algebraic equations, which was invented in the Song and Yuan period in China but by this period had been preserved only in Korea (Hong Chŏngha, date unknown, Book 3, 74a–75a).

While no doubt satisfying to his Korean readers, Hong did not want to make his story simply one of a patriotic confrontation in which Korean mathematicians defeated an arrogant imperial astronomer. In the later part of his account, for instance, the imperial astronomer taught the two Koreans a new method for finding the length of one side of a regular pentagon inscribed in a circle. It was trigonometry. The Koreans confessed that 'the Eastern Kingdom has never seen this method (東國未有此法),' and asked him to give them an instant lesson on it. The imperial astronomer recommended that they read the *Jihe yuanben* 幾何原本

(Elements of Geometry) and the *Celiang quanyi* 測量全義 (Complete Meaning of Measurement), works on Western geometry and surveying techniques complied respectively by Matteo Ricci and Giacomo Rho in the late Ming period. The Koreans also asked the imperial astronomer to evaluate their level of attainment in mathematics, to which He Guozhu replied favorably, though condescendingly, 'You have already mastered seven or eight of the computational techniques, and the remaining two or three will be easy to master. You gentlemen should try harder' (Hong Chŏngha, date unknown, Book 3, 77a–78a).

Hong's story did not end here. At the end of his account, he had the imperial astronomer betray his interest in the counting rods with which the Korean mathematicians solved the algebraic equation, confessing that 'we don't have these counting tools in the Middle Kingdom (中國無如此算子).' Hong gave his counting rods to the imperial astronomer as a gift. He Guozhu also recorded the two Korean mathematicians' names to 'make them known to the Great Country (大國)' (Hong Chŏngha, date unknown, Book 3, 81b).

All in all, Hong portrayed his conversation with He Guozhu as that of an encounter between the mathematical traditions of the Eastern Kingdom and the Middle Kingdom. Each country had its own mathematical tradition, one centering on the celestial origin method and the counting rod calculation and the other relying on Western mathematics. To the mathematics of his own country, Hong even appeared to give a distinctive name: the 'Eastern Computation' (*tongsan* 東算). This term appeared in the title of a manuscript, which was considered by historians as an earlier draft of Hong's work written in around 1718, the *Draft of the Eastern Computation* (Tongsan ch'o 東算草). Without a preface or any other clarification by the author, we do not know which technical and cultural meanings Hong wanted to give to this name. Based on Hong's portrayal of his contact with the Qing astronomer, however, I can suggest the following two points.

First, the so-called Eastern Computation, as coined by Hong, should not be understood in the modern nationalist sense. Although the Eastern Computation certainly referred to the techniques and methods that were being practiced in Korea, Hong knew very well that their computational methods, including the celestial origin method, had their origin in the great mathematicians of the Song and Yuan periods in China. When opposing Eastern Computation against Chinese (or Qing-Western) mathematics, Hong could very well have intended to highlight that the essence of Chinese classical mathematics, which had been completely lost in China, was now preserved intact in Korea. If this was his intention, then the identity strategy of Korean mathematicians paralleled that found in Ch'oe Sŏkchŏng's preface to Matteo Ricci's world map, in which he promoted Chosŏn as the heir of the scientific legacy of Ming China and the entire Middle Kingdom lineage.

Second, Hong did not imagine the community of Eastern mathematicians as culturally closed. On the contrary, Hong structured the story of his mathematical encounter with the Qing astronomer as a process of reciprocal give and take exchanging what they had for what they lacked. Why did Hong place this story at the end of his treatise on Eastern Computation? This shows, I suggest, that Eastern Computation, as Hong envisioned it, was an open-ended entity that was to be enriched and continually updated as the Chosŏn mathematicians learned and incorporated the new mathematics of Qing China into their own work. Similar to the changes that occurred in the field of astronomy, mathematics in Korea also underwent a fundamental change in the eighteenth century with the introduction of the new Qing-Western mathematics in the mid-eighteenth century, which was synthesized in the *Shuli jingyun* 數理精蘊 (the Essence of Numbers and Their Principles), the Qing mathematics text compiled together with the *Lixiang kaocheng*.

6 Conclusions

So far, this chapter has explored the scientific exchanges between Korea and Qing China in the late seventeenth and early eighteenth centuries, reaching a culmination with He Guozhu's visit to Chosŏn in 1713. These scientific exchanges took place in an environment characterized by overwhelming political inequality between the two, with marked disparities in the cultural resources each possessed. While this imbalance often framed scientific exchanges between actors from the two states as either a unidirectional flow of advanced scientific knowledge from the center to the periphery or the empire's unilateral collection of information from its tributary state, the evidence suggests that the actual exchange between the two was quite reciprocal.

The specific forms of reciprocity varied from field to field. In the field of astronomy, where the political hierarchy of the empire and the tributary state directly translated into the hierarchy of knowledge, exchanges took the form of one-way transmission, and both He Guozhu and Hŏ Wŏn understood their exchanges as a process by which the advanced observational astronomy of the Qing was transmitted to Korea. The elite scholar-officials such as Ch'oe Sŏkchŏng only added an ideological variation to this one-way flow by insisting that the astronomy they embraced was a legacy of the Ming dynasty, not the Qing. In the case of geography, the Qing sought to collect geographical information from Chosŏn, taking advantage of their overwhelming power. However, Chosŏn actors were also able to explore and collect the Qing's geographical information in their own manner. As a result, there was no fundamental difference in the quality of geographical information available to the actors of the two states. In the case of mathematics, according to the accounts of Chosŏn mathematicians, the encounter was a fairly symmetrical and reciprocal exchange of Western mathematics held by the Qing for classical Chinese mathematics held by Chosŏn.

In conclusion, the political tension and conflicts between the Qing and Chosŏn did not prevent the scientific exchange between them from being reciprocal. While Korean actors seized these scientific exchanges with the Qing as unique opportunities to construct their own scientific tradition with distinct ideological and scientific characteristics, their efforts never led to the creation of a radically different, self-sufficient national tradition. Instead, their science remained interconnected with, and thus commensurable with, the Qing's imperial science.

Acknowledgements I am grateful to the organizers of the conference, 'Needham's Dialogical Vision: Understanding Science as a Multi-Civilizational Outcome' and the editors of this volume, especially Professors Arun Bala and Raymond W.K. Lau, for giving me the opportunity to present my paper and then kindly guiding the process of its revision. I would also like to thank Professor Ku Beomjin of Seoul National University for his valuable comments on an earlier version of this paper.

References

Cams, Mario. 2017. *Companions in Geography: East-West Collaboration in the Mapping of Qing China (c.1685–1735)*. Leiden: Brill.

Chang, Ping-Ying. 2023. *The Chinese Astronomical Bureau, 1620–1850: Lineages, Bureaucracy and Technical Expertise*. New York: Routledge.

Elman, Benjamin A. 2005. *On Their Own Terms: Science in China, 1550–1900*. Cambridge: Harvard University Press.

Harrison, Carol E., and Ann Johnson. 2009. Introduction: Science and National Identity. *Osiris* 24 (1): 1–14.

Hŏ T'aeyong. 2009. *Chosŏn hugi chunghwaron-kwa yŏksa insik* 朝鮮後期 中華論과 歷史認識 [Sino-centrism and Historiography in Late Chosŏn Period]. Seoul: Ak'anet.

Hong Chŏngha. date unknown. *Kuil chip* 九一集 [Treatise on Nine and One], 3 vols. A Manuscript Copy Preserved in the Kyujanggak Institute for the Korean Studies, Seoul National University.

Jami, Catherine. 2011. *The Emperor's New Mathematics: Western Learning and Imperial Authority During the Kangxi Reign (1662–1722)*. Oxford: Oxford University Press.

Jartoux, Pierre. 1713. The Description of a Tartarian Plant, call'd *gin-seng*; with an Account of Its Virtues. In a Letter from Father Jartoux, to the Procurator General of the Missions of India and China. Taken from the Tenth Volume of Letters of the Missionary Jesuits, printed in Paris in octavo, 1713. *Philosophical Transaction* 28: 237–247 https://royalsocietypublishing.org/doi/10.1098/rstl.1713.0025. Accessed 17 May 2023.

Seonmin, Kim. 2007. Ginseng and Border Trespassing Between Qing China and Chosŏn Korea. *Late Imperial China* 28 (1): 33–61.

Kim Seulki. 2017. Sukchongdae Kwansanggam-ŭi sihŏnnyŏk haksŭp: Ŭryunyŏn yŏksŏ sagŏngwa kŭe taehan Kwansanggam-ŭi taeŭng-ŭl chungsimŭro (Kwansanggam's Learning of *Shixian li* in King Sukchong's Reign: Chosŏn Astronomers' Response to the Year Ŭryu Calendar Affair). *Hanguk kwahaksa hakhoeji* 韓國科學史學會誌 [Korean Journal for the History of Science] 39 (3): 435–464.

Kim Seulki. 2023. Chŏngjo 13-nyŏn-ŭi Chŏnmunhak p'ŭrojektŭ [The Astronomy Project of the Chosŏn Government in 1789]. A Paper Presented in the Colloquium for the History of Science in Chosŏn Korea, Seoul National University, 6 October 2023.

Li Hua-tzu. 2008. *Cho Ch'ŏng kukkyŏng munje yŏn'gu* 朝ø國境問題ɔ究 [A Study on the Issues Concerning the Chosŏn-Qing Border]. Seoul: Jimmundang.

Lim Jongtae. 2012. Learning Western Astronomy from China: Another Look at the Introduction of the *Shixian li* Calendrical System into Late Joseon Korea. *The Korean Journal for the History of Science* 34 (2): 197–217.

Jongtae, Lim. 2013. Journeys of the Modest Astronomers: Korean Astronomers' Missions to Beijing in the Seventeenth and Eighteenth Centuries. *Extrême-Orient Extrême-Occident* 36: 81–108.

Lim Jongtae. 2016. The Tributary Relations Between the Chosŏn and Ch'ing Courts to 1800. In *The Cambridge History of China, Vol. 9: The Ch'ing Empire to 1800. Part II*, ed. Willard J. Peterson, 146–196. Cambridge: Cambridge University Press.

No Taehwan. 2005. *Tongdo sŏgi ron hyŏngsŏng kwajŏng yŏngu* 東道西器論形成過程ɔ究. [Emergence of the Idea of Eastern Way and Western Utility in Late Chosŏn Korea]. Seoul: Ichisa.

Pae Usŏng. 1998. *Chosŏn hugi kukt'ogwan-kwa ch'ŏnhagwan-ŭi pyŏnhwa* 朝鮮後期國土觀과 天下觀의 變化 [Changing Perception on the National Territory and World in Late Chosŏn Korea]. Seoul: Ilchisa.

Pibyŏnsa tŭngnok 備邊司謄錄 [Records of the Border Defense Council], Electronic Version by Kuksa p'yŏnch'an wiwonhoe. https://db.history.go.kr/item/level.do?itemId=bb

Sŏ Hosu, Sŏng Chudŏk, and Kim Yŏng, eds. 1796, *Kukcho yŏksang ko* 國朝曆象考 (Compendium of the Calendrical Astronomy in Chosŏn Dynasty) [Korean Translation by Moon Joong-Yang, Yi Ŭnhŭi. Seoul: Somyŏng cg'ulp'an, 2004].

Sukchong sillok 肅宗實錄 [Veritable Records of King Sukchong's Reign], Electronic Version by Kuksa p'yŏnch'an wiwonhoe. http://sillok.history.go.kr

Sŭngjŏngwŏn ilgi 承政院日記 [The Daily Records of Royal Secretariat of Chosŏn Dynasty], electronic version by Kuksa p'yŏnch'an wiwonhoe. http://sjw.history.go.kr

Yi Ŭnsŏng. 1985. Taerisŏkche Sinbŏp chuipyŏng ilgu-wa Sohyŏn Seja [The Plane Sundial According to the New Western Method and the Crown Prince Sohyŏn]. *Tongbang hakchi* 48: 85–124.

7

Practical and 'Precise' World Geographical Knowledge Developed in Premodern Chinese and Islamic Worlds Through Multi-civilizational Connections and Contact

Hyunhee Park

1 Introduction

Through a lifelong effort epitomized by his magnificent series *Science and Civilisation in China* (hereinafter SCC), Joseph Needham significantly reshaped the history of science and technology by underscoring the vital contributions of non-European civilizations (Finlay 2000, pp. 265–303). Volume III of this series, focusing on 'Mathematics and the Sciences of the Heavens and the Earth,' stands as a testament to this, introducing an unprecedented discourse on Chinese geography and cartography alongside related disciplines like mathematics, astronomy, and geology. This volume illuminates a tradition of scientific precision in Chinese practices that found parallels in Islamic societies, challenging the traditional Eurocentric narrative of technological progress in world history.

H. Park (✉)
History Department, CUNY John Jay College and The Graduate Center, New York, NY, USA
e-mail: hpark@jjay.cuny.edu

A. Bala et al. (eds.), *Multicivilizational Exchanges in the Making of Modern Science*,
https://doi.org/10.1007/978-981-97-3541-9_7

In SCC Volume III's Chapter 22 titled 'Geography and Cartography,' Needham and his colleagues compare Chinese technological achievements with those of contemporary European and Islamic societies, adopting a comparative and global historical perspective (Needham and Wang 1959, pp. 497–590). They meticulously analyze a wide array of significant maps and geographical treatises spanning from ancient to the early modern period in China, a period also marked by the introduction of new European influences through Jesuit missionaries. This comprehensive examination leads them to suggest that China's cartographic tradition, intertwined with aspects of religious cosmography, was as advanced, if not more so, than those found in Europe and many other global societies. This chapter builds upon these foundational insights, endorsing and expanding upon the two central themes presented in Volume III. It aims to provide a more nuanced and comparative analysis of the evolution of geographical knowledge, exploring the significant impacts and interconnections of the premodern Chinese and Islamic worlds on the broader scope of global geographical understanding.

2 Inter-civilizational Exchanges and the Evolution of Cartography: Revisiting Volume III of SCC

Volume III of SCC not only showcases the comparative technological advancements but also reveals the complex web of cultural and intellectual exchanges that have shaped the cartographic traditions of various civilizations. The first central theme of the volume emphasizes the potential influences of one civilization on another, often facilitated by bilateral exchanges. This theme is explored through a series of detailed examples within the volume. However, at the time of its publication, limited studies were available, leading the authors to suggest rather than conclusively demonstrate the importance of these knowledge transfers through inter-societal exchange. Notably, the volume presents cases indicating the possible influence of Arab geographic and cartographic techniques on Chinese practices, and in some instances, even on European cartography

(Needham and Wang 1959, pp. 561–564). With the surge in relevant studies in recent decades, it becomes imperative to revisit and reassess these influences in light of new evidence.

The second central theme, intricately linked with the first, examines the development of practical and precise geographic knowledge in China, significantly shaped by its extensive multi-civilizational connections and contacts throughout history.[1] This aspect of the volume's discourse, now bolstered by recent scholarly work, calls for a deeper exploration. The exchange of geographic knowledge with the Islamic world—referred to as the 'Arab world' in Needham's volume—is particularly notable. By the medieval period, these Islamic societies had developed some of the world's most practical and precise geographic knowledge. While Needham's volume explores the religious and cosmographical aspects of these societies' worldviews, it emphasizes the scientific drive behind China's quest for expanded, real-world knowledge. This emphasis on scientific inquiry in premodern China, in conjunction with the valuable insights provided by the Islamic world, underlines the importance of these inter-civilizational exchanges in shaping the history of cartography.

It is important to note that, since this 1959 publication of SCC Volume III first discussed Chinese geography and cartography from a comparative perspective, the field of the history of geography and cartography in general has enjoyed an unprecedented boom worldwide. New studies in different countries have advanced new methods and introduced new sources in greater quantity, variety, and detail. Among the most significant events in this international trend has been the establishment of the bi-annual conference on the History of Cartography (ICHC), in which not only scholars but also map collectors and enthusiasts from all over the world actively participate. Developments like this in turn led to major publications in the field, including the journal

[1] Needham had to omit the discussions advanced by Sinologists by then because of the space limit (Needham and Wang 1959, 497). In the context of this chapter, 'practical' refers to aspects of geographical knowledge that prioritize the actual doing or experience of mapping and spatial understanding over theoretical or conceptual ideas. This definition emphasizes the functionality and applicability of geographical information in real-world scenarios, as opposed to purely symbolic or religious representations. It highlights the focus on empirical accuracy and utility for navigation and territorial administration, ensuring that the geographical information is directly applicable to practical needs such as travel, trade, and governance.

Imago Mundi and the series *History of Cartography*. In particular, *History of Cartography, a massive multi-volume series* published since 1987 and edited by John B. Harley and David Woodward, attempts to cover extensively the cartographic traditions of non-Western societies in its three large volumes, and its Open Access platform sets important foundations for further developments in comparative research.[2] Scholars have continued to advance the comparative approach in conferences and other venues worldwide.[3]

In the realm of recent advancements in the history of cartography, a crucial shift has been noted: a move away from the singular, modern Eurocentric standards toward an appreciation of the rich diversity of geographic imagery from multiple cultural contexts. This trend, responding to Harley's call in the 1980s for a 'humanistic turn' in cartography and geography, has illuminated previously unrecognized forms of cultural diversity in world societies, including those in Asia.[4] However, this broader and more flexible approach, while enriching our understanding of diverse geographic representations, sometimes inadvertently neglects a critical aspect of geographical and cartographical development in premodern Asian societies. This aspect is the development of practical and precise world geographical knowledge, which was intricately tied to cross-cultural contacts. Notably, this phenomenon is epitomized by the historical Silk Roads, which facilitated not only the movement

---

[2] Volume 1 covers Europe; Part 1 of Volume 2 the Islamic World and South Asia; Part II of Volume 2 East and Southeast Asia; and Part 3 of Volume 2 the rest of the world (Harley and Woodward 1987–1998). Many ancient maps of the non-Western world received greater attention through these volumes.

[3] For example, the Geography Research Institute of Yunnan University in China translated all the volumes of Harley and Woodward's *History of Cartography* into Chinese. It also held an international academic conference in 2019 that assembled many experts in the history of cartography to initiate collaborative research and seek further global cross-cultural research methods for future endeavors in the field.

[4] Matthew H. Edney (2007) offers a comprehensive overview of the field's expansion and evolution since the 1970s, highlighting both traditional concerns and new directions in cartographic scholarship. Further illustrating this trend are works like those of D. K. Yee (1994a, 1994b, 1994c), which reassess Chinese cartography in its own right, the book by and Martin Lewis and Karen E. Wigen (1997), which stimulates a reevaluation of large-scale spatial constructs by challenging traditional geographical views and revealing the subjectivity in maps and spatial categorizations, ultimately advocating for new frameworks that enhance our understanding of human diversity and foster a more conscious approach to defining and perceiving our diverse world.

of people, goods, and ideas but also played a significant role in the scientific and technological advancements discussed in the Needham volumes, including those in geography and cartography. It is this intersection of transnational, multi-civilizational connections and contacts that was pivotal in shaping the geographic understanding of the world in premodern times, an aspect that must be given due attention in contemporary cartographic studies.

To establish a balance amid the evolving landscape of the global history of geography and cartography, this chapter emphasizes the significance of scientific methods in geographical study that have brought to light lesser-known yet more practical and perhaps more 'accurate' forms of geographical knowledge developed in premodern Asia. These forms, having emerged in societies outside the commonly studied European context, benefited immensely from the thriving networks of cross-cultural interactions, known as the Silk Road. The study of foreign geographical knowledge as amassed by the Chinese across various historical periods has long been a focal point in Sinology.[5] Numerous distinguished Sinologists and Sinology centers across Asia have consistently dedicated efforts to gathering, preserving, and disseminating research findings related to geographic and cartographic knowledge.[6]

Recently, the scope of scholarship has expanded to include an in-depth understanding of the role of geographic knowledge transfer in the fabric of global history, thereby enhancing our comprehension of these intricate intercultural exchanges. Nonetheless, many studies in the history of cartography, particularly those adopting global perspectives,

---

[5] The significance of Chinese literature in Sino-foreign relations is underscored through various translations and annotated works of primary sources. These works collectively highlight the long-standing importance of Sinology in understanding and documenting foreign lands, as evidenced by the works on Zhao Rukuo's *Zhufan zhi* (Description of the Foreign Lands), authored in the early thirteenth century (Hirth and Rockhill 1911; Yang 1996; Fujiyoshi 1991).

[6] A notable example of significant cartographic achievement is the Kangnido (混一疆理歷代國都之圖, The Map of Integrated Lands and Regions of Historical Countries and Capitals), created in Korea in 1402. This map reflects the extensive Afro-Eurasian connections and contacts of the thirteenth and fourteenth centuries, symbolizing the global interconnectedness characteristic of the Mongol era. Its significance, particularly for the place names in western Eurasia, has garnered sustained interest among historians and geographers (Sugiyama, 2007). For further analytical studies on the Kangnido, see Muraoka (2020) and Kenzheakhmet (2021); Park (2018) provides a recent analysis of the Kangnido map's content and historical context, along with a concise overview of major studies on the map.

have predominantly focused on European cartographic developments from the early modern period. Additionally, some works have concentrated on Asian cartographic changes within the context of eighteenth and nineteenth-century colonization.[7] When addressing Asian cartographies, the prevalent emphasis has been on the influence of the Jesuits, their introduction of European cartographic techniques to Asia from the sixteenth century onward, and the subsequent innovative synthesis of these techniques in Asian cartography (Cams 2017). This focus, however, often overshadows the vibrant development of geography in Asian countries, especially the scientific knowledge accrued through their extensive multi-civilizational connections and contacts. There is a clear need for more comprehensive studies in this area to achieve a balanced representation in the field of cartographic history.

This gap in scholarship is particularly striking when juxtaposed with Western perceptions, which often depict China as 'incurious' about foreign lands and cultures. This caricature is even echoed by Joseph Needham, typically an advocate for Chinese scientific contributions, who once remarked, somewhat out of character: 'The greater population of China was self-sufficient...generally content with only occasional exploration, essentially incurious [my emphasis] about those far parts of the world' (Needham 2005, quoted in Cohen 2010, p. 135). However, contrary to this prevailing stereotype, a wealth of historical evidence demonstrates China's profound interest in foreign societies. The research of scholars like Jongtae Lim and Ryuji Hiraoka vividly illustrates China's extensive exchanges with neighboring East Asian countries, such as Korea and Japan (Hiraoka and Watanabe 2015; Lim chap 6 of the present volume). These interactions, however, were not confined to its immediate vicinity. China also forged significant economic and intellectual ties with more distant civilizations, particularly those of the Islamic world. These connections underscore a broader narrative of China's active engagement

[7] For an illustration of this focus on Asian cartography in the context of colonialism, see Matthew Edney's (1999) monograph, *Mapping an Empire: The Geographical Construction of British India, 1765–1843*, which details how Britain utilized modern survey techniques in India to shape and legitimize the spatial image of its empire, thereby playing a crucial role in its colonial endeavors.

and curiosity in the wider world, challenging the notion of its insularity and lack of interest in global affairs.

A closer examination of the Silk Road unveils the distinctive nature of Chinese interactions with the Islamic world during the premodern era, contrasting sharply with its relationships with neighboring East Asian countries and later with European powers during their age of global expansion. The exchanges with Islamic societies, particularly those in the West, were instrumental in broadening the horizons of Chinese scientific geography. They infused Chinese cartographic traditions with a depth and detail that were unparalleled, borne from the rich tapestry of knowledge woven through these premodern networks. This historical interplay raises an intriguing question: How did the fusion of Chinese and Islamic geographical knowledge impact key European discoveries, such as the Americas and the sea route around the Cape? These events, which George Saliba considers pivotal in the rise of modernity, may well have been influenced, directly or indirectly, by the confluence of insights from the East. This hypothesis invites a reevaluation of the traditional narrative of European exploration, suggesting a more integrated global history where the contributions of Asian civilizations are duly recognized in shaping the world as we know it.

In exploring these questions, we find valuable guidance in the perspectives of scholars like Prasenjit Duara, who challenge the traditional view of scientific and technological history as a linear, evolutionary progression. Duara (Chap 2 of the present volume) proposes the concept of 'circulatory histories,' highlighting the dynamic and fluid nature of historical developments. This framework recognizes that advancements in science and technology do not occur in isolation; they flow through various regions, interacting with and adapting to local contexts, leading to their continuous transformation.[8] By adopting this perspective, we gain a more nuanced and comprehensive understanding of how scientific knowledge and practices have been shared, modified, and evolved across different civilizations. This approach, emphasizing the interconnected

[8] For the role of distilled liquors in the transfer and localization of food technology, see Park 2021.

web of global scientific progress, leads us seamlessly into a closer examination of specific instances where such circulatory histories are evident, including in the realm of geography and cartography.

3 Precision and Exchange in Geography and Cartography: The Evolution of Chinese Mapping Practices and Cross-cultural Influences

A striking example of such circulatory histories can be seen in the scientific practices of Chinese geography and cartography, as highlighted in the Needham volume. Since ancient times, one notable practice has been the implementation of a grid system. *The Tracks of Yu*, one of the earliest maps to depict all of China, employs this grid as a unit of measurement, with each square representing 100 *li* (Needham and Wang 1959, p. 547). This map is a testament to the advanced mapmaking techniques of Chinese cartographers, who achieved a level of accuracy unparalleled in their era. The development of this grid system is often credited to Pei Xiu (224–271 CE), known as 'the father of geography,' who established systematic principles for creating precise maps, subsequently adopted by later Chinese geographers.[9] Furthermore, the Needham volume sheds light on various documentary records, illustrating how Chinese governments compiled accurate and practical geographical knowledge for governance purposes (Needham and Wang 1959, pp. 517–521). This extensive and systematic approach to topographical documentation, as noted by Sinologists, likely surpasses that of any other nation in both scope and detail, underscoring the sophistication of Chinese cartographic practices in historical context (Needham and Wang 1959, 517).[10]

[9] Few maps of China proper or foreign regions, including Pei Xiu's maps, during the first millennium survived. However, evidence including the earliest surviving maps hint at his influence on later maps (Park 2012, p. 35).

[10] Also see Alexander Wylie's monograph, *Notes on Chinese Literature: With Introductory Remarks on the Progressive Advancement of the Art* (1867). See also Hargett (1996, pp. 405–442) for an in-depth analysis of Song Dynasty local gazetteers. Recent scholarship has further explored

While the Needham volume, due to its space constraints and the limited evidence available at the time of publication, could only give modest attention to the broader development of world geography, recent studies have significantly expanded our understanding in this area. These studies uncover more evidence of the transfer and exchange of geographical and cartographic knowledge during the premodern period. Through comparative examinations of various geographical treatises and maps, they have revealed a rich tapestry of historical connections and contacts among societies, which played a pivotal role in the production and dissemination of maps and geographic treatises. The exchange of knowledge between the Chinese and Islamic worlds, both dominant civilizations during the medieval period (approximately 700–1500), brings into focus a critical aspect of their interconnected history.[11]

As major contributors to the network of multi-civilizational connections across Afro-Eurasia, China and the Islamic world enjoyed periods of significant economic, institutional, and cultural development, largely driven by extensive international trade. This vibrant interplay of interactions and exchanges did not merely foster economic growth; it also catalyzed the accumulation of practical and accurate geographic knowledge within both civilizations. Their geographic understanding extended beyond their immediate regions, encompassing territories and peoples far beyond their traditional borders. This deepened knowledge base reflects the profound impact of their exchanges and highlights the need for a more inclusive view of the evolution of global geographic knowledge.

While both China and the Islamic world developed unique geographical and cartographic traditions that differed from modern scientific practices, their methodologies were sometimes significantly influenced

---

the production of geographical knowledge in imperial China, particularly highlighting the role of local gazetteers (*difang zhi* 地方志). As discussed by Alexis Lycas, Masato Hasegawa, and Shih-Pei Chen (Lycas et al. 2022, pp. 161–165), historians of imperial China have led the way in analyzing the geographical production of places, especially through the cultural and historiographical significance of local gazetteers during the early medieval period. This recent research underscores the systematic and comprehensive nature of local topographical writings in China.

[11] For pioneering works that examine the contacts between China and the Islamic Middle East, see Park's (2012) monograph, *Mapping the Chinese and Islamic Worlds*, and D. D. Leslie's (1986) *Islam in Traditional China: A Short History to 1800*.

by their mutual interactions. A recent collaborative research project examining medieval cartography, which included Latin-Christian, Arab-Islamic, and Chinese maps, brought to light the different emphases placed on large-scale structures in geographical understanding across these societies.[12] For example, Chinese maps, as revealed in the project, tended to focus less on physical depiction and more on symbolizing political and cultural unity. In contrast, Latin-Christian and Arab-Islamic traditions demonstrated a broader scope, encompassing detailed representations of the entire known world. This comparative study underscores the diversity and significance of mapping traditions within distinct cartographic cultures, revealing how each was shaped by their unique historical contexts and exchanges. These insights suggest the importance of recognizing the exchange of geographic information and techniques between China and the Islamic world, as it was instrumental in refining the cartographic practices of both civilizations. Each contributed unique perspectives and methodologies, leading to the development of sophisticated and interconnected mapping traditions.

Understanding these interconnections is crucial, as they laid the foundation for a more intricate and globally integrated approach to cartography. This deeper recognition allows for a comprehensive exploration of the individual and collective contributions of these civilizations to the field of cartography, enhancing our appreciation of their role in shaping the geographical understanding of the world. Such a perspective not only acknowledges the diversity of cartographic legacies but also reinforces the importance of cross-cultural influences in the evolution of geographic knowledge.

While the earliest surviving Chinese maps from the eleventh century do not depict the wider world, they reflect the accumulation of practical and remarkably accurate geographical knowledge, much of which is found in the vast array of their geographical treatises. This depth of knowledge was largely derived from the accounts of travelers, including those who journeyed to the 'Western regions' (lands west of China)

12 The collaborative study by Christoph Mauntel, Klaus Oschema, Jean-Charles Ducène, and Martin Hofmann attempted to identify the different traditions' marks of emphasis, each created in a specific cultural context and used as part of a strategy to arrange space in accord with it (Mauntel et al. 2018, pp. 295–367).

since ancient times. These travelers' reports provided a rich source of information for Chinese scholars, who documented detailed knowledge about places in Central Asia, and even lands further west like Iran and Rome, based on indirect sources, in official histories and other literary forms. A notable example is Zhang Qian (died 113 BCE), the envoy who pioneered the overland Silk Route, whose explorations informed the descriptions of Western Regions (Xiyu) in the classic *Records of the Grand Historian* (*Shiji*) by Sima Qian (145–circa 86 BCE). Sima Qian, in his historical works, provides detailed information about the physical distances between various cities in China and the Western regions. Building upon these textual descriptions, sinologists D. D. Leslie and K. H. J. Gardiner have reconstructed maps of Central and Western Asia, representing these regions with relatively accurate spatial precision based on the accounts from the official histories of the Han dynasty. This suggests that even without the aid of detailed maps, people of the time could still envision these territories accurately using the textual descriptions alone.[13]

This rich tradition of documenting foreign lands continued unbroken since Sima Qian first wrote about the Western peoples. The depth and accuracy of this literature grew substantially after China's Western contacts peaked during the Silk Road era.[14] Under the reunified Sui (581–617) and Tang (617–907) dynasties, China intensified its international connections, leading to a significant enhancement of its geographical understanding. During this period, figures like Pei Ju (557–627) ventured to Central Asia, creating works like *Maps of and Notes about the Western Regions* (Xiyu tuji 西域圖記). Although most of his work is

[13] The article by D.D. Leslie and K.H.J. Gardiner, 'Chinese Knowledge of Western Asia During the Han' (1982, pp. 254–308) gives critical introduction of main texts containing information about places in Central and West Asia during the Han period and provides different hypotheses for identifying Chinese names of place made by earlier sinologists including E. Chavannes and Fujita Toyohachi.

[14] For several specific cases, see articles in the special issue of *The Journal of Asian History* 49 (2015), especially Hyunhee Park's 'Information Synthesis and Space Creation' (pp. 119–140).

lost, its preface in the *History of the Sui* offers a glimpse into its content.[15] It reveals that the work contained detailed illustrations and extensive information, highlighting the diverse political, economic, and cultural aspects of the Silk Road and the Western Regions. This documentation showcases the advanced and increasingly accurate geographical understanding possessed by the Chinese during this period. As this knowledge accumulated, it not only facilitated further Chinese exploration into the wider world but also contributed to the further development of remarkably accurate and practical geographic knowledge about these expansive regions. The so-called world maps of China, dating back to the Song dynasty, while focusing primarily on China, began to include detailed names of foreign places along their margins, such as Dashi—referring to the Islamic world at the time (Park 2012, pp. 35–37). These names represented actual locations, moving beyond the imaginary places often depicted in earlier maps, and demonstrated the expanding geographical awareness and accuracy in Chinese cartography.

Alongside the depiction of foreign places at the margins of some maps focusing on the wider world, the expansion of Chinese geographic knowledge was significantly enhanced by the inclusion of detailed information such as distances in various genres of geographic literature.[16] This enrichment occurred as Chinese dynasties forged diplomatic connections with various countries. The sources of this accurate and practical geographic knowledge were diverse, ranging from Chinese travelers to international visitors, including merchants and diplomatic envoys who had been coming to China since ancient times. Notably, during the eighth and ninth centuries, merchants from West Asia navigated well-established maritime routes to reach China, where they interacted with the receptive officials of the Tang and Song dynasties. These encounters, facilitated by the dynasties' welcoming policies toward foreign merchants, played a crucial role in the enhancement of geographical understanding in China. These exchanges, representing a significant era of knowledge accumulation, highlight the dynamic interplay of domestic

[15] For a biography of Pei Ju 裴矩, see Wei Zheng (2000), *Sui Shu* (History of the Sui) 67, pp. 1577–1585.

[16] For a succinct examination of how information about Western countries, accumulated over centuries, was synthesized in a thirteenth-century Buddhist work, see Park (2015, pp. 133–136).

and international influences in shaping China's geographic knowledge. Such significant developments in the history of Chinese cartography and geography warrant an in-depth exploration of their key elements and impacts, particularly through the lens of cross-civilizational interactions. This exploration will delve into how Chinese, Islamic, and European mapping practices collectively influenced and shaped the evolution of global cartographic traditions.

4 Cross-civilizational Cartographic Exchange: The Confluence of Chinese, Islamic, and European Mapping Practices

The influx of foreign merchants, especially in the period following the rise of Islam and the subsequent expansion of Islamic empires across Afro-Eurasia, significantly bolstered maritime trade networks linking the Middle East to East Asia. These trade routes were crucial in driving the economic and cultural growth of the Islamic world, with West Asian merchants playing a central role. A key example of this interconnectedness is the work of Tang geographer Jia Dan (729–805), who authored 'The Route to the Foreign Countries across the Sea from Guangzhou.' This comprehensive navigational guide from Guangzhou to the Abbasid caliphate (750–1258) illustrates the depth of interaction and knowledge exchange during this era. Its contents, parallel to the emerging Arabic geographical works of the mid-ninth century, underscore the significant role played by Islamic Middle Eastern merchants in shaping a more systematic and realistic geographic understanding (Park 2012, pp. 31–34). This period marked not only a critical transfer of advanced geographic knowledge from the Islamic world to China but also symbolized a pivotal moment in the evolution of accurate and practical geographical knowledge, bridging the knowledge gap between East and West.

Inheriting the rich legacy of ancient Greek and Roman geographical knowledge, Islamic scholars during the medieval period crafted maps of the Afro-Eurasian oikoumene with remarkable accuracy. These maps, informed by the information brought back by traders who regularly journeyed to China, reflect an enhanced understanding of the vast region. A notable example is the prototype of the tenth-century-world maps by the Balkhī School, as featured in the Needham volume (Needham and Wang 1959, p. 562). These maps, though rudimentary in design, provided a relatively accurate representation of Eurasia and northern Africa as understood by Muslim geographers of the era. Building upon Greek and Roman cartographic principles, and positioned centrally in Afro-Eurasia, these Muslim geographers produced some of the most accurate geographical treatises and maps of the world available during that era, surpassing even those developed in China.

When contrasting this development with the state of cartography in Europe during the early and high Middle Ages (500–1300), a stark difference becomes evident. European societies, isolated from the vibrant Asian trade routes and largely disconnected from contemporaneous Asian civilizations, relied on maps such as the 'T-in-the-O' (T–O) map, the predominant mappa mundi of the period. This map, steeped in mythological views, depicted the Eastern hemisphere as an idyllic paradise, reflecting less the geographical reality and more the prevailing worldviews (Needham and Wang 1959, pp. 529–530).[17] However, when comparing these European maps with those from the Chinese and Islamic worlds, it becomes clear that the latter was developed in a context of more scientific geographic knowledge, fostered by extensive multi-civilizational connections and exchanges. This observation challenges the relativist interpretations that seek to counter Eurocentric perspectives by acknowledging diversity in geographic understanding. Such interpretations risk falling into an alternative Eurocentric trap that undervalues the advancements in practical and precise geographic knowledge achieved in the premodern Chinese and Islamic worlds. By carefully examining the geographic works from these civilizations, we gain a more balanced understanding that recognizes their contributions as closely aligned with the standards of modern scientific geography.

---

[17] For a more recent, detailed analysis of various forms of medieval world maps in the T–O style, see David Woodward's article in *The History of Cartography: Volume One* (Woodward 1987, pp. 318–353).

The Needham volume hints at a potential breakthrough in Chinese geographic and cartographic practices during the thirteenth and fourteenth centuries, suggesting that they were influenced directly by the Islamic world through the unprecedented contacts and exchanges under Mongol dominance. Recent studies have brought to light many new pieces of evidence supporting this assertion. A notable example involves Muslim scholars from the Il-Khanate, a Mongol regime in Central Asia and Iran closely affiliated with China's Mongol-run Yuan dynasty. These scholars brought Islamic world maps to the Mongol court in China, which depicted a more comprehensive view of the Afro-Eurasian landmass, thus playing a crucial role in the development of Chinese maps that, for the first time, included regions beyond China's traditional borders (Park 2012, pp. 99–105). One surviving map from this era, part of a project undertaken by the Yuan government to enhance its geographic database (a project briefly mentioned but not extensively explored in the Needham volume), is particularly illustrative of this influence. This map places south at the top, aligning with Islamic mapping traditions and features a grid function that mirrors modern longitude and latitude systems, pinpointing various locations within the vast Mongol empire, including sites in Central and West Asia. The presence of these features indicates a significant integration of Islamic cartographic techniques into Chinese practice, reflecting a blend of traditions that contributed to a more accurate representation of the world in some Chinese maps of the time.[18]

This blending of Chinese and Islamic cartographic traditions during the Mongol era is further evidenced in maps drawn based on these Mongol-era maps in later periods; a primary example is the *Kangnido* (produced in 1402), which is a remarkable embodiment of cross-cultural cartographic influences, blending Mongol-period maps with Korean sources to depict regions extending from Korea and China to Arabia, Africa, and Europe, underscoring its unique role in medieval cartography.[19] Recent scholarly works exploring the connections between

[18] This contrasts with traditional Chinese grid systems that show distances (Park 2013, pp. 125–158).

[19] See this chapter's footnote 6 for major works on this.

Europe and Asia suggest that this period was crucial for the transmission of Asian geographic knowledge to Europe. This exchange significantly influenced European global exploration and the development of scientifically accurate cartographic works. For example, European geographic accounts and world maps from the thirteenth and fourteenth centuries, like Marco Polo's (1254–1324) account and the Fra Mauro map (produced between 1457 and 1459), show expanded worldviews, reflecting Europe's increasing connections to Asia from the late thirteenth century onward.[20] These developments mark a pivotal moment in the history of cartography, where the interplay of Chinese, Islamic, and European mapping practices contributed to a more comprehensive and accurate understanding of the world.

The height of interactions between China and the Islamic world during the Mongol era was marked by extraordinary instances of cross-cultural exchange and exploration, with significant active participation from the Chinese side. One notable figure was the Chinese globetrotter Wang Dayuan (fl. 1311–1350), who, inspired by the historical accounts of Sima Qian, claimed to have ventured to West Asia and Africa. His travels culminated in the creation of a geography rich with new, accurate, and practical insights about these regions, providing testimony to the flourishing international trade of his era (Park 2012, pp. 115–118). About a century later, Ma Huan, a Muslim descendant in China, followed in these footsteps during his travels with Zheng He's (1371–1433) expeditions, journeying to Mecca on pilgrimage and providing eyewitness validation of Wang Dayuan's reports (Park 2012, pp. 161–162, 176–179). Likewise, merchants and scholars from the Islamic Middle East, as well as some European adventurers leveraging the re-established extended Eurasian trade networks during the Mongol period, traveled extensively. They reached far-flung destinations

[20] Recent analysis by Cattaneo of the Fra Mauro map (2022, 270–271) highlights its depiction of late Mongol-era sea routes, reflecting Venice's engagement with Eastern trade and the influence of narratives from travelers like Marco Polo. This map, showcasing detailed navigation routes in the Indian Ocean and trade circuits from China to Persia, represents one of the earliest comprehensive portrayals of these routes on a world map. Additionally, for an in-depth examination of the significant impact of this map on medieval European maps, see Marianne O'Doherty's (2011, pp. 30–57) article, 'A Peripheral Matter? Oceans in the East in Late-Medieval Thought, Report, and Cartography.'

including China, sharing their contemporary geographic knowledge and thereby influencing a wide range of cultures and societies. This up-to-date understanding of the world they disseminated was a product of, and contributed to, the rich tapestry of knowledge shaped by those actively engaged in multi-civilizational connections and exchanges.

5 Conclusion

This chapter highlights the vital contributions of China and other Asian countries to the global advancement of geographical knowledge, emphasizing the practical and 'precise' nature of their cartographic achievements. The geographers and cartographers of the medieval Chinese and Islamic cultural worlds developed extensive and remarkably accurate geographic knowledge, relying on sophisticated literary and cartographic data amassed through their multi-civilizational connections and exchanges. While Needham's first volume introduced the concept of these exchanges, more recent studies have built upon this foundation, providing concrete evidence that further reinforces the early findings. These explorations into the past not only enrich our understanding of historical cartographic practices but also underscore the global nature of geographic knowledge development, challenging the notion of any single civilization's dominance in the scientific arena. The collective contributions of various civilizations, particularly those of Asia, played a pivotal role in shaping the world's geographical understanding, a legacy that continues to resonate in contemporary geographic and cartographic studies.

While mapping familiar territories have always been relatively easier, depicting distant lands historically presented a significant challenge, often leading to imaginative or speculative portrayals. However, the height of interactions between China and the Islamic world during the medieval period signified a notable shift in this trend. This era saw intrepid explorers, including the renowned European traveler Marco Polo, the Chinese globetrotter Wang Dayuan, and the Muslim explorer Ma Huan (1380–1460) accompanying Zheng He, venturing far beyond their native lands. While not all these travelers may have reached the destinations they claimed, the information they gathered, whether firsthand or secondhand, provided up-to-date and practical geographic knowledge of a world increasingly interconnected even before the era of early modern global exploration. The wealth of knowledge about regions such as West Asia, Africa, and beyond, gathered and shared along travel routes frequented by numerous explorers from diverse backgrounds, culminated in the creation of detailed and accurate geographical works. Their insights, gathered from far beyond their native lands, were instrumental in shaping international trade and cross-cultural interactions, underscoring the profound impact of these explorations on the global understanding of geography.

In considering the postmodernist trend of relativism, this paper recognizes the importance of appreciating diverse cartographic practices across different societies.[21] However, it also cautions against overlooking the evolutionary development of geographic knowledge as a result of these extensive exchanges. During the medieval era, societies in Asia were not mere passive recipients but active contributors to the acquisition and application of scientific knowledge. This active involvement in cross-cultural trade and exchanges facilitated a practical understanding of the world, demonstrating the significant role these societies played in the evolution of scientific and geographic understanding.

Finally, this paper advocates for a more inclusive and nuanced understanding of the global narrative, one that recognizes the foundational role of Asian societies in shaping our world's scientific heritage. This paper emphasizes that the conception of scientific and technological geographic knowledge was not an exclusive domain of the West, especially not since the advent of modern times.[22] Rather, it was intricately and robustly constructed in Asian societies during the Middle Ages, subsequently influencing and inspiring European advancements. As discussions about

---

[21] For example, Zayde Antrim (2018) argues that history provides the best means to overcome the persistently Eurocentric perspective on mapping the Middle East and establish resilient alternative geographic perspectives for the future. Taking this approach, she builds her case on the region's rich repository of precolonial, pre-national indigenous mapping traditions that offered users conceptualizations of terrain that appropriately addressed local concerns.

[22] Geoffrey C. Gunn (2018) demonstrates how the Westerners (Europeans) developed more accurate understanding of Asia for almost 2000 years since the antiquity, a situation that has left them in what the author originally calls a 'Ptolemaic trap' or 'Ptolemaic template.'

world history increasingly focus on the modern era, it is essential to broaden our historical lens to include the rich tapestry of scientific and technological exchanges during the premodern era. In particular, the Mongol period, often relegated to a historical footnote, was a crucible for cross-cultural exchanges that shaped the trajectory of global history leading into the modern age. By delving into the plethora of geographic and cartographic works produced during this era, especially between the Chinese and Islamic worlds, we can reevaluate the significance of practical geographic knowledge in world history. Such an exploration enables a comprehensive reassessment of premodern exchanges and their enduring impact on scientific developments.

References

Antrim, Z. 2018. *Mapping the Middle East*. London: Reaktion Books.

Cams, M. 2017.*Companions in Geography: East-West Collaboration in the Mapping of Qing China (c.1685–1735)*. Leiden: Brill.

Cattaneo, A. 2022. Connected Histories. The Mongol Empire and the Creation of New Worldviews in the Fifteenth Century: Fra Mauro's Mappa Mundi (Venice, c. 1450) and the Honil Gangni Yeokdae Gukdo Ji Do (Hanseong, c. 1480). In *The Mongol Empire in Global History and Art History*, ed. A. Dunlop. Cambridge, MA: Harvard University Press.

Cohen, H.F. 2010. *How Modern Science Came into the World: Four Civilizations, One 17th-Century Breakthrough*. Amsterdam: Amsterdam University Press.

Edney, M.H. 1999. *Mapping an Empire: The Geographical Construction of British India, 1765–1843*. Chicago: University of Chicago Press.

Edney, M. H. 2007. Recent Trends in the History of Cartography: A Selective, Annotated Bibliography to the English-Language Literature. *Coordinates: Online Journal of the Map and Geography Round Table of the American Library Association, Series B* 6.

Finlay, R. 2000. China, the West, and World History in Joseph Needham's 'Science and Civilisation in China.' *Journal of World History* 11 (2): 265–303.

Fujiyoshi, M. 1991. *Shoban shi* 諸蕃志 [Description of the Foreign Lands] *by Zhao Rugkuo* 趙汝适 *(1170–1228)*. Osaka: Kansai Daigaku.

Gunn, G.C. 2018. *Overcoming Ptolemy: The Revelation of an Asian World Region.* Lanham, MD: Lexington Books.

Hargett, J.M. 1996. Song Dynasty Local Gazetteers and Their Place in the History of Difangzhi Writing. *Harvard Journal of Asiatic Studies* 56 (2): 405–442.

Harley, J.B. and Woodward, David, eds. 1987–1998. *The History of Cartography*, vols. 7. Chicago: University of Chicago Press.

Hiraoka, R., and A. Watanabe. 2015. A Jesuit Cosmological Textbook in 'Christian Century' Japan: De Sphaera of Pedro Gomez (Part II). *SCIAMVS* 16: 125–223.

Hirth, F.F., and W.W. Rockhill. 1911. *Chau Ju-Kua: His Work on the Chinese and Arab Trade in the Twelfth and Thirteenth Centuries, entitled Chu-fan-chi.* St. Petersburg: Printing Office of the Imperial Academy of Sciences.

Kenzheakhmet, N. 2021. *Eurasian Historical Geography as Reflected in Geographical Literature and in Maps from the 13th to the Mid-17th Centuries.* Großheirath: Ostasien Verlag.

Leslie, D.D. 1986.*Islam in Traditional China: A Short History to 1800.* Belconnen, ACT: Canberra College of Advanced Education.

Leslie, D.D., and K.H.J. Gardiner. 1982. Chinese Knowledge of Western Asia During the Han. *T'oung Pao,* LXVIII (4–5): 254–308.

Lycas, A., Hasegawa, M., and Chen, S. 2022. The Production of Geographical Knowledge in Medieval China. *East Asian Science, Technology, and Medicine* 54 (2): 161–165.

Mauntel, C., Oschema, K., Ducène, J-C., and Hofmann, M. 2018. Mapping Continents, Inhabited Quarters and the Four Seas. Divisions of the World and the Ordering of Spaces in Latin-Christian, Arabic-Islamic and Chinese Cartography in the Twelfth to Sixteenth Centuries. A Critical Survey and Analysis. *Journal of Transcultural Medieval Studies* 5 (2): 295–367.

Muraoka H. 村岡倫, ed. 2020. *Saiko no sekai chizu o yomu: Kon'itsu kyōri rekidai kokuto no zu kara miru riku to umi* 最古の世界地図を読む: 混一疆理歴代国都之図から見る陸と海 [Analyzing the Oldest World Map]. Kyoto: Hōzōkan 法藏館.

Needham, J. 2005. *The Grand Titration: Science and Society in East and West.* London: Routledge.

Needham, J., and Wang, L. 1959. *Science and Civilisation in China. Vol. 3: Mathematics and the Sciences of the Heavens and the Earth.* Cambridge: Cambridge University Press.

O'Doherty, M. 2011. A Peripheral Matter? Oceans in the East in Late-Medieval Thought, Report, and Cartography. *Bulletin of International Medieval Research* 16: 30–57.

Park, H. 2012. *Mapping the Chinese and Islamic Worlds: Cross-Cultural Exchange in Pre-Modern Asia.* New York: Cambridge University Press.

Park, H. 2013. Cross-Cultural Exchange and Geographic Knowledge of the World in Yuan China. In *Eurasian Influences on Yuan China*, ed. M. Rossabi. Singapore: Institute of Southeast Asia Studies.

Park, H. 2015. Information Synthesis and Space Creation: The Earliest Chinese Maps of Central Asia and the Silk Road, 1265–1270. *The Journal of Asian History* 49: 119–140.

Park, H. 2018. The World Map Produced in Korea in 1402 and Its Possible Sources from the Islamic World. *The Journal of Asian History* 52 (2): 209–234.

Park, H. 2021. *Soju: A Global History*. Cambridge: Cambridge University Press.

Sugiyama, M. 杉山正明. 2007. Tōzai no sekaizu ga kataru jinrui saisho no dai chihei 東西の世界図が語る人類最初の大地平 [The First Portrait of the World Depicted in the World Maps in the East and the West]. In *Daichi no Shōzō—Ezu・Chizu ga kataru sekai* 大地の肖像—絵図・地図が語る世界 [Portrait of the Earth: The World Described by Pictorial Maps and Geographical Maps], ed. J. Fuji'i, A. Kinda, and M. Sugiyama, 54–83. Kyoto: Kyoto Daigaku Gakujutsu Shuppamkai.

Wei Zheng 魏徵, et. al. 2000. *Sui Shu* 隋書 [The History of Sui]. Beijing: Zhonghua shuju.

Woodward, D. 1987. Medieval Mappaemundi. In *The History of Cartography: Volume One: Cartography in Prehistoric, Ancient, and Medieval Europe and the Mediterranean*, ed. J.B. Harley and D. Woodward, 286–370. Chicago: University of Chicago Press.

Wylie, A. 1867. *Notes on Chinese Literature: With Introductory Remarks on the Progressive Advancement of the Art*. Shanghai: American Presbyterian Mission Press; London: Trübner.

Yang, B., ed. 1996. *Zhufan zhi jiaoshi* 諸蕃志校释 [Description of the Foreign Lands, with Annotations and Footnotes] *by Zhao Rugkuo* 趙汝适 *(1170–1228).* Beijing: Zhonghua Shuju.

Yee, D.K. 1994a. Reinterpreting Traditional Chinese Geographical Maps. In *The History of Cartography: Volume Two, Book Two: Cartography in the Traditional East and Southeast Asian Societies*, ed. J.B. Harley and D. Woodward, 35–70. Chicago: University of Chicago Press.

Yee, D.K. 1994b. Taking the World's Measure: Chinese Maps between Observation and Text. In *The History of Cartography: Volume Two, Book Two: Cartography in the Traditional East and Southeast Asian Societies*, ed. J.B. Harley and D. Woodward, 96–127. Chicago: University of Chicago Press.

Yee, D.K. 1994c. Traditional Chinese Cartography and the Myth of Westernization. In *The History of Cartography: Volume Two, Book Two: Cartography in the Traditional East and Southeast Asian Societies*, ed. J.B. Harley and D. Woodward, 170–202. Chicago: University of Chicago Press.

Part III

Natural Sciences in Dialogue

8

Liberating Mathematics from Civilizations

George Gheverghese Joseph

1 Introduction

Over the last three decades there has emerged increasing recognition of the role that mathematical exchanges across civilizations played in the growth of mathematical ideas.[1] It is well recognized that since the Bronze

[1] This paper began as an attempt to draw together the muti-civilizational exchanges that nurtured the growth of mathematical knowledge. However, over time, I found myself health-wise unable to complete the chapter. Instead I thought it would be a good idea to draw together dialogical connections I discussed in the course of writings about the mathematics of different civilizations in my book *The Crest of the Peacock: Non-European Roots of Mathematics* into one narrative by exploring how they integrate together. This approach has some drawbacks that have to be admitted. First I have been unable to address more recent work in the areas I discuss. Second, I have had to largely see dialogue in terms of non-European contributions without adequtely covering the Greek and European contributions to the mathematics of other civilizations since my original study was focussed on the non-European mathematical traditions. Nevertheless, I think the study does reflect the broad general pattern of mathematical exchanges across civilizations and cultures. I also wish to thank Arun Bala for his invaluable help in not only kindly assisting me in the process by putting together dialogical connections discussed in this chapter that are drawn from my book, but also the numerous fruitful exchanges we had which helped to weave this material into a continuous and coherent narrative.

G. G. Joseph (✉)
University of Manchester, Manchester, UK
e-mail: beavergrease@yahoo.com; beavergrease@gmail.com

A. Bala et al. (eds.), *Multicivilizational Exchanges in the Making of Modern Science*,
https://doi.org/10.1007/978-981-97-3541-9_8

Age about 5000 years ago ideas have circulated across the Eurasian region between the agrarian civilizations that emerged in the river valleys of the Tigris-Euphrates, the Nile, the Indus, and the Hwang-Ho. Over time these exchanges became more intensive and extensive as the connections between these civilizations developed through empires, religions, and commerce. Although the global turn in the history of mathematics has led to many studies of mathematical connections between cultures there have been few studies of how the flow of mathematics over the region shaped in crucial ways the growth of mathematical ideas. This chapter will begin by delineating these flows and examining how the dynamics of mathematical development has been influenced by the interaction and merging of mathematical ideas and traditions across the Afro-Eurasian region over time. It will also trace their culminating impact on the emergence of modern mathematics in recent years.

Since the global turn in history most writers have tended to focus on histories of mathematics which trace the development of ideas in Mesopotamia, Egypt, Greece, India, and China. The only exception to such accounts is Arabic-Islamic civilization that is seen to be the receptacle of both Greek and Indian mathematical ideas which later came to be transmitted to Europe. However, it is possible to discern major turning points in mathematics which occurred when ideas flowed and merged across separate civilizations. Intra-civilizational histories that confine themselves to the growth of mathematics within civilizations, without regard to influences from the outside, do not allow us to see clearly these historical processes and mathematical reorientations resulting from intercultural fusions. This is what I explore in this chapter.

2 Between Egypt and Mesopotamia

The Egyptians developed a hieroglyphic decimal numeral system that was well-developed quite early in the third millennium BCE. It led to the construction of innovative algorithms for basic arithmetical computations. These developments emerged especially in the two centuries around 1800 and 1600 BCE. It is also generally known that ancient Egyptians were already resolving problems involving linear equations with a variety of geometrical objects, including the surface of a hemisphere that involved calculating their surface areas and well their volumes (Lumpkin 2000). In the past these had been seen as originating only with Greek mathematics. However, there is little knowledge about how the Egyptians discovered the methodology for dealing with such problems. We have to assume that they had rational approaches of some sort, although we can be sure that if could not have been based on reasoning with explicit axioms (Joseph 2011, p. 120).

While Egypt developed around the River Nile the other early civilization of the period with mathematical achievements was Mesopotamia which emerged between the two great rivers of the Tigris and Euphrates. Unlike Egypt which based its developments in numbers on a decimal place values system, Mesopotamian numbers were grounded on a sexgesimal place-value system with base 60. This continues to influence modern day numbering of time intervals with 60 seconds in one minute, and 60 minutes in an hours, and measurements of 360 degrees in a circle. All of these served numerical because 60 is a highly composite number with factors that are 1, 2, 3, 4, 5, 6, 10, 12, 15, 20, 30, 60 which greatly simplified numerical computations with fractions.

The greatest achievements of its mathematics took place during 1800—1600 BCE that has come to be known as the First Babylonian Period. These have been found written in cuneiform on clay tablets covering a wide range of mathematical topics such as fractions, algebra, quadratic, and cubic equations. They also the value of the square root of 2 computed to accuracy of three significant sexagesimal digits. In modern terms this is six significant decimal digits. Indeed, in their numerical and algebraic achievements the Mesopotamians were a great deal more advanced than the Egyptians in that period. It is known as Babylonian mathematics because the greatest advances were made under Babylonian rule.

However, Mesopotamian achievements in geometry were also considerable. They had knowledge of what later came to be known as Pythagorean triples. Although it is generally supposed that they did not have knowledge of the general theorem encompassing these Neugebauer argues that 'it was known during the whole duration of Babylonian mathematics that the sum of the squares on the lengths of the sides of a right triangle equals the square of the length of the hypotenuse' (Neugebauer, 1969, p. 36). They also did much work on similar triangles, with formulas that involved the ratios of sides of such triangles. Their work on both similar triangles and Pythagorean triples constitutes examples of their considerable accomplishments and contributions to geometry (Joseph 2011, p. 173).

Although researchers in the past had considered the Egyptian and Mesopotamian traditions to have developed separately, to a large extent shaped by their quite distinct decimal and sexagesimal base number systems, more detailed studies have revealed their mutual influences on each other. This is not surprising given their close proximity in both historical time and geographical locations (Joseph 2011, p. 121). There is increasing evidence that even before the Hellenistic period of the close economic and political contacts between them. One example is the spread of Mesopotamian algebra and geometry to Egypt and parallel developments of geometry and algebra in the two regions (Parker 1972; Freiberg 2005; Joseph 2011, pp. 16–17). Indeed there is a great continuity in the two traditions, with mutual influences from one to the other, that spanned over three millennia. It culminated with the period of Alexandrian mathematics in the beginning of the first millennium when they came to be fused together under Ptolemaic rule of the Hellenistic Era.

3 Babylonian-Egyptian Diffusions to Greece

The two river valley civilizations also had a profound impact on Greek civilization in general, as well as in mathematics. Egyptian influence on the Greeks prior to the Alexandrian Age was most crucial in the earlier half of the first millennium BCE. At the time there were widespread Greek colonies scattered all over the Mediterranean, with trading networks that furnished many cross-channels of interchange. This is evident in the resemblances between the mathematics of Egypt and Greece, especially in number theory. The Greek literature also provides acknowledgement by the Greeks themselves of their debt to

Egypt, with many of them claiming that mathematics originated in Egypt. It has been confirmed by both contemporaries and later historians that pioneering Greek mathematicians, including Thales, Pythagoras, and Eudoxus journeyed to Egypt in search of knowledge. This led Katz and Lumpkin to conclude:

> Not only is Pythagoras supposed to have studied in Egypt, but so is Thales, the supposed father of Greek geometry. Also, Oenopides. Herodotus, Heron of Alexandria, Diodorus Siculus, Strabo, Socrates (through Plato), and Aristotle—all say that geometry was first invented by the Egyptians and then passed on to the Greeks. The question always seems to be, in this regard, what we mean by geometry. If, by geometry, we mean an axiomatic treatment with theorems and proofs in the style of Euclid, then it is clear that this was a Greek invention. But mathematicians have always known that, in general, one does not discover theorems by the axiomatic method. One discovers theorems by experiment, by trial and error, by induction, etc. Only after the discovery is there a search for a rigorous 'proof' (Katz and Lumpkin 1995, pp. 10–11. See also Katz 2007).

Equally significant is the great debt of Greece to Mesopotamia in the early stages of its mathematics and astronomy. The campaigns of the Assyrian king Sargon II (722–705 BCE), and Ashurbanipal's later occupation of Egypt including the meeting he had with Gyges of Lydia toward the middle of the seventh-century BCE, all corroborate Mesopotamian and Greek contacts. It was then that Anaximander of Miletus (610–546 BCE) introduced the mathematical and astronomical instrument of the gnomon from Babylon into Greece. Later, following the release of the Jews from captivity in 538 BCE by the Persian King Cyrus the Great, they carried Babylonian learning westward. Over the next few centuries, as the Persians reigned in the East, including many Greek cities and provinces such as Lydia, Macedonia, and Thrace, as well as the South in Egypt and Libya, philosophy, literature, and mathematics began their seminal development in the independent Greek city states, especially Athens.

After the Persian Empire was conquered by Alexander of Macedonia, and the three regions of Greece, Egypt, and Mesopotamia were linked together under Greek rule (even though his empire disintegrated into several parts) there emerged some of the most significant advances in mathematics. It led to a flowering of Greek creativity over the next few centuries (Joseph 2011, p. 16). It has also led to a far greater appreciation of the fructifying links between the Egyptian, Mesopotamian, and Greek mathematics. It has been suggested by Friberg that we can find not only 'unexpected links between Egyptian and Babylonian mathematics' but also 'amazing traces of a Babylonian origin in Greek mathematics.' Friberg notes that such traces are evident in the knowledge of Babylonian 'metric algebra'—a methodology combining geometry, metrology, and the solution of quadratic equations—in the work of several famous Greek mathematicians such as Euclid, Heron, Ptolemy, Diophantus, and Archimedes. He gives eighteen pieces of evidence in eighteen chapters to substantiate his claim. These involve comparing Greek written work with the mathematics in both Babylonian and pre-Babylonian mathematical clay tablets his conjecture (Friberg 2005, 2007b). He has chapters comparing and contrasting Babylonian geometry with similar topics dealt with by Greek mathematicians including Theon of Smyrna, Theodorus of Cyrene, Heron, and Ptolemy. Friberg's thirteenth chapter explores the impact of Babylonian mathematical methods with Diophantus's *Arithmetica*. His studies highlight the interdependence between the three mathematical traditions of Greece, Egypt, and Mesopotamia (Joseph 2011, pp. 185–186).

Of particular significance is the link between Babylonian methods and Euclid's *Elements* which is also addressed by Friberg in several chapters. It is reasonable to suppose that the seed of Euclid's work on similar triangles drew directly or indirectly from Mesopotamian discoveries in the same area. What Euclid did, but the Mesopotamians did not, is to frame it within the distinctive Greek approach of axiomatic deductive logic, and giving it a significance that is hard to underestimate (Joseph 2011, pp. 171–172).

It was the coming together in multicivilizational exchanges of the mathematical traditions of Mesopotamia, Egypt, and Greece during the third-century BCE and the third-century CE that led to one of the greatest periods in mathematics. Euclid, Archimedes, Apollonius, and Diophantus all produced their seminal works in this era. It was to an extent made possible by the pre-Euclidean mathematics of Egypt and Mesopotamia infused with the powerful methodology of Euclid's deductive axiomatic orientation (Joseph 2011, pp. 17–18). It also brought together Classical Greek mathematics, with its geometric and deductive tradition, and the empirical and algebraic of Egypt and Mesopotamia. 'This synthesis can be seen exemplified in the works of Archimedes, Ptolemy, Diophantus, Pappus, and Heron' (Joseph 2011, pp. 120–122).

Nevertheless, there is a prevalent inclination to understand Alexandrian mathematics merely as a direct extension of only previous Greek mathematics, notwithstanding the unique character of its mathematics as shown in the notable achievements of Archimedes, Heron, Diophantus, and Pappus. An important explanation for this oversight is the view expressed by Kline that the Egyptians and Mesopotamians 'barely recognized mathematics as a distinct discipline.' This view invokes reasons such as that they had no general rules, offered no 'mathematical proofs, lacked the capacity for mathematical abstraction, failed to distinguish clearly between exact and approximate results, and did not have a notion of mathematics as a discernible activity to be studied for its own sake' (Joseph 2011, p. 181). This is to simply impose the distinctive Greek view of mathematics upon the discipline as a whole! It is also to ignore how much of the Greek mathematical legacy is built upon discoveries first made in Egypt and Mesopotamia.

We have to conclude that there were significant influences that flowed between the mathematics traditions of the Old Babylonian mathematics of Mesopotamia and that of the Middle Kingdom of Egypt. Indeed even studies of later texts such as the Cairo demotic mathematical papyrus of the third-century BCE show that significant themes and methods from Late Babylonian mathematics were known to Ptolemaic Egyptians. We can see that two traditions existed in parallel during the lifetime of Euclid lived in Alexandria—one a practical Greek-Egyptian mathematics (which Friberg calls 'metric algebra') and the other a theoretical (Euclidean) 'Greek mathematics' (Joseph 2011, p. 185).

4 From Mesopotamia to India

Since there is some evidence of links with the Indus Valley and Mesopotamian civilizations which were flourishing at the same period during the Bronze Age we might expect mathematical exchanges between them. Archaeological evidence indicates cultural and commercial contacts between Mesopotamia and the Indus Valley. With a two-way traffic, but contacts became closer once the Persian Empire came to include portions of India west of the Indus, and Mesopotamia and Egypt under its suzerainty.

Although there is little evidence of early Egyptian mathematical transmissions to India, the Mesopotamian influence is recognizable. This enabled India from the very beginning to play an important role in the transmission and diffusion of ideas. While there is no direct evidence of mathematical exchange between the two cultural areas, the oldest Indian astronomical and astrological text—V*edanga Jyotisa*—reveals this in the close parallels with astronomical calculations it deploys to identify the longest and shortest day, as well as the list of twenty-eight *nakshatras* (constellations) also found in early Indian Vedic texts and Mesopotamian approaches and ideas (Joseph 2011, p. 18).

5 Greece to India Transmissions

Following the conquests of the Persians under Darius around 513 BCE of regions to the West of the Indus River, there developed a period of turbulence in Indian for six centuries. Broken only by a century and a half of relative peace under the Mauryan rule, India was subjected to invasions by the Greeks, the Sakas, the Pahlavas, and the Kusanas. It was also an era of increasing contact between India and the West. The Persian Empire which linked to Greece in the West and India in the East and also brought a mingling of Greek and Indian soldiers, along with Mesopotamians and Egyptians, in its armies saw an era of exchanges of knowledge as well. Greek writings now begin to refer to Indians as indoi. We find suggestive parallels emerging in Indian and Pythagorean philosophy, and some Greek sources speak of Pythagoras as travelling to

India in his search for knowledge. Perhaps the most systematic attempt to delineate these Greco-Indian exchanges has been made by McEvilley who argues that both Greek and Indian thought had crucial impact on each other. He traces many of the ideas of the Presocratics to Indian origins and also suggests that this was followed later by seminal influences of Greek thought on Buddhist logic and dialectic (McEvilley 2001).

There is ample evidence that the Greeks also continued to carry Mesopotiamian influence into Indian astronomy during the Hellenistic period. The astronomical texts known as the Siddhantas from the first millenium BCE show the impact of the mathematics and astronomy of the Ptolemaic and Seleucid dynasties upon Indian science. Later impact of Greek knowledge, especially in medicine, has been found in the great learning centre Jund-i-Shapur in Persia between 300 and 600CE. Indeed, with the demise of Alexandria as a centre of learning, and the closure of Plato's Academy. Many scholars sought refuge in Jund-i-Shapur which became an arena of contact for Greek and Indian scholars (Joseph 2011, pp. 18–19). The sixth-century astronomical text Panchasiddhantika written in Sanskrit by the astronomer Varahamihira describes the approach of five traditions—Surya Siddhanta (Treatise of the Sun), Vasishtha Siddhanta (treatise of Vasishtha), Paitāmaha Siddhanta (Treatise of Pitamaha), Romaka Siddhanta (Treatise of Romans), and Paulisa Siddhanta (Treatise of Paul). The last two are likely to be treatises from the West and of Greek origins.

6 Indian Mathematics Travels to Arabic Civilization

The most important exchanges of mathematical ideas in the second half of the first millennium were between the Islamic world, Greece, and India. In the case of Greece these included Euclid's elements which became a fundamental text of Islamic mathematics, as well as number theory which were further extended and developed by Islamic mathematicians. In the case of Indian mathematics three main influences can be identified. First and most important was the spread of Indian numerals with their associated algorithms, which was refined further in the Islamic world and then transmitted to Europe. The second was the transmission of Indian trigonometry, especially the use of the sine function. Indeed, both the diffusion of Indian numerals, algebra, and trigonometry are well documented in studies, as is the transmission of Greek geometry, and Islamic extensions to it. Finally the mathematicians of the Islamic world drew on the solutions of equations in general, and of indeterminate equations in particular. (Joseph 2011, pp. 19–20).

However, such transmissions were not merely one-way. After the establishment of Mughal rule in North India during the medieval period Arabic mathematics became more widespread in North India. Indeed from the fifteenth century we can find two mathematical traditions operating side-by-side in India—a Sanskrit-derived mainstream tradition of Indian mathematics especially exemplified in South India by the Kerala School, and an Arabic tradition associated with the madrassahs or Muslim schools mainly in the north (Joseph 2011, p. 21).

Some attempt at fusing the traditions did occur, especially in the area of astronomy. Mahendra Suri, an astronomer in the court of Firoz Tughlaq, published his work *Yantraraja* in 1370. This introduced into the Sanskrit *Siddhanta* tradition the ideas and practices from Persian and Islamic astronomy. Efforts along these lines of integration within astronomy—its instruments and mathematics—persisted into the seventeenth century and provided the basic materials for training with the Ptolemaic system. Ultimately the two traditions resisted fusion so that over time there developed two separate schools of mathematical astronomy. This is reminiscent of the sixth century where Varahamihira described five schools of mathematical astronomy of which two were foreign. Often under tolerant rulers they were accommodated together in practice as was the case when they were under the patronage of an enlightened rulers such as the Mughal emperor Akbar, or the Hindu Raja Jai Singh of Rajasthan. Indeed the large masonry astronomical instruments at Delhi, Jaipur, Ujjain, and Varanasi (Benares) are reminders of the latter's interest in observational astronomy on a scale remarkable in India (Joseph 2011, pp. 510–512).

7 Arabic Mathematics Enters Europe, China, and India

Islamic scholars and their contributions also had significant impact of the growth of mathematics in early modern Europe. In this regard they were more than mere custodians and transmitters of ancient Greek and Indian mathematics (Joseph 2011, p. 20). One major influence was Al-Khwarizmi's Algebra, which played a key role in the development of the subject in Europe after its translation into Latin. This was mediated largely by the Latin translations of al-Khwarizmi's work by Robert of Chester and Gerard of Cremona in the twelfth century. It was about the same time that Fibonacci translated al-Khwarizmi's work which he came across either in his youth during his learning and travels in North Africa, or his travels as an adult across Egypt, Syria, Greece, and Sicily. Indeed the final fifteenth chapter of Fibonacci's Liber Abaci (1202) features notably al-Khwarizmi's work on the theory of quadratic equations (See Sigler 2003).

Another major influence was Abu Kamil (c. AD 900). Known as the 'Egyptian Calculator' his commentary al-Khwarizmi's work systematically discussed and developed the fundamental rules of algebraic operations, solution of equations, and accept the use of irrational numbers in solutions and coefficients of equations. His techniques were later imported into Europe by Fibonacci, whose seminal influence on European mathematics is hard to overestimate. Another great influence was Ibn al-Haytham, who is better known for his optics and its influence on Renaissance art, but who introduced along with it a geometric theory of the solutions to equations.

But mention must also be made of the two great mathematicians Thabit and Omar, whose work though less well known in early modern Europe, also influenced Arabic mathematics with their rigorous geometrical derivation of results. It was only in the 1850s after Woepcke's translation of his *Algebra* did Western historians of mathematics began to refer to Omar's work. However, Kasir (1931, pp. 6–7) argues that even a century earlier there was evidence that European mathematicians were aware of his studies. In the case of Thabit ibn Qurra he did not seem to have a direct impact on European mathematics, although we can find some circumstantial evidence which is quite suggestive that he did. Some of the texts translated by Gherardo of Cremona (c. 1175), who brought more Islamic science into Europe than anyone else at the time, may have included discoveries drawn from Thabit's texts. It is also possible that John Wallis's *Treatise on Angular Sects.* (1685) may have drawn upon Thabit's generalization of the Pythagorean theorem. Indeed the likelihood of such a conclusion increases when we consider the creation of places for professors of Arabic in major universities of Europe conversant with the latest mathematical and astronomical discoveries of the Islamic world, and their role in studying and disseminating the continuous stream of Arabic manuscripts then flowing into a receptive in the seventeenth century (Joseph 2011, pp. 486–487).

These Arabic mathematical contributions include Thabit's rule for generating amicable numbers (sometimes attributed to Fermat and Euler) and the Horner-Ruffini method for solutions of higher-order equations. We also cannot ignore the considerable debt that Nicolaus Copernicus (1473–1543) owed to the Islamic mathematician-astronomers Nasir al-Din al-Tusi (1201–1274) and Mu'ayyad al-Din al-'Urdi (d. 1266) which played key roles in his mathematical framing of his revolutionary vision of a heliocentric universe (Saliba 2007, pp. 193–232).

There is also some suggestive evidence of Islamic and Chinese exchanges in mathematics. For example, Islamic mathematics may have drawn inspiration from Chinese methods for the solution of numerical equations of higher order, and Chinese may have learnt from Islamic trigonometry and arithmetic. Under Mongol rule in the Yuan Dynasty Islamic astronomers had been invited into the Astronomical Bureau in China. It was a time when the Mongol Empire included both China in the East and the Muslims in Central Asia. Later Muslim rulers such as Hulegu Khan and Ulugh Beg were patrons and practitioners of science who brought Chinese and Islamic mathematicians and astronomers together, notably in the Maragha observatory.

Islamic mathematicians such as al-Kashi (c. 1400) seem to have been influenced by Chinese work in their numerical methods of solving equations of higher order, such as quadratics and cubics. Conversely there is a strong likelihood that Islamic scholars may have played a role in introducing trigonometric ideas to Chinese mathematicians (Joseph 2011, pp. 20–21).

8 India and China Exchanges

The spread of Buddhism into China beginning about the first-century CE, and the travels of Chinese pilgrims into India during the Tang Dynasty when Buddhism became dominant in China, led to many scientific and cultural exchanges between the two countries over several centuries. Many Indian works on astronomy, mathematics, and medicine were translated into Chinese. Indian astronomers were invited into the Chinese Astronomical Bureau from about 600 CE, and there is evidence that they also introduced into China the principles of Indian astronomy built on quite distinctively different cosmological ideas. We find use of the method of *kuttaka* in India and of *qiuyishu* in China being used in the solution of indeterminate equations, though the direction of transmission in this regard is not clear.

9 From India to Europe

While Indian mathematical transmissions to Europe through the mediation of Islamic civilization is widely acknowledged, there is also the possibility that there may have been a direct transmission of Indian mathematics. Indeed, the number of Indian mathematical ideas made in India before the Europeans arrived by sea to India in 1498, and later emerging in Europe is quite remarkable. It includes the method of construction of 24 sine values used by Aryabatha in 499 CE which was later deployed by Regiomontanus in 1533, the radian measure of Aryabatha adopted by European mathematicians in the late nineteenth century, and the algebraic technics of summation of finite series used by Aryabatha also attributed to the Italian mathematician Bombelli in the sixteenth century. Mention can also be made of the prime factoring method involving difference of squares discovered by Narayana in 1356 and used by Fermat in the seventeenth century, as well as the continued fractions formulas used by Bhaskara II in the twelfth century and later by John Wallis in 1665 (Joseph 2011, pp. 436–7).

Most notable are the discoveries of the Kerala School in South India after the Portuguese established themselves following the arrival of Vasco da Gama in 1498. The school had made advances through a series of mathematicians such as Madhava, Nilakantha, and Jyesthadeva in the study of infinite series for π and certain trigonometric functions. These discoveries in mathematical analysis were made 250 years before European mathematicians, beginning with Madhava, before they were rediscovered by Leibniz, Newton, and Gregory—the pioneers of infinitesimal calculus. It raises the possibility that they were somehow transmitted to Europe since the European achievements followed the Indian ones, and the Indian mathematicians who made these discoveries worked in the area where the Portuguese landed. Hence it is quite possible that knowledge of the Indian achievements may have come into Europe and stimulated similar discoveries there (Joseph 2011, p. 21).

In the above cases there is little documentary evidence to establish transmission by way of acknowledgements or translations. But it is possible to establish circumstantial evidence for transmissions by identifying methodological, algorithmic, and epistemological similarities. Especially significant are similarities involving duplications of erroneous results or approximations from the contributing culture. While correct results may have been independently discovered, it is more likely that matching errors are due to copying. Moreover, the presence of an accessible corridor of communication between the two cultures involved, as well as an appropriate chronology of priority—both established by the arrival of Europeans with mathematically trained Jesuit scholars in India after the discoveries in India, does strengthen the circumstantial case (Joseph 2011, pp. 435–437).

A similar case in reverse transmission may be seen as happening with the Indian mathematician Srinivas Ramanujan. Without formal training in pure mathematics, he made significant discoveries in number theory, infinite series, continued fractions, and mathematical analysis. His early work was done in isolation, and altogether he completed thousands of results, especially in identities and equations. These were precisely the areas in which Indian mathematicians had made contributions in the past. Also like earlier Indian mathematicians he seemed to treat his results as generalizations of experience and intuition rather than demonstrable by axiomatic deduction. His major inspiration that took him beyond the Indian tradition was a library copy of *A Synopsis of Elementary Results in Pure and Applied Mathematics* by G. S. Carr's listing of 5,000 theorems of modern mathematics that he assiduously studied and developed further.

His achievement is a classic example of how discovery and proof in mathematics can be undertaken by different persons. It exemplifies why we cannot reduce the context of justification of mathematics (proof) to the context of discovery (finding a theorem). As we saw earlier when the Greeks declared that the Egyptians discovered geometry, they were the results that they were referring to and not the methods of proof. This is often overlooked by those who deny Egyptian mathematical discoveries simply because they did not deduce them from first principles or axioms.

10 Journeys Between China, Japan, and Europe

There is considerable evidence that during the medieval period there was a substantial flow of not only products such as silk and ceramics from China to Europe but also technologies (Lach 1965, Needham 1954). This had begun in the first-century CE and continued into the fifteenth- and sixteenth-centuries CE. We can lists some of these—the square-pallet chain pump, metallurgical blowing engines operated by water power, the wheelbarrow, the sailing carriage, the wagon mill, the crossbow, the technique of deep drilling, the so-called Cardan suspension for a compass, the segmental arch-bridge, canal lock-gates, numerous inventions in ship construction (including watertight compartments and aerodynamically efficient sails), gunpowder, the magnetic compass for navigation, paper, printing, and porcelain (Needham 1954, pp. 240–241). It is reasonable to assume that these transfers of technologies were also accompanied by mathematical ideas and algorithms discovered in China including indeterminate analysis, the 'Chinese remainder theorem, algorithms for extracting square and cube roots, and solutions of cubic and higher-order equations by Horner's method as now called' (Joseph 2011, pp. 21–22).

Chinese influence is often underestimated by ignoring the role of the Jesuits as two-way mediators of the exchange of knowledge between China and Europe. There is a tendency to only see them as transferring European knowledge to China—exactly how they presented themselves since their trump card for their missionary mission was that European superiority in science, especially mathematics and astronomy, was also a mark of the superiority of the religion they brought to China. But we should not rule out the possibility that Chinese mathematical ideas could have been transmitted to Europe even though there is no direct evidence that either William Horner was aware of the Chinese solution. Indeed, the influence itself may be in the form of a stimulus effect—while the actual method may not have been transferred the knowledge that such a method existed may have stimulated its rediscovery in Europe. This would explain why Horner's Italian rival and contemporary Paolo Ruffini also arrived at the same solution (Joseph 2011, p. 282).

In two areas, also discussed in other chapters in this volume, the Chinese impact may have been seminal—the solution to the mathematics of equal temperament and the discovery of non-linear algebra and determinants. The first concerns the contribution of the musician Chu Tsai-Yu (1536–1611) who first to solve the mathematical problem of 'equal temperament'. His ideas could have been transmitted from Ming China to Europe by Jesuit missionaries. An equal temperament is a system of musical tuning where there is an identical frequency ratio between each pair of adjacent notes. Equal temperaments are designed to approximate some form of 'just intonation' wherein the frequencies of notes are related by ratios of integers. In such cases, an interval, generally an octave, is divided by a series of equal *steps* of frequency ratios. The most common tuning system in Western music involves dividing the octave into twelve equal parts (semitones), customarily by starting with a standard pitch of 440 Hz It is what is generally termed the twelve-tone equal temperament scale.

One of the first to propose an equal temperament was Vincenzo Galilei, the father of the more renowned Galileo Galilei. He advocated equal temperament in a treatise in 1581. But it was Chu Tsai-Yue who discovered the first correct solution to the problem of achieving equal temperament in 1584. His discovery was repeated and published in Europe about thirty years Marin Mersenne (1588–1648) and Simon Stevin (1548–1620). This raises the question of whether Chu's solution may have been spread through Jesuits or any other agency from China to Europe. We can also ask, as with Kerala mathematics, whether we are dealing here with direct transmission or merely stimulus effects of the knowledge that a solution to the problem of equal temperament had been found in China reaching Europe via the Jesuits.

Another influence of Chinese mathematics is the spread of linear algebra from China into the West in the early modern era. In a highly novel and stimulatingly cogent study Roger Hart makes a convincing case that linear algebra originally developed in China and spread throughout Eurasia spreading West to Europe and East to Japan. Its origins and methods can be seen in Chinese mathematical treatises from the first-century CE until the seventeenth century. The earliest record of its arrival in Europe as a set of linear algebra techniques is in the thirteenth-century CE. What is now recognized as the most important algorithm of linear algebra–Gaussian elimination—had already circulated to Europe by the sixteenth century.

However, the discipline is still seen as originating in the West and appearing de novo in Europe in the sixteenth-seventeenth centuries when there was an increasingly expanding trade and intellectual exchanges between China and Europe. Hart demonstrates that textual analysis shows that this is a false notion—we can trace through close studies the many parallels between Chinese and European methods and how the latter is a transformation and transmission of the computing techniques of Chinese mathematical practices.

The period of the early modern era was also a time when European mathematics came to be connected by the Jesuits with India, China, and Japan. We have seen that these connections may have facilitated the transmission of some Indian and Chinese mathematical ideas to Europe. But it may also have inspired new mathematical developments in Asia. Particularly noteworthy is the case of Seki Takakazu (alias Seki Kowa, 1642–1709). Many of his discoveries were in advance of, and parallel to, European discoveries at the time. They suggest that the simple meeting of the Chinese and European traditions stimulated discoveries on both ends of the Eurasian continent. Seki Takakazu made contributions in advance of, or He took further the Chinese work on solving numerical equations of higher order using the Horner-Ruffini method, discovered determinants ten years ahead of Leibniz, identified the conditions of existence for polynomials of positive and negative roots, polynomials, carried out pioneering work with continued fractions, and discovered one year before Bernoulli the numbers named after the latter. He also calculated the value of π to nine decimal places by applying a geometrical method with a polygon with 2^{17} correct sides (Shigeru and Rosenfeld 1997, Joseph 2011, pp. 299–300).

11 Liberating Mathematics from Civilizations

In the past the history of mathematics has been written largely with a view to tracing the evolution of mathematical knowledge within civilizations. It has only incidentally focused on the way mathematical knowledge has been enriched, fructified, and transformed by exchanges across civilizations. Such exchanges have not only advanced our knowledge of numbers, algebra and geometry but also the methodological and epistemological orientations to the subject. It offers a new perspective on mathematics following other historical approaches in the last three decades. These shifted attention from the study of civilizations as autochthonous entities to understanding them as part of the larger ecosystem of interacting civilizations—especially in the Afro-Eurasian World zone since the Bronze Age. Its title 'Liberating Mathematics from Civilizations' mirrors this connected, dialogical, circulatory, or entangled histories of civilizations and their mathematical traditions.

Such an approach raises the problematic issue of separating transmissions across civilizations from independent discoveries within civilizations. The same problems arise even in the writing of the history of science. Thomas Kuhn (1962) has touched on this concern with regard to individual scientists. How do we know who discovered oxygen? Was it

Priestley who first isolated the gas, or Lavoisier who identified it as a separate component of air? These are tangled concerns in both the history of scientific discovery across individuals and across civilizations. They have generated much debate but this chapter has not made any effort to enter these complex and knotty issues.

However, in this chapter we have adopted the view that the proper approach is for those claiming independent discovery to establish their case against the odds. Here we have followed Joseph Needham who wrote concerning establishing transmissions from China to Europe:

> Of course, there may have been some degree of independence in the European advances. Even when we have good reason to believe in a transmission from China to the West, we know very little of the means by which it took place. But as in all other fields of science and technology the onus of proof lies upon those who wish to maintain fully independent invention, and the longer the period elapsing between the successive appearances of a discovery or invention in two or more cultures concerned, the heavier that onus generally is (Needham 1970, p. 70).

We have given Needham's criterion a wide scope of applicability here in considering transmissions across civilizations and cultures. Even if this criterion itself is open to dispute it paves the way to explore how far a multicultural exchanges approach liberating mathematics from civilizations can illuminate the history of the discipline.

References

Friberg. 2005. *Unexpected Links Between Egyptian and Babylonian Mathematics.* Singapore: World Scientific.

Friberg. 2007a. *A Remarkable Collection of Babylonian Mathematical Texts.* New York: Springer.

Friberg. 2007b. *Amazing Traces of a Babylonian Origin in Greek Mathematics.* Singapore: World Scientific.

Joseph, George Gheverghese, and (with Nelson, David and Williams, Julian). 1993. *Multicultural Mathematics: Teaching Mathematics from a Global Perspective.* Oxford: Oxford University Press.

Joseph, George Gheverghese. 2009. *Kerala Mathematics: History and Its Possible Transmission to Europe*. Delhi: B.R. Publishing Corporation.

Joseph, George Gheverghese. 2011. *The Crest of the Peacock: Non-European Roots of Mathematics*. Princeton: Princeton University Press.

Joseph, George Gheverghese. 2016. *Indian Mathematics: Engaging With the World from Ancient to Modern Times*. Singapore: World Scientific.

Joseph, George Gheverghese. 2018. *A Passage to Infinity: Medieval Indian Mathematics from Kerala and its Impact*. New Delhi: SAGE Publications Pvt. Ltd.

Kasir, Daoud S. 1931. *The Algebra of Omar Khayyam*. New York: Teachers College, Columbia University, January 1.

Katz, V., and B. Lumpkin. 1995. The Relationship of Egyptian to Greek Mathematics. *History and Pedagogy of Mathematics Newsletter* 35: 10–13.

Katz, Victor, ed. 2007. *The Mathematics of Egypt, Mesopotamia, China, India, and Islam: A Sourcebook*. Princeton: Princeton University Press.

Kuhn, Thomas. 1962. *The Structure of Scientific Revolutions*. Chicago: University of Chicago Press.

Lach, D.F. 1965. *Asia in the Making of Europe*, 2 vols. Chicago: university of Chicago Press.

Lumpkin, B. 2000. The Mathematical Legacy of Ancient Egypt: A Response to Robert Palter. In *Debating Black Athena*, ed. Bernal Martin. NC: Duke University Press.

McEvilley, Thomas C. 2001. *The Shape of Ancient Thought*. Comparative Studies in Greek and Indian Philosophies. Allworth.

Needham, Joseph. 1954. *Science and Civilisation in China, Volume 1: Introductory Orientations*. Cambridge: Cambridge University Press.

Needham, Joseph. 1970. *Clerks and Craftsmen in China and the West*. Cambridge: Cambridge University Press.

Neugebauer, Otto. 1969. *The Exact Sciences in Antiquity*, 2 ed., vol. 9. New York: Dover Publications.

Parker, R.A. 1972. *Demotic Mathematical Papyri*. Brown Egyptological Studies 7. Providence: Brown University Press.

Saliba, George. 2007. *Islamic Science and the Making of the European Renaissance*. Cambridge, MA: MIT Press.

Shigeru and Rosenfeld. 1997. Seki Kowa. In *Encyclopedia of the History of Science, Technology and Medicine in Non-Western Cultures*, ed. H. Selin. Dordrecht: Kluwer Academic Publishers.

Sigler, L.E. 2003. *Fibonacci's Liber Abaci: A Translation into Modern English of Leonardo Pisano's Book of Calculation*. New York: Springer.

9

Ancient Chinese Origins of Modern Western Science; or, The Early History of Linear Algebra

Roger Hart

Linear algebra is one of the mathematical foundations of modern science. Today, together with calculus, it is one of the most important core courses in modern mathematics at the university level. The main problem of linear algebra is the solution of systems of *n* linear equations in *n* unknowns. There are two main solutions: Gaussian elimination, the most general solution, and the most important of all matrix algorithms; and determinants, which offer an elegant solution for simple systems of linear equations for very small *n*. Linear algebra has applications in a broad range of sciences including physics and engineering, economics and finance, statistics and probability, control theory and robotics, computer graphics, cryptography, and network theory. Through the 'matrix mechanics' developed by Werner Heisenberg, Max Born, and Pascual Jordan in the 1920s, modern linear algebra is one of the foundations of quantum mechanics, which led to the Third Industrial Revolution and Digital Age. In the twenty-first century, linear

R. Hart (✉)
Texas Southern University, Houston, TX, USA
e-mail: rhart@rhart.org

A. Bala et al. (eds.), *Multicivilizational Exchanges in the Making of Modern Science*,
https://doi.org/10.1007/978-981-97-3541-9_9

algebra continues to play a fundamental role in a wide range of sciences, including artificial intelligence, machine learning, quantum information sciences, and the Second Quantum Revolution.

Traditional accounts assert that linear algebra is exclusively Western in origin. Early hagiographies attribute the discovery of Gaussian elimination to Carl Friedrich Gauss (1777–1855) and the discovery of determinants to Gottfried Wilhelm Leibniz (1646–1716). Subsequent research has given considerable cause to question these traditional attributions: Gaussian elimination can be found in the earliest extant Chinese mathematical treatise, *Nine Chapters on the Mathematical Arts* (*Jiuzhang suanshu* 九章算術, c. first-century CE); during the period of rapidly increasing trade between Europe and China, Gaussian elimination suddenly 'simultaneously' appears *ex nihilo* in several European treatises in the sixteenth century, beginning with the French monk Johannes Buteo's (c. 1492–c. 1572) *Logistica* (1559). Similarly, we are to believe that determinants were 'simultaneously' discovered *ex nihilo* by Leibniz, a Sinophile, and the Japanese mathematician Seki Takakazu 關孝和 (c. 1642–1708), who studied from Chinese treatises. Yet despite all this, linear algebra has still been held to be exclusively Western in origin; in the received historiography, the emergence of linear algebra in Europe is without history or precedent before the sixteenth century.

Research I have completed demonstrates that linear algebra developed first in China and later spread throughout Eurasia, including to Europe in the west and Japan in the east. More specifically, my first research monograph, *Chinese Roots of Linear Algebra* (Johns Hopkins University Press, 2011), demonstrates that the essentials of key methods used today in modern linear algebra—augmented matrices, row operations, row reductions, Gaussian elimination, and determinantal-style calculations—can be found in the arcane arts known as *fangcheng* 方程, practiced by anonymous adepts, and recorded in extant Chinese texts beginning in the first-century CE in China. My second research monograph, *Imagined Civilizations: China, the West, and Their First Encounter* (Johns Hopkins University Press, 2013), demonstrates that some of these practices circulated to Europe by the early thirteenth century. My first two research monographs fundamentally revise our understanding of the circulation of knowledge during this period, demonstrating that the most significant mathematics of this period circulated from China to Europe, and not vice versa. In this chapter I summarize these results, focusing on the mathematical evidence, using material I have previously published in *Chinese Roots of Linear Algebra*, *Imagined Civilizations*, and my website http://rhart.org/algebra/, together with new additional evidence available only after the publication of my research monographs.

In terms of this conference, then, the best currently available evidence I know for 'understanding science as a multi-civilizational outcome' is this research I published a decade ago demonstrating that linear algebra developed first in China and then circulated to Europe. To the question, 'How and why were China, and other major Asian civilizations, able to articulate knowledge that subsequently came to enrich modern science', the early history of linear algebra offers one compelling answer; and given this evidence, it is simply impossible to deny that that modern science is a 'multi-civilizational outcome'. Linear algebra is, however, hardly unique in this respect: this conference presents important contributions to the ongoing project of tracing the global circulation of science. In the terminology of this conference, then, we must conclude that modern science has always been 'dialogical'.[1]

1 Outline of the Argument

This chapter is divided into two parts. The key evidence I present here is technical mathematics recorded in extant mathematical treatises. Part I presents a summary of the following key facts from extant Chinese mathematical treatises:

[1] All quotations in this paragraph are from announcements for this conference. For a background on these debates, see Bala (2006, 2009, 2012), Bala and Duara (2016), and Lau (forthcoming).

- Gaussian elimination: the key elements of linear algebra—augmented matrices, row operations, row reductions, and Gaussian elimination—can be found in extant Chinese mathematical treatises from the first-century CE, along with dozens of later mathematical treatises throughout imperial China through the seventeenth century.
- Determinants: Calculations of determinants, limited to special classes of matrices, were apparently known in China by the first-century CE; the earliest extant explicit record of a determinantal calculation is 1261 CE; the earliest extant explicit record of a complete determinantal solution is 1661 CE.

Part II presents evidence of the global circulation of these practices throughout Eurasia:

- Europe: the earliest extant record demonstrating the circulation of linear algebra techniques from China to Europe is 1202 CE.
- Japan: linear algebra circulated from China to Japan no later than 1663 CE.
- Gaussian elimination: The earliest extant record of Gaussian elimination in Europe is the French monk Johannes Buteo's (c. 1492–c. 1572) *Logistica* (1559); the writing of these problems in the form of augmented matrices, the row operations, row reductions, Gaussian elimination, and results at each step are the same as those found in Chinese texts.

2 Part I. Linear Algebra Recorded in Chinese Mathematical Treatises

Earliest Extant Records of Matrices and Gaussian Elimination, ca. First-Century CE

The earliest known extant records of matrices and what we now recognize as Gaussian elimination are found in the *Nine Chapters on the Mathematical Arts* [*Jiuzhang suanshu* 九章算術],[2] in the eighth of those chapters, which is titled '*Fangcheng*' 方程. The *Nine Chapters*, the earliest transmitted Chinese text devoted specifically to mathematics, is traditionally dated c. 150 BCE, but modern scholarship places the likely date of composition during the first-century CE. In Chapter 8, '*Fangcheng*', 18 problems are recorded as word problems, without diagrams. These problems range from systems with two conditions in two unknowns to systems with five conditions in five unknowns; there is also one problem, problem 13, with five conditions in six unknowns, which I will refer to as the 'well problem'. The *Nine Chapters* stipulates that all 18 problems are to be solved by the *fangcheng* procedure (*fangcheng shu* 方程術), one variant of what we now call Gaussian elimination.[3] This section will illustrate how the *fangcheng* procedure, as reconstructed from written records preserved in the *Nine Chapters*, was a mathematical practice that used simple two-dimensional visual patterns on a counting board to produce solutions to what we would now call systems of linear equations with n conditions in n unknowns, using Gaussian elimination. And as we will see in following sections, some of these problems also have determinantal-style solutions, which are recorded in later texts and are transmitted to Europe by the thirteenth century.

Textual Records Versus Mathematical Practices

[2] *Jiuzhang suanshu* 九章算術 [Nine chapters on the mathematical arts], in *SKQS* (1983–1986). For the best analysis of the mathematics, see Shen et al. (1999); for a translation and study focused on textual questions, see Chemla and Guo (2004).

[3] Gaussian elimination has many variants; it also may be understood at various levels of abstraction. For definitions of Gaussian elimination, see Strang (1993, pp. 12–13) and Stewart (1973, p. 113, 1998, p. 148). In particular, Stewart notes: 'There are four closely related ways to approach Gaussian elimination: 1. Gaussian elimination as the elimination of variables in a linear system, 2. Gaussian elimination as row operations on a linear system, 3. Gaussian elimination as a transformation of a matrix to triangular form by elementary lower triangular matrices, 4. Gaussian elimination as the factorization of a matrix into the product of lower and upper triangular factors'.

First, however, we must distinguish between written records of the *fangcheng* procedure preserved in extant Chinese mathematical treatises and the *fangcheng* procedure itself as a mathematical practice. Problems in the *Nine Chapters* are recorded in one-dimensional narrative form.

For example, problem 18 is as follows:

> Now it is given: 9 *dou*[4] of hemp, 7 *dou* of wheat, 3 *dou* of legumes, 2 *dou* of beans, and 5 *dou* of millet, [together] are valued at 140 coins; 7 *dou* of hemp, 6 *dou* of wheat, 4 *dou* of legumes, 5 *dou* of beans, and 3 *dou* of millet, [together] are valued at 128 coins; 3 *dou* of hemp, 5 *dou* of wheat, 7 *dou* of legumes, 6 *dou* of beans, and 4 *dou* of millet, [together] are valued at 116 coins; 2 *dou* of hemp, 5 *dou* of wheat, 3 *dou* of legumes, 9 *dou* of beans, and 4 *dou* of millet, [together] are valued at 112 coins; 1 *dou* of hemp, 3 *dou* of wheat, 2 *dou* of legumes, 8 *dou* of beans, and 5 *dou* of millet, [together] are valued at 95 coins. It is asked: What is the value of one *dou* [for each of the grains]? Answer: 7 coins for one *dou* of hemp; 4 coins for one *dou* of wheat; 3 coins for one *dou* of legumes; 5 coins for one *dou* of beans; 6 coins for one *dou* of millet. Procedure: Follow *fangcheng*; use the procedure for positives and negatives.

The earliest extant editions of the *Nine Chapters* provide no two-dimensional diagrams. Without prior knowledge of this mathematical practice on two-dimensional counting boards, these problems, recorded in this one-dimensional narrative form, are simply unintelligible.

It should also be noted that these problems are highly contrived, mathematical practices without any known applications, presented in problems that are entirely pretextual. For example here, ordinarily, we calculate the total price for combinations of items based on the prices of the individual items: for each item, we multiply the number of units by the cost of that item per unit and then add the subtotals together to find the total price. This is simple everyday practical mathematics. In this problem, in contrast, we are given the prices for five sets of different combinations of five items and then asked to employ complicated linear

[4] A unit of volume, equal to ten *sheng* 升. Currently, one *dou* is equal to one decaliter.

Table 1 Chinese rod numerals for one through nine, and ten through ninety; zero is denoted by a blank space, as indicated here for numbers 10 through 90

1	2	3	4	5	6	7	8	9
𝍠	𝍡	𝍢	𝍣	𝍤	𝍥	𝍦	𝍧	𝍨
10	20	30	40	50	60	70	80	90
𝍩	𝍪	𝍫	𝍬	𝍭	𝍮	𝍯	𝍰	𝍱

algebra computations to find the solution to this system of five conditions in five unknowns, thereby determining the price of each individual item.

Counting Rod Calculations

As a mathematical practice, the *fangcheng* procedure was performed on a two-dimensional counting board with counting rods. Numbers were recorded using differing arrangements of counting rods. The Chinese numeral-rod system is not difficult to understand: the first position is for numbers 0–9 (0 is denoted by an empty space); 1–5 are denoted by the corresponding numbers of rods placed vertically, and a horizontal rod denoting 5 is placed above one to four vertical rods to denote 6–9. The next position to the left is for tens, which are placed horizontally, then hundreds, and so on. Red counting rods denote positive numbers and black counting rods denote negative numbers. A summary of the representation of numbers by counting rods is shown in Table 1.

Thus, for example, the number 721 is arranged on the counting board as 𝍦𝍪𝍠.

Augmented Matrices

Fangcheng problems were arranged using counting rods in a rectangular array, in a manner identical to the augmented matrix familiar from modern linear algebra, if we allow for differences in writing, as will be demonstrated below using problem 18.

STEP 0. The following diagram shows how problem 18 is laid out using counting rods on the counting board:

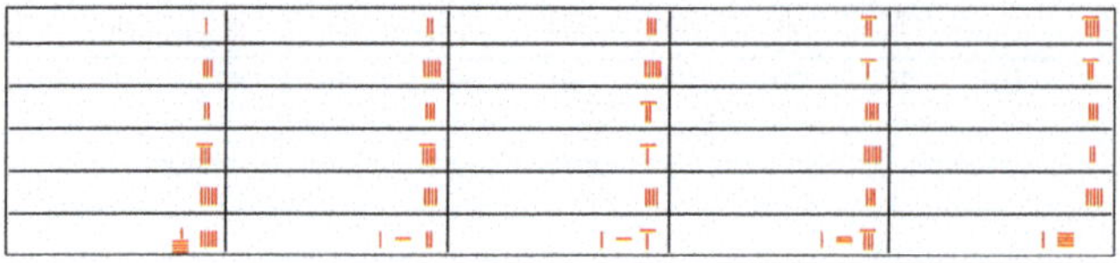

Written in modern mathematical notation as an augmented matrix, we have the following:

$$\begin{bmatrix} 9 & 7 & 3 & 2 & 5 & 140 \\ 7 & 6 & 4 & 5 & 3 & 128 \\ 3 & 5 & 7 & 6 & 4 & 116 \\ 2 & 5 & 3 & 9 & 4 & 112 \\ 1 & 3 & 2 & 8 & 5 & 95 \end{bmatrix} \tag{1}$$

The only difference between this and the modern mathematical notation is the orientation of the array, which corresponds to differences in writing (writing in imperial China proceeded from top to bottom, then right to left; modern English is written from left to right, then top to bottom). Augmented matrices, rotated clockwise 90°, correspond exactly to the counting rod diagrams.

This problem can also be written out more explicitly as a system of n linear equations in n unknowns, which is probably the form most familiar to a general audience of modern readers:

$$9x_1 + 7x_2 + 3x_3 + 2x_4 + 5x_5 = 140 \tag{2}$$

$$7x_1 + 6x_2 + 4x_3 + 5x_4 + 3x_5 = 128 \tag{3}$$

$$3x_1 + 5x_2 + 7x_3 + 6x_4 + 4x_5 = 116 \tag{4}$$

$$2x_1 + 5x_2 + 3x_3 + 9x_4 + 4x_5 = 112 \tag{5}$$

$$x_1 + 3x_2 + 2x_3 + 8x_4 + 5x_5 = 95. \tag{6}$$

It should be noted, however, that in modern linear algebra, the simpler augmented matrix (1), which corresponds to the counting board representation, is much preferred over the more cumbersome use of x_n to denote the unknowns (that is, in this case, x_1, x_2, $x3$, x_4, x_5) in Eqs. (2–6).

Elementary Row Operations

If *fangcheng* problems were displayed using what we would now call an augmented matrix, the solution proceeds by what we would now call elementary row operations—multiplication of a row by a nonzero constant and subtracting one row from another. Solving *fangcheng* problems requires as a prerequisite a knowledge of counting board operations for addition, subtraction, multiplication, and division of integers (including negative numbers). It also requires a knowledge of the corresponding operations for fractions, although fractions rarely appear. Given knowledge of these operations, as a mathematical practice, the *fangcheng* procedure is perhaps best understood as simple visual patterns on the counting board that are applied repeatedly in order to solve the problem. Extant treatises provide specific names for only two of these patterns:

- Cross-multiplication (*biancheng* 徧乘), the multiplication of an entire column of entries by one entry of another column, or in modern terms, the multiplication of one entire row by a nonzero constant;
- Term-by-term subtraction (*zhi chu* 直除, which literally means 'directly subtract'), the term-by-term subtraction of the entries of one column from those of another. In modern terms, this is subtracting one row from another.

In addition, there are other simple visual patterns, which are not provided names that are even more difficult to explain in words. These patterns are illustrated below.

Multiplication of a Row by a Number

The *fangcheng* procedure begins with what we would in modern terms call multiplication of an entire row by a number, as shown below.

STEP 1. Cross-multiplication (*biancheng* 徧乘). Following the *fangcheng* procedure, the first step is to cross-multiply, that is, to multiply each entry in the second column by the first entry of the first column. More specifically, the column (7, 6, 4, 5, 3, 128) is multiplied by 9, as is indicated in the following diagram[5]:

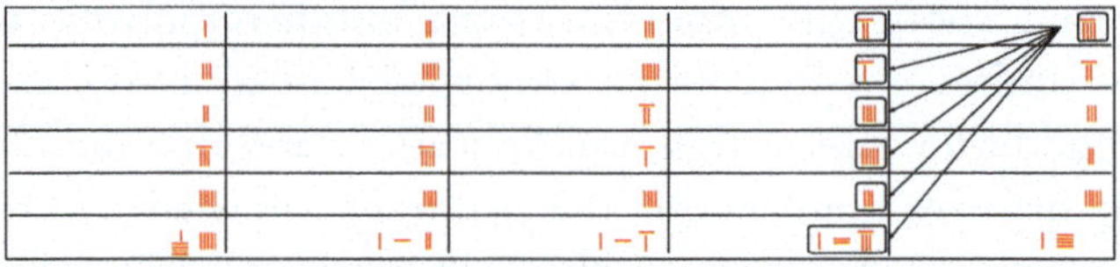

Written in modern mathematical notation, this is simply the augmented matrix modified by what is called a row operation, that is, by multiplying the entire second row by the pivot in the first row, as shown in the augmented matrix below:

$$\begin{bmatrix} \boxed{9} & 7 & 3 & 2 & 5 & 140 \\ \boxed{7} & \boxed{6} & \boxed{4} & \boxed{5} & \boxed{3} & \boxed{128} \\ 3 & 5 & 7 & 6 & 4 & 116 \\ 2 & 5 & 3 & 9 & 4 & 112 \\ 1 & 3 & 2 & 8 & 5 & 95 \end{bmatrix}$$

[5] The counting board, as shown in this diagram, and all of the following diagrams, was used to display the results of the calculations. In practice, the addition, multiplication, subtraction, or division of two individual entries might have been calculated in the head; but it is also possible that recording intermediary results might have been necessary. Written records of *fangcheng* practices do not describe these details.

STEP 2. Cross-multiplication. The second step is another cross-multiplication, which is similar to the previous step. We multiply the entire first column by the first entry in the second column, that is, we multiply (9, 7, 3, 2, 5, 140) by 7.

Subtracting One Row from Another

The next step in the *fangcheng* procedure is what we would call in modern terms subtracting one row from another.

STEP 3. Term-by-term subtraction (*zhi chu* 直除). In the third step, we subtract, term-by-term, the entries in the first column from those in the second column, that is, $(63, 54, 36, 45, 27, 1152) - (63, 49, 21, 14, 35, 980)$, which, on the counting board, again follows a very simple pattern, as can be seen from the diagram below:

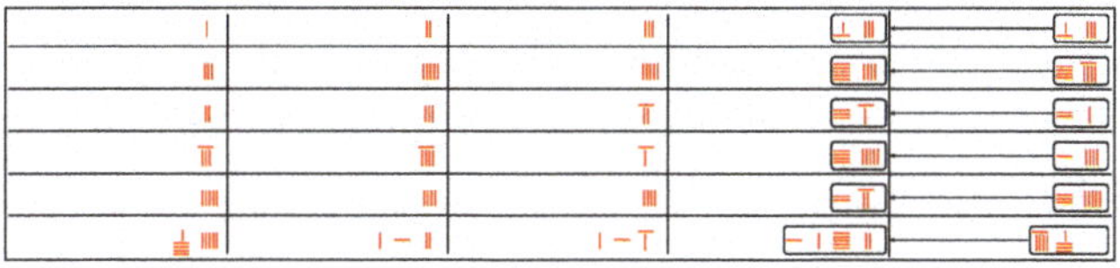

On the corresponding augmented matrix, this pattern is as follows:

$$\begin{bmatrix} \boxed{63} & \boxed{49} & \boxed{21} & \boxed{14} & \boxed{35} & \boxed{980} \\ \downarrow & \downarrow & \downarrow & \downarrow & \downarrow & \downarrow \\ \boxed{63} & \boxed{54} & \boxed{36} & \boxed{45} & \boxed{27} & \boxed{1152} \\ 3 & 5 & 7 & 6 & 4 & 116 \\ 2 & 5 & 3 & 9 & 4 & 112 \\ 1 & 3 & 2 & 8 & 5 & 95 \end{bmatrix}$$

Elimination of Entries

This results in what in modern terms we would call the elimination of entries. That is, the result $(63, 54, 36, 45, 27, 1152) -$

(63, 49, 21, 14, 35, 980) = (0, 5, 15, 31, −8, 172) is represented on the counting board as shown in the following diagram[6]:

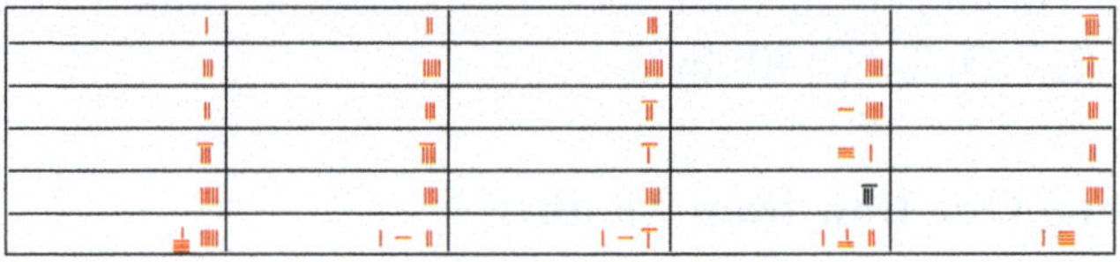

In modern mathematical notation, this result is as follows:

$$\begin{bmatrix} 9 & 7 & 3 & 2 & 5 & 140 \\ 0 & 5 & 15 & 31 & -8 & 172 \\ 3 & 5 & 7 & 6 & 4 & 116 \\ 2 & 5 & 3 & 9 & 4 & 112 \\ 1 & 3 & 2 & 8 & 5 & 95 \end{bmatrix}$$

This is precisely row reduction of the augmented matrix, the first step of Gaussian elimination, where we have eliminated the first entry in the second row.

Upper-Triangular Form

STEPS 4–30. Continued cross-multiplication and term-by-term subtraction for the remaining columns. We then eliminate the entries in the remaining columns that lie above the diagonal entries, proceeding as we did for the first and second columns. That is, by simply continuing to use the same pattern of cross-multiplication and term-by-term subtraction, we eliminate the entries in the remaining columns, arriving at the following:

[6] At this point, as shown in the diagram, the original values of the first column (9, 7, 3, 2, 5, 140), are restored. Again, extant treatises do not fully explain these details.

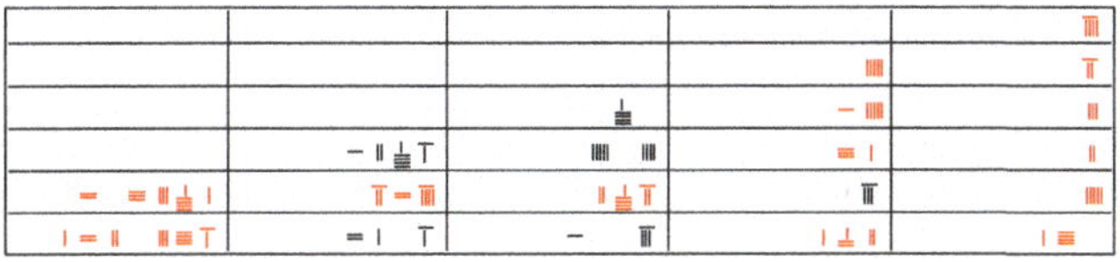

Using modern mathematical notation, in the augmented matrix, we have the following:

$$\begin{bmatrix} 9 & 7 & 3 & 2 & 5 & 140 \\ 0 & 5 & 15 & 31 & -8 & 172 \\ 0 & 0 & -90 & -504 & 297 & -1008 \\ 0 & 0 & 0 & -1296 & 729 & -2106 \\ 0 & 0 & 0 & 0 & 203391 & 1220346 \end{bmatrix}$$

This is just the upper-triangular form familiar from modern linear algebra. We have completed the first half of the solution, namely, forward substitution. It should be noted that, just as is the case for modern Gaussian elimination, this procedure can be applied to an array of any size, that is, any system of n conditions in n unknowns.

Algorithm for Elimination

The *fangcheng* procedure recorded in the *Nine Chapters*, which is the same as Gaussian elimination, can be written in modern mathematics as shown in Algorithm 1.

Algorithm 1: *Fangcheng* Procedure

```
procedure ELIMINATION(A, b)
    for k ← 1 to n − 1 do                                  ▷ For columns 1 to n − 1
        if A[k,k] = 0 then                                 ▷ If pivot a_kk is zero
            [Not explicitly addressed]                     ▷ Problem 1 gives no instructions
        else                                               ▷ Pivot a_kk ≠ 0
            for i ← k + 1 to n do                          ▷ For each row below the pivot a_kk
                a[i] ← A[i,k]                              ▷ Save the value of the entry a_ik
                for j ← k to n + 1 do                      ▷ For the remaining entries in the row
                    A[i,j] ← A[k,k] * A[i,j]               ▷ Multiply the row by the pivot a_kk
                end for j
                for j ← k to n + 1 do                      ▷ For the remaining entries in the row
                    A[i,j] ← A[i,j] − a[i] * A[k,j]        ▷ Subtract entries multiplied by a_ik
                end for j
            end for i
        end if
    end for k
return A′, b′                  ▷ Transformed augmented matrix (in upper-triangular form)
end procedure
```

It should be noted here that there are numerous cases for which extant texts do not provide guidance, such as zero pivots, as noted above.

Conclusion

The earliest transmitted Chinese mathematical treatise, the *Nine Chapters*, is the earliest known extant record of what we now in modern mathematics call Gaussian elimination—augmented matrices, row operations, row reductions, leading to upper-triangular form. The *Nine Chapters* then employs a highly-sophisticated integer-preserving back-substitution algorithm to solve the problem.[7]

Distinctive Matrices with Determinantal Solutions in the *Nine Chapters*

The *Nine Chapters* contains several problems of a distinctive form of augmented matrix, in which the absence of entries simplifies what we

[7] For detailed analysis of the integer-preserving back substitution in the *Nine Chapters*, see Hart (2011, pp. 93–108). Integer-preserving Gaussian elimination was a topic of research in the early days of computers: see Bareiss (1966).

would now call determinantal calculations. When I use the term 'determinants' in this study of the *early* history of linear algebra, I am referring to solutions found for specific cases that are equivalent to those found by determinants, and not a general or complete theory of determinants: historically speaking, specific cases preceded later generalization under the rubric of a complete theory of determinants. Here I follow conventional usage in research on the history of mathematics, and more specifically, 'determinantal methods' (*Determinantenmethode*) in the typology in Tropfke's classic work *Geschichte der Elementarmathematik*.[8] I will use the term 'determinantal' broadly to refer to schematic methods similar to Cramer's Rule involving the multiplication of individual entries. I will use the term 'determinantal calculation' when such a method does not yield solutions to the unknowns (see section below on 'Earliest Extant Determinantal Calculations, 1261 CE'); I will use the term 'determinantal solution' when such a method does yield complete solutions to the unknowns (see section below on 'Earliest Extant Determinantal Solutions, 1661 CE).

Determinantal calculations and solutions are explicitly recorded only in later treatises, not in the *Nine Chapters*, but the high proportion of these problems is evidence that these solutions were known at the writing of the *Nine Chapters*. More specifically, in modern terminology, these problems have the following form:

$$\begin{bmatrix} k_1 & l_1 & 0 & \cdots & 0 & b_1 \\ 0 & k_2 & l_2 & \ddots & \vdots & b_2 \\ \vdots & \ddots & \ddots & \ddots & 0 & \vdots \\ 0 & \cdots & 0 & k_{n-1} & l_{n-1} & b_{n-1} \\ l_n & 0 & \cdots & 0 & k_n & b_n \end{bmatrix} \tag{7}$$

[8] More specifically, solutions to systems of n linear equations in m unknowns are categorized as follows: (1) 'Reduction to an equation with one unknown'; (2) 'Special procedures'; (3) 'Elimination methods'; (4) 'Determinantal methods' (*Determinantenmethode*). Tropfke et al. (1980, pp. 388–389). Determinantal methods stand in contrast to other approaches to solving systems of linear equations, such as elimination by row reductions, in which solutions are found through a series of successive operations.

As we will see below, problems of this form are very distinctive and cannot be found in modern mathematical treatises on linear algebra. Yet among the eighteen problems in '*Fangcheng*', Chapter 8 of the *Nine Chapters*, there are five problems, including the well problem, that are variants of the distinctive form given by augmented matrix (7), for which the absence of entries permits relatively simple determinantal solutions. A brief conspectus of these problems is presented in Table 2.

For each of these problems, following visual simple patterns, it is possible to compute the values for all of the unknowns; and from the

Table 2 From among the eighteen problems in '*Fangcheng*', Chapter 8 of the *Nine Chapters*, five problems are variants of augmented matrix (7), for which the absence of entries allows simple determinantal solutions

Problem	Counting board	Augmented matrix	Notes
3	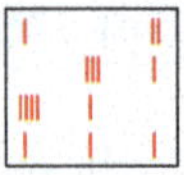	$\begin{bmatrix} 2&1&0&1 \\ 0&3&1&1 \\ 1&0&4&1 \end{bmatrix}$	Variant with three conditions in three unknowns
12		$\begin{bmatrix} 1&1&0&40 \\ 0&2&1&40 \\ 1&0&3&40 \end{bmatrix}$	Variant with three conditions in three unknowns
13	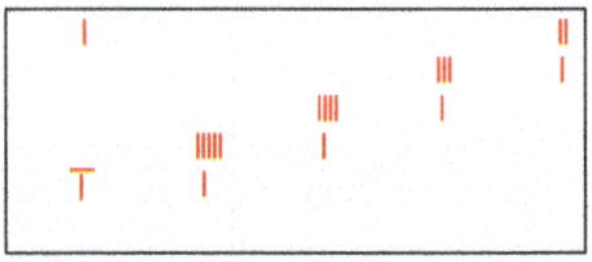	$\begin{bmatrix} 2&1&0&0&0&y \\ 0&3&1&0&0&y \\ 0&0&4&1&0&y \\ 0&0&0&5&1&y \\ 1&0&0&0&6&y \end{bmatrix}$	Five conditions in six unknowns
14	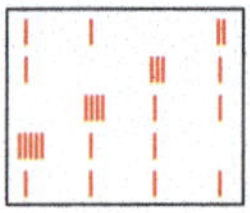	$\begin{bmatrix} 2&1&1&0&1 \\ 0&3&1&1&1 \\ 1&0&4&1&1 \\ 1&1&0&5&1 \end{bmatrix}$	More complex variant with four conditions in four unknowns
15	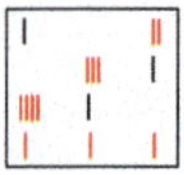	$\begin{bmatrix} 2&-1&0&1 \\ 0&3&-1&1 \\ -1&0&4&1 \end{bmatrix}$	Variant with three conditions in three unknowns, with negative coefficients

point of view of the original *fangcheng* procedure (a variant of Gaussian elimination, described above) presented in the *Nine Chapters*, there is nothing noteworthy about any of these problems that would explain their inclusion in the text.

Later Chinese mathematical treatises continue to record numerous problems that are variants of the distinctive form given in augmented matrix (7): often as many as 10–20% of the problems are similar. It was because of the importance of these problems in Chinese mathematical treatises that I chose an example of a *fangcheng* problem with nine conditions in nine unknowns of the form of augmented matrix (7) for the cover of my book *Chinese Roots of Linear Algebra.*

Earliest Extant Determinantal Calculations, 1261 CE

The earliest known calculations of what we would now in modern terms call the determinant of the matrix of coefficients have the following features:

- Determinantal calculations are for a special subclass of problems of the form of augmented matrix (7). In modern terms, these problems are n equations in $n + 1$ unknowns, where the $(n + 1)^{\text{th}}$ unknown is $y = b_1 = b_2 = \ldots = b_n$.
- The exemplar for this class of problems is the 'well problem', problem 13 of Chapter 8 of the *Nine Chapters*.
- These problems are solved by first employing a nonsensical and, from a modern point of view, mathematically spurious calculation to determine the $(n + 1)^{\text{th}}$ unknown y, as will be explained below.

The well problem is presented in the *Nine Chapters* in one-dimensional narrative form as follows:

> Given are five families who share a well. The deficit of 2 of A's well-ropes is the same as 1 of B's well-ropes. The deficit of 3 of B's well-ropes is the same as 1 of C's well-ropes. The deficit of 4 of C's well-ropes is the same as 1 of D's well-ropes. The deficit of 5 of D's well-ropes is the same as 1 of E's well-ropes. The deficit of 6 of E's well-ropes is the same as 1 of

> A's well-ropes. Each then obtains the well-rope that makes up the deficit, and all reach the water. Problem: how deep is the well, and how long is each of the well-ropes?[9]

The well problem, displayed on the counting board, is as shown in problem 13 in Table 2 on the previous page. More generally, it is an exemplar for problems with n equations in $n + 1$ unknowns of the following form:

$$\begin{bmatrix} k_1 & l_1 & 0 & \cdots & 0 & y \\ 0 & k_2 & l_2 & \ddots & \vdots & y \\ \vdots & \ddots & \ddots & \ddots & 0 & \vdots \\ 0 & \cdots & 0 & k_{n-1} & l_{n-1} & y \\ l_n & 0 & \cdots & 0 & k_n & y \end{bmatrix} \tag{8}$$

The earliest record of these determinantal calculations is Jia Xian's 賈憲 (fl. 1023–1063) commentary on the *Nine Chapters* (c. 1050 CE). Jia Xian's commentary is no longer extant, but excerpts are preserved as quotations in Yang Hui's 楊輝 (c. 1238–c. 1298) *Nine Chapters on the Mathematical Arts, with Detailed Explanations* (*Xiang jie jiuzhang suanfa* 詳解九章算法, c. 1261).[10]

In the earliest received version of the *Nine Chapters*, the $(n + 1)^{\text{th}}$ unknown is simply stipulated to be 721, with no explanation provided. In Jia Xian's commentary, an explicit calculation is provided to find the $(n + 1)^{\text{th}}$ unknown: the diagonal terms are multiplied together, the remaining terms are multiplied together (that is, the super-diagonal and the term in the corner), and then the two products are added, yielding,

$$2 \cdot 3 \cdot 4 \cdot 5 \cdot 6 + 1 \cdot 1 \cdot 1 \cdot 1 \cdot 1 = 721.$$

Setting the $(n + 1)^{\text{th}}$unknown $y = 721$ transforms the problem to a linear system of five conditions in five unknowns, which then has a

---

[9] For complete analysis of this problem, see Hart (2011, Chapter 7), 'The Well Problem'.

[10] In *ZKJDT* (1993).

unique solution, which can be found by determinantal methods or Gaussian elimination. Problems of this form were recognized to be a category of problems (*bi lei* 比類, parallel category); an additional example with three equations and four unknowns is presented and solved.

A Nonsensical, Mathematically Spurious Calculation

The calculation presented here for the $(n + 1)^{\text{th}}$ unknown y is nonsensical and, from a modern perspective, mathematically spurious. That is, the unknown depth of the well y depends, nonsensically, on the following: (1) the number of ropes used; (2) the number of ropes borrowed; and (3) the parity (odd or even) of the number of families. More specifically, the unknown depth of the well y is found not by measuring the well's depth, but instead is calculated by the following nonsensical procedure:

> *Ropes used*: First, we multiply together the number of ropes used by each family;
> *Ropes borrowed*: Next, we multiply together the number of ropes borrowed by each family;
> *Parity*: Finally, because the number of families is odd, we add the product of the ropes borrowed to the product of the ropes used to obtain the depth of the well (if the number of families is even, we would then subtract).

According to this nonsensical procedure, then, a change in the number of ropes used, ropes borrowed, or parity of the number of families results in a change in the depth of the well!

This calculation is not just nonsensical, from a modern perspective, it is also mathematically spurious, as shown by dimensional analysis: multiplying together dimensionless numbers (ropes used and ropes borrowed), then adding or subtracting the subtotals depending on the parity of the number of families, gives a dimensionless result that is supposed to be the depth of the well, which must have the dimension of length. As we will see below, this same nonsensical, mathematically spurious calculation is recorded in Fibonacci's works.

Earliest Extant Determinantal Solutions, 1661 CE

The earliest known extant record of a step-by-step determinantal solution is preserved in Fang Zhongtong's 方中通 (1634–1698) *Numbers and Measurement, an Amplification* (*Shu du yan* 數度衍, 1661),[11] used to solve the well problem. This procedure has no analogue in modern mathematics and shows the complexity of these solutions. For the sake of brevity, I present only brief examples of the calculations; for the complete solution, see Hart (2011, pp. 132–144).

'Set the Negatives' and Calculate the 'Dividend'

The first part of the calculation, called 'set the negatives' (*li fu* 立負), follows simple visual patterns. The second part of the calculation follows simple visual patterns of cross-multiplication and addition, which are difficult to explain in words, used to calculate the 'dividend'. The following step is one example of this visual pattern:

STEP 15. For the last step of this second part of the calculation, the visual pattern of cross-multiplication and addition is again repeated one final time. Specifically, we perform the following calculation: (1) we begin with the entry located in row 6 column 1, which at the beginning of this calculation, from the previous step, was originally 10815. (2) We then multiply this entry by the entry in row 4 column 2, which is 5, giving 10815·5, and replace the entry in row 6 column 1 with the temporary result 54075. (3) We then multiply the entry in row 6 column 2, 721 by the entry in row 1 column 4, −1, giving 721·−1 = −721. (4) Finally, we subtract the result of the previous step, −721, from the temporary entry currently in row 6 column 1, 54075, namely 54075−(−721), and then replace the temporary entry in row 6 column 1 with the result, 54796.

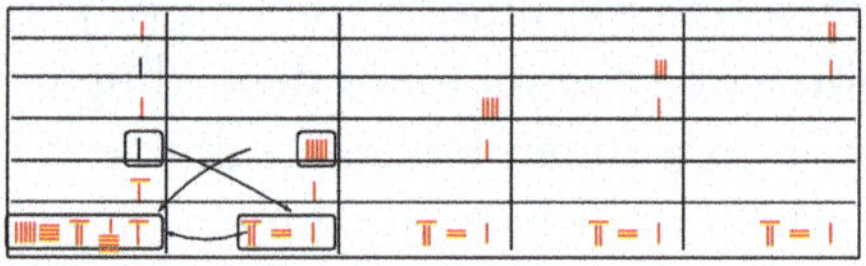

[11] In *SKQS*.

Written out, in symbolic terms, for general *n*, the result is as follows:

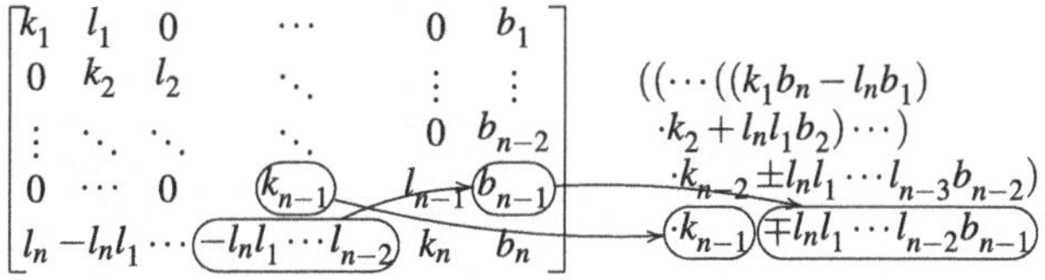

In symbolic terms, for general *n*, the result calculated is as follows:

$$((\ldots((((k_1b_n - l_nb_1)k_2 + l_nl_1b_2)k_3 - l_nl_1l_2b_3)k_4 + l_nl_1l_2l_3b_4)\ldots)$$
$$\cdot k_{n-2} \pm l_nl_1l_2\ldots l_{n-3}b_{n-2})k_{n-1}$$
$$\mp l_nl_1l_2\ldots l_{n-2}b_{n-1}$$

Calculating the 'Divisor'

The third part of the calculation is to calculate the 'divisor' (*fa* 法) by multiplying the elements of the diagonal of the coefficient matrix, then adding (or subtracting, depending on whether the number of unknowns is odd or even) the product of the elements on the super-diagonal (sub-diagonal on the counting board) together with the element l_n. The result is simply what we would now call the determinant of the coefficient matrix.

Dividing the 'Dividend' by the 'Divisor'

The final part of the calculation is a simple division, dividing the 'dividend' we have calculated by the 'divisor' we have calculated.

STEP 17. The final step, a simple division, gives the value of the *n*th unknown, which can be expressed in general symbolic terms as follows:

$$x_n = \frac{\begin{array}{c}((\ldots((((k_1b_n - l_nb_1)k_2 + l_nl_1b_2)k_3 - l_nl_1l_2b_3)k_4 + l_nl_1l_2l_3b_4)\ldots) \\ \cdot k_{n-2} \pm l_nl_1l_2\ldots l_{n-3}b_{n-2})k_{n-1} \\ \mp l_nl_1l_2\ldots l_{n-2}b_{n-1})\end{array}}{(k_1k_2k_3\ldots k_{n-1}k_n \mp l_nl_1l_2l_3\ldots l_{n-1})}, \quad (9)$$

where again $\pm$ is $+$ for n even and $-$ for n odd.

This procedure is entirely general, valid for any $n \geq 3$, and all of the n unknowns can be found by similar calculations.

3 Part II. Global Circulation of Linear Algebra throughout Eurasia

These practices circulated over the centuries throughout Eurasia, to Europe in the west and Japan in the east. The diffusion of these practices to Europe was likely effected by merchants, missionaries, and travelers, following the routes by which commerce, art, and religion circulated in what is increasingly understood to be the 'global Middle Ages', to use Geraldine Heng's helpful term.[12] This comports with the considerable existing evidence of the global circulation of mathematics.[13] This explains the almost simultaneous appearance in the seventeenth century of a complete theory of determinants in the works of Gottfried Wilhelm Leibniz (1646–1716), a Sinophile, and the Japanese mathematician Seki Takakazu 関孝和 (1642–1708), who had studied Chinese mathematics. The considerable variety of problems, preserved in extant Chinese treatises, shows that determinants were not suddenly discovered *ex nihilo,*

[12] See the Global Middle Ages Project (G-MAP) website, http://globalmiddleages.org; Heng (2022, 2018).

[13] For several examples, see Vogel (1983), Chemla (1994, 1997), Dold-Samplonius et al. (2002), Katz and Imhausen (2007), Rowe and Horng (2015), and Brentjes (2010).

but emerged from over one thousand years of explorations of increasingly complex patterns of determinantal-style computations. This also explains the otherwise inexplicable appearance of Gaussian elimination, identical to Chinese methods, in the work of the French monk Johannes Buteo (c. 1492–c. 1572) and others in the sixteenth century.

Circulation of Distinctive Determinantal Solutions to Europe, 1202 CE

The arcane determinantal-style calculations and solutions described above for problems of the form of augmented matrix (8) circulated across the Eurasian continent and are recorded in the works of Leonardo Pisano (c. 1170–c. 1250), more commonly known today by the name Fibonacci. Most tellingly, the same nonsensical, mathematically spurious calculations for the solution of the $(n + 1)^{\text{th}}$ unknown are recorded in Fibonacci's work.

Problems of the more general form given by augmented matrix (7) are already so distinctive that they can serve as a 'fingerprint', a kind of unique identifier. In my research for *Chinese Roots of Linear Algebra*, I searched through thousands of matrices recorded in modern mathematical treatises[14] and found only one example of a problem that was similar (a problem with three equations in three unknowns). Because these problems are so distinctive, I then searched for similar problems in European mathematical treatises from the seventeenth and eighteenth centuries, including the extant mathematical writings of Gauss and Leibniz. Unable to find similar problems, at the time, I felt that the most that I could definitively conclude is that '[t]he essentials of the methods used today ... in modern linear algebra *were not first discovered* by Leibniz or by Gauss'.[15] It turns out that I was looking for problems of the form given in

[14] Strang (2006), Serre (2002), Zhang (1999), Stewart (1998), Horn and Johnson (1985), Horn and Johnson (1991), Cullen (1990), Stewart (1973), Hoffman and Kunze (1971), Muir (1906–1923).

[15] Hart (2011, p. 192), emphasis in italics added.

augmented matrix (7) in seventeenth- and eighteenth-century European texts, at least four centuries after they were transmitted to Europe.[16]

Even more distinctive still are problems of the subclass given by augmented matrix (8). The solutions to these problems are without analog in modern mathematics, and in fact these solutions are esoteric enough that they have not been adequately analyzed in previous studies of Fibonacci's mathematics.[17] To put it another way, it was because of my familiarity with Chinese *fangcheng* problems and their solutions that the problems and solutions I found in Fibonacci made sense.

Here I will briefly present as an example one problem of the form of the subclass of distinctive augmented matrices (8), 'Another Problem on Five Men', and its solution (as Table 3 on page XX shows, this is but one among several examples). In Fibonacci, this problem is recorded as a word problem. In modern mathematical terms, it is a system of five conditions in six unknowns, which can be written as follows:

$$\begin{aligned} &x_1 + \frac{2}{3}x_2 = y,\ x_2 + \frac{4}{7}x_3 = y,\ x_3 + \frac{5}{11}x_4 = y, \\ &x_4 + \frac{6}{13}x_5 = y,\ x_5 + \frac{8}{19}x_1 = y \end{aligned} \tag{10}$$

If we write this as an augmented matrix, to facilitate comparison with the *fangcheng* problems, we can immediately see that it is similar in form to that given by augmented matrix (8):

$$\begin{bmatrix} 1 & \frac{2}{3} & 0 & 0 & 0 & y \\ 0 & 1 & \frac{4}{7} & 0 & 0 & y \\ 0 & 0 & 1 & \frac{5}{11} & 0 & y \\ 0 & 0 & 0 & 1 & \frac{6}{13} & y \\ \frac{8}{19} & 0 & 0 & 0 & 1 & y \end{bmatrix}. \tag{11}$$

[16] For further evidence of mathematical practices first recorded in Chinese texts that can be found in Fibonacci, see Ji (2015).

[17] Tropfke et al. (1980), Sesiano (1985), Spiesser (2000), and Hannah (2007, 2010).

Table 3 Several examples from Fibonacci's *Liber Abaci* of problems of the form given in augmented matrix (8) (the last problem listed here, 'On Five Men Who Bought a Horse', is solved in *Liber Abaci* by a variant of false position)

Title of problem	Problem written as an augmented matrix	Notes
'On the Purchase of a Horse by Three Men, When Each One Takes Some Bezants from the Others'. Sigler (2002, pp. 338–339)	$\begin{bmatrix} 1 & \frac{1}{3} & 0 & y \\ 0 & 1 & \frac{1}{4} & y \\ \frac{1}{5} & 0 & 1 & y \end{bmatrix}$	Three conditions in four unknowns
'Another Problem on Three Men According to the Above written Method'. Sigler (2002, p. 341)	$\begin{bmatrix} 1 & \frac{2}{3} & 0 & y \\ 0 & 1 & \frac{4}{7} & y \\ \frac{5}{9} & 0 & 1 & y \end{bmatrix}$	Three conditions in four unknowns
'On the Same with Four Men'. Sigler (2002, pp. 341–342)	$\begin{bmatrix} 1 & \frac{1}{3} & 0 & 0 & y \\ 0 & 1 & \frac{1}{4} & 0 & y \\ 0 & 0 & 1 & \frac{1}{5} & y \\ \frac{1}{6} & 0 & 0 & 1 & y \end{bmatrix}$	Four conditions in five unknowns
'On Five Men Who Bought a Horse'. Sigler (2002, p. 458)	$\begin{bmatrix} 1 & \frac{1}{2} & 0 & 0 & 0 & y \\ 0 & 1 & \frac{1}{3} & 0 & 0 & y \\ 0 & 0 & 1 & \frac{1}{4} & 0 & y \\ 0 & 0 & 0 & 1 & \frac{1}{5} & y \\ \frac{1}{6} & 0 & 0 & 0 & 1 & y \end{bmatrix}$	Five conditions in six unknowns

To solve the problem, Fibonacci writes the problem out in a two-dimensional form as a series of fractions arranged contiguously, as follows,

The details of the taking are written in order thus: $\frac{8}{19}\,\frac{6}{13}\,\frac{5}{11}\,\frac{4}{7}\,\frac{2}{3}$.[18]

The operations he uses to solve the problem are not those familiar from modern mathematics.

First, the $(n + 1)^{\text{th}}$ unknown is found as follows:

> And all of the numbers that are under the fraction are multiplied together; there will be 57057. *As the number of men is odd*, the product of the numbers that are over the fractions is *added* to this, that is, the 2 times

[18] Sigler (2002, p. 345) and Boncompagni (1857, p. 234), available online through Linda Hall Library Digital Collections, https://catalog.lindahall.org/permalink/01LINDAHALLINST/19lda7s/alma991855313405961.

> the 4 times the 5 times the 6 [times the 8]; there will be 58977, which is had for the price of the horse.[19]

The above calculation notes that when 'the number of men is odd' the products of the two sets of numbers (the numerators and the denominators) are to be added. That is, the price of the horse is given by

$$3 \cdot 7 \cdot 11 \cdot 13 \cdot 19 + 2 \cdot 4 \cdot 5 \cdot 6 \cdot 8 = 58977. \tag{12}$$

The problem is then solved for each unknown in succession. The calculation for the first unknown is explained in Fibonacci as follows:

> And as the first man's bezants are had, the upper number of the fraction of the takings is subtracted from the lower number of the same fraction, that is the 2 from the 3; there remains 1, and it is multiplied by the 7; there will be 7, to which you add the product of the 2 and the 4; there will be 15, which you multiply by the 11; there will be 165, from which you subtract the product of the 2 and the 4 and the 5; there remains 125, which you multiply by the 13; there will be 1625, to which you add the product of the 2 and the 4 and the 5 and the 6; there will be 1865, which is multiplied by the 19; there will be 35435, and the first man has this many.[20]

Written out using modern mathematical notation, this is the following calculation,

$$((((3 - 2) \cdot 7 + 2 \cdot 4) \cdot 11 - 2 \cdot 4 \cdot 5) \cdot 13 + 2 \cdot 4 \cdot 5 \cdot 6) \cdot 19 = 35435. \tag{13}$$

Each of the remaining unknowns is calculated in a similar manner.

As with *fangcheng* problems, these problems were solved using simple visual patterns in two dimensions, as shown in the example below of the calculations to find the first unknown.

---

[19] Sigler (2002, p. 345), emphasis in italics added, interpolation based on the mathematics added; Boncompagni (1857, p. 234), emphasis and interpolation added.

[20] Sigler (2002, p. 345) and Boncompagni (1857, pp. 234–235).

STEP ($2n-2$). For the final steps, we continue to employ simple visual patterns. The next-to-last step is to multiply the upper terms of all but the last of the fractions, from right to left, giving $k_1 k_2 \dots k_{n-1}$. We alternate between adding and subtracting these terms, we will add this term if n is odd, and subtract if n is even, as follows:

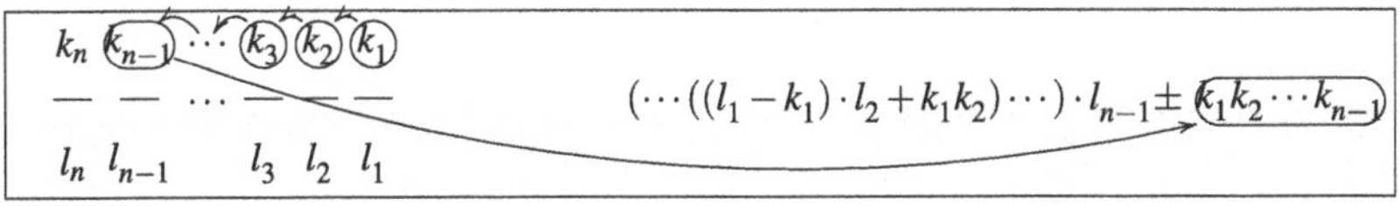

STEP ($2n-1$). The final step is as follows:

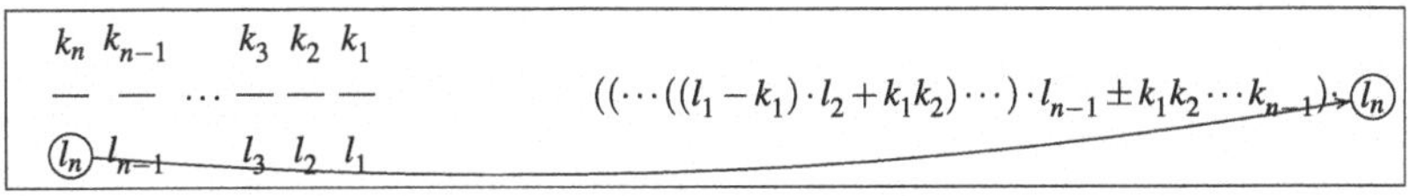

The result is

$$((\dots(((l_1-k_1)\cdot l_2+k_1k_2)\cdot l_3-k_1k_2k_3)\dots)\cdot l_{n-1}\pm k_1k_2\dots k_{n-1})\cdot l_n$$

This solution is entirely general, valid for any $n \geq 3$, and all of the n unknowns can be found by similar calculations. This is only one example of several similar problems of systems of n conditions in $n+1$ unknowns of the form of augmented matrix (8) recorded in Fibonacci's *Liber Abaci*, as shown in Table 3. A complete explanation of the solution of these problems can be found in Hart (2013, pp. 173–80).

Most telling of all, in this series of problems, the nonsensical, mathematically spurious calculation first found in Chinese texts to determine the $(n+1)^{\text{th}}$ unknown is recorded by Fibonacci to determine the $(n+1)^{\text{th}}$ unknown: we multiply together the dimensionless numbers of the upper part of the fractions, then multiply together the dimensionless numbers of the lower part of the fractions, and then add or subtract based on whether the number of men is odd or even, to give us the total currency in bezants. Changing the fractions of borrowing changes the sum total currency in bezants; similarly, whether the borrowing is between an odd or even number of men determines whether the second term is subtracted or added, changing total currency in bezants.

Circulation Throughout Eurasia

These problems, equivalent in modern terms to systems of n conditions in $n + 1$ unknowns of the form of augmented matrix (8), circulated throughout Eurasia. Later European mathematical treatises recording these problems include the following: Fibonacci's two minor works, *Flos* (1225) and *Epistola ad Theodorum* (n.d.); German mathematician Christoff Rudolff's (1499–1545) *Behend vnnd Hubsch Rechnung durch die kunstreichen regeln Algebre* (1525); German monk and mathematician Michael Stifel's (1487–1567) *Arithmetica Integra* (1544); and Portuguese scholar Pedro Nunes' (1502–1578) *Libro de algebra en arithmetica y geometria* (1567).[21]

Similar problems also circulated eastward to Japan and are recorded in some of the earliest extant Japanese mathematical treatises. For example, Muramatsu Shigekiyo's 村松茂清 (1608–1695) *Sanso* 算俎 [Calculation altar] (n.p., 1663) records several problems under the heading 'excess and deficit' (*eichū* 盈朒), with problems which, written in modern mathematical notation, are of the form of the augmented matrix (7), and are solved by determinantal methods, including the following[22]:

$$\begin{bmatrix} 4 & 6 & 0 & 0 & 0 & 90 \\ 0 & 12 & 3 & 0 & 0 & 111 \\ 0 & 0 & 7 & 15 & 0 & 138 \\ 0 & 0 & 0 & 8 & 5 & 80 \\ 6 & 0 & 0 & 0 & 14 & 184 \end{bmatrix}$$

---

[21] I am particularly indebted here to research completed by historian of medieval mathematics Albrecht Heeffer, who is assembling an online database titled 'Sources in the history of algebra before 1600', available at http://logica.ugent.be/albrecht/math.php, which 'currently contains 197 manuscripts, books, and reprints, and 1259 problems'. See the following: For *Flos*, see Heeffer, 'Conspectus of problems from: *Flos*', http://logica.ugent.be/albrecht/problems.php?code=FLF; for *Epistola ad Theodorum*, see Sesiano (1985, pp. 123–124), Rudolf (1525), and Stifel (1544), available through Linda Hall Library Digital Collections online at https://catalog.lindahall.org/permalink/01LINDAHALL_INST/19lda7s/alma99737173405961; Nunes (1567).

[22] For a translation into English, see Horiuchi (2010, pp. 48–52); for translation into modern Japanese, see Satō (1987, pp. 89–91). See also Knobloch et al. (2013) and Ogawa et al. (2018). For more information on the circulation of Chinese mathematical treatises in Japan, see Shigeru (2000).

We must thus conclude that determinants do indeed have an early history: a complete theory of determinants did not simply suddenly appear through *ex nihilio* simultaneous discovery in Japan by Seki and in Europe by Leibniz; instead, the complete theories of determinants by Seki and Leibniz were the result of centuries of experiments with increasingly complex varieties of special cases.

Earliest Extant Records of Gaussian Elimination in Europe, 1560 CE

Linear algebra problems and Gaussian elimination solutions similar to those recorded in Chinese treatises first appear *ex nihilo*—without other precedent or explanation—in France in the sixteenth century and in Switzerland in the seventeenth century. The earliest known extant record of augmented matrices, row operations, and Gaussian elimination is the French monk Johannes Buteo's (c. 1492–c. 1572?) *Logistica* (Lyon, 1560). Similar problems, with a variety of solutions, can also be found in the following works: the Italian physician, astrologer, and mathematician Girolamo Cardano's (1501–1576) *Ars magna* (1545); the French poet Jacques Peletier du Mans' (1517–1582) *L'Algebre* (Lyon, 1554), which attributes the problems to Cardano; the French algebraist Guillaume Gosselin's (c. 1536–c. 1604) *De Arte Magna* (Paris, 1577); and the Swiss mathematician Johann Rahn's (1622–1676) *Teutsche Algebra* (Zurich, 1659). There is little secondary scholarship on these figures, with the exception of Cordano.[23]

[23] The best sources are Cifoletti (1992) and Gosselin (2016). On Cardano, see Cardano (1968), Giglioni (2021), and Grafton (1999).

Buteo's *Logistica*

The earliest known extant record of Gaussian elimination in Europe is Buteo's *Logistica*, which records the following three problems[24]:

$$\begin{bmatrix} 3\ 1\ 1\ 42 \\ 1\ 4\ 1\ 32 \\ 1\ 1\ 5\ 40 \end{bmatrix}, \quad \begin{bmatrix} 2\ 1\ 1\ 34 \\ 1\ 3\ 1\ 51 \\ 1\ 1\ 4\ 68 \end{bmatrix}, \quad \begin{bmatrix} 2\ 1\ 1\ 1\ 34 \\ 1\ 3\ 1\ 1\ 36 \\ 1\ 1\ 4\ 1\ 52 \\ 1\ 1\ 1\ 6\ 78 \end{bmatrix} \tag{14}$$

These three problems, written in the form of augmented matrices, are very similar to those recorded in Chinese texts. If we take the first problem as an example, Buteo first records the following conditions:

$$1A,\ \tfrac{1}{3}B\ \tfrac{1}{3}C\ [14 \qquad 1B,\ \tfrac{1}{4}A,\ \tfrac{1}{4}C\ [8 \qquad 1C,\ \tfrac{1}{5}A,\ \tfrac{1}{5}B\ [8$$

Buteo then records the four following steps, using Gaussian elimination to find the solution, as transcribed below from left to right (superscripts in the original indicate the number by which the entire row from the above three conditions was multiplied):

$3A.\ 1B.\ 1C\ [42\|\ 1^3$	$3A.\ 12B.\ 3C\ [96$	$3A.\ 3B.\ 15C\ [120$	$22B.\ 154C\ [858$
$1A.\ 4B.\ 1C\ [32\|\ 2^4$	$3A.\ 1B.\ 1C\ [42$	$3A.\ 1B.\ 1C\ [42$	$22B.\ 4C\ [108$
$1A.\ 1B.\ 5C\ [40\|\ 3^5$	$11B.\ 2C\ [54$	$2B.\ 14C\ [78$	$150C\ [750]$

Available evidence suggests that Buteo copied these three problems and solutions from sources with (possibly distant) Chinese origins[25]: (1) The writing of these problems in the form of augmented matrices, the row operations, row reductions, Gaussian elimination, and results at each step is precisely the same as that found in Chinese treatises. (2) The only known extant records of Gaussian elimination from anywhere in the world from before the sixteenth century are Chinese mathematical treatises. (3) Conversely, all records of linear algebra problems from

[24] Buteo (1559).

[25] Buteo himself (falsely) attributes algebra to Euclid, attacking both Arab sources and abacists. See Cifoletti (1992, pp. 47, 271).

elsewhere in the world from before the sixteenth century are limited to extremely special cases, with solutions limited to extremely special calculations. (4) Buteo's work, and those of his contemporaries, evinces only the most limited acquaintance with linear algebra—more exactly, all the problems recorded in Buteo (which are also recorded Cardano and Peletier) represent a single special case, for which Buteo inexplicably employs a completely general solution.

Comparison with enormous sophistication found in the *Nine Chapters*, composed over 1500 years earlier, best demonstrates the implausibility that Gaussian elimination was somehow independently discovered Buteo or his contemporaries. Evidence preserved in the *Nine Chapters* demonstrates considerable familiarity with a wide range of problems, presumably the result of centuries of calculations through widespread use of the two-dimensional counting board. Most tellingly, Chapter 8 of the *Nine Chapters* evinces a clear understanding that Gaussian elimination (*fangcheng* method) is a general solution for systems of n conditions in n unknowns: (1) It is used to solve all 18 problems. (2) In particular, it is used to solve numerous special problems for which there are special solutions (e.g., determinantal methods). (3) It is shown to be a solution when there are fractions, whether in the given coefficients or solutions. (4) It is shown to be a solution when there are negative numbers, whether in the given coefficients or solutions, or in the calculations. (5) It is used in problem 17 to solve a completely general problem with 4 conditions in 4 unknowns. (6) It is used in problem 18 to solve a completely general problem with 5 conditions in 5 unknowns. In fact, the *Nine Chapters* records evidence of considerably more sophisticated calculations, including the following: (1) The highly-sophisticated integer-preserving back substitution in the *Nine Chapters*.[26] (2) The fact that the solution of problem 18 requires the use of at least 77 counting rods.[27] If the *Nine Chapters* shows that Gaussian elimination emerged as a general solution from likely centuries of experimentation with a considerable number of limited specific cases with specific solutions, there is nothing in sixteenth European treatises indicating anywhere near this kind of sophistication.

[26] See Hart (2011, chapters 5–8).

[27] Hart (2012).

In sum, it is simply not plausible that Gaussian elimination, the most general solution to *n* conditions in *n* unknowns, was discovered independently *ex nihilo* in Europe on the basis of one or several extremely limited specific cases. In any case, even though the earliest known extant record in Europe of Gaussian elimination is Buteo's *Logistica*, credit for the discovery of Gaussian elimination has never been attributed to Buteo, or for that matter, to any of his contemporaries, Cardano, Peletier, Gosselin, or Rahn. Traditional accounts alleging exclusive Western origins for Gaussian elimination have thus now arrived at a dead end, on the one hand unwilling to accept indisputable evidence of global circulations mathematics (arguably one of the most transmissible forms of knowledge), and on the other hand unwilling and indeed unable to credit discovery of the most important algorithm in linear algebra to Buteo or his contemporaries.

4 Conclusions

Linear algebra is one of the mathematical foundations of modern science: it has broad applications in a wide range of modern sciences; it is the mathematical framework for much of twentieth-century quantum mechanics; and it plays a fundamental role in twenty-first-century sciences, including artificial intelligence, quantum information sciences, and the Second Quantum Revolution.

Linear algebra first developed in early China and later spread throughout Eurasia, to Europe in the west and Japan in the east. Extant mathematical treatises provide a *terminus ante quem* for the following key developments:

- The earliest known extant written record of the essential elements of what we now call linear algebra—augmented matrices, row operations, row reductions, and Gaussian elimination, together with determinantal calculations and solutions—is the 18 problems and solutions recorded in '*Fangcheng*', Chapter 8 of the *Nine Chapters*, composed no later than the first-century CE.

- Evidence from the *Nine Chapters* shows that by the first-century CE, Gaussian elimination (*fangcheng* method), a general solution to n conditions in n unknowns, had been discovered: it is used to solve all 18 problems, including numerous special problems with special solutions found by determinantal methods; it is shown to be a solution for problems with fractions and with negative numbers; it is used to solve one completely general problem with 4 conditions in 4 unknowns and one completely general problem with 5 conditions in 5 unknowns; it is used together with a completely general highly-sophisticated integer-preserving algorithm for back substitution.
- Evidence from the *Nine Chapters* shows that by the first-century CE, determinantal calculations were used to solve a special class of systems of n conditions in $n + 1$ unknowns: a value for the $(n + 1)^{\text{th}}$ unknown was assigned by calculating what we would now call a determinant. The earliest written record of such a determinantal calculation is from Jia Xian's commentary on the *Nine Chapters*, c. 1050 CE, which is no longer extant; excerpts are quoted in Yang Hui's *Nine Chapters on the Mathematical Arts, with Detailed Explanations*, which is extant, dating from 1261 CE.
- Evidence from the *Nine Chapters* shows that by the first-century CE, determinantal solutions were employed for special types of *fangcheng* problems. The earliest extant written record of such a determinantal solution is preserved in Fang Zhongtong's *Numbers and Measurement, an Amplification*, dated 1661 CE.
- Arcane determinantal calculations and solutions—so distinctive that they can serve as 'fingerprints'—circulated across the Eurasian continent by the thirteenth-century CE. The earliest known extant European record is Fibonacci's *Liber Abaci*, dated 1202 CE. Most tellingly, nonsensical and mathematically spurious calculations for the $(n + 1)^{\text{th}}$ unknown first recorded in Chinese treatises are recorded in *Liber Abaci*. The earliest extant record of the transmission of similar problems to Japan is Muramatsu Shigekiyo's *Calculation Altar* (1663). These distinctive problems continued to be recorded in later European and Japanese mathematical treatises.
- Gaussian elimination appears suddenly in Europe *ex nihilio* first in the solutions to three problems recorded in the French monk Johannes

Buteo's *Logistica* (1559). Available evidence suggests that these three problems and their solutions are from sources with (possibly distant) Chinese origins: the writing of these problems in the form of an augmented matrix, the row operations, row reductions, Gaussian elimination, and results at each step are the same as those found in Chinese texts; the only known extant records of Gaussian elimination from anywhere in the world from before the sixteenth century are Chinese mathematical treatises; it is not plausible that Buteo and his contemporaries, who were acquainted with only one or several extremely special problems, somehow discovered a general solution; indeed, there is literally no argument for an independent European discovery of Gaussian elimination—credit for the discovery of Gaussian elimination has never been attributed to Buteo or any of his contemporaries, Cardano, Peletier, Gosselin, or Rahn.

The early history of linear algebra then fills in gaps in the historical record extending from the first-century CE up through the seventeenth century. During this period of widespread use of the counting board for calculations in China, anonymous and likely illiterate adepts employed simple two-dimensional visual patterns to solve what we would now call systems of n linear equations in n unknowns. These two-dimensional mathematical practices on the counting board were non-literary: these practices were not calculated, taught, or transmitted through texts; these practices are only very rarely recorded in texts, often in the pursuit of patronage; indeed, recording these two-dimensional practices in one-dimensional narrative form renders them almost unintelligible. What is most astonishing is the considerable sophistication of these problems and calculations.

In modern linear algebra, determinants provide elegant solutions to systems of n linear equations in n unknowns for very small n. The apparent *ex nihilio* discovery of determinants by Seki in Japan and by Leibniz in Europe is explained by centuries of experiments with determinantal methods (*Determinantenmethode* in Tropfke's typology) for limited special cases. The *Nine Chapters*, the earliest extant transmitted treatise in China specifically on mathematics, provides clear evidence that determinantal calculations and solutions were known by the first-century

CE. The key is the display of entries of the system in a two-dimensional form. Later extant treatises record increasingly explicit details for both determinantal calculations and solutions. These problems circulated to Europe no later than the thirteen century and are recorded in later treatises; these calculations are also recorded in Japanese mathematical treatises of the seventeenth century. It must be conceded, however, that we may never obtain more complete information on Seki's sources or Leibniz's sources. During this early period, written texts must be viewed as a recording technology—written texts are not the medium of linear algebra practices but do on exceedingly rare occasions record them. In this sense, we might consider as a homely analogy wildlife cameras: wildlife cameras certainly provide indisputable evidence, yet wildlife cameras record only a minuscule fraction of wildlife activity.

In modern linear algebra, Gaussian elimination is the most important matrix algorithm, providing a general solution to systems of n linear equations in n unknowns for any n. It was first discovered no later than two millennia ago, and perhaps even earlier. Again, the key is the display of entries of the system in the form of a two-dimensional augmented matrix; simple visual patterns are repeatedly employed to find the solution. Gaussian elimination in the *Nine Chapters* is essentially the same as taught today in undergraduate mathematics courses. In the sixteenth century, during a period of increasing global trade between Europe and China, Gaussian elimination first appears in Europe ex nililio in Buteo's *Logistica*, recorded in a form identical to that found in Chinese treatises. Again, although we may never obtain more complete information on the sources used by Buteo and his contemporaries, the only possibility is sources with (possibly distant) Chinese origins.

In sum, to return to the ideas framing this conference, the early history of linear algebra makes it impossible to deny that modern science is a 'multi-civilizational outcome'. We must conclude that modern science has always been 'dialogical'. The early history of linear algebra is only one contribution to the larger ongoing project to trace the multi-civilizational global circulations of science.

Acknowledgements I would like to express my sincere gratitude to the Joseph Needham Foundation for Science & Civilisation (JNFSC, Hong Kong), the Needham Research Institute (Cambridge), together with the advisory committee, conference organizers, and especially Dr. Peter Lee, Chairman of JNFSC, Dr. Arun Bala, Prof. Raymond W. K. Lau, and Prof. MEI Jianjun 梅建军. I would also like to thank conference participants for their outstanding contributions, interesting discussions, and helpful suggestions. Finally, I would like to thank the National Endowment for the Humanities for funding supporting this research, and Linda Hall Library, one of the world's foremost independent science, technology, and engineering research libraries, and especially Dr. Benjamin Gross, Vice President for Research and Scholarship, for an ideal environment for research and writing. Parts of this paper were presented to China Forum, Jesus College, Cambridge University, November 16, 2022; I would like to thank participants for their helpful comments. Any views, findings, conclusions, or recommendations expressed in this chapter do not necessarily represent those of the National Endowment for the Humanities or Linda Hall Library.

References

Chinese Collectanea

SKQS Yingyin Wenyuan ge Siku quanshu 影印文淵閣四庫全書 [Complete Collection of the Four Treasuries, Photolithographic Reproduction of the Edition Preserved at the Pavilion of Literary Erudition]. 1773–1782. Edited by Ji Yun 紀昀 (1724–1805) and Lu Xixiong 陸錫熊. 3461 titles in 1500 vols. Reprinted from the collection of the Guoli gugong bowuyuan 國立故宮博物院, Taipei: Taiwan shangwu yinshuguan 臺灣商務印書館. Digitized full text, searchable edition. Hong Kong: Chinese University Press, 1983–1986.

ZKJDT Zhongguo kexue jishu dianji tonghui: Shuxue juan 中國科學技術典籍通彙:數學卷 [Comprehensive Collection of the Classics of Chinese Science and Technology: Mathematics Volumes]. Edited by Guo Shuchun 郭书春. 5 vols. Zhengzhou: Henan jiaoyu chubanshe 河南教育出版社, 1993.

Primary Sources in East Asian Languages

Fang Zhongtong 方中通 (1634–1698). 1661. *Shu du yan* 數度衍 [Numbers and Measurement, an Amplification]. Reprint. In *SKQS.*

Jiuzhang suanshu 九章算術 [Nine Chapters on the Mathematical Arts]. Commentary by Liu Hui 劉徽 (fl. 263 CE), subcommentary by Li Chunfeng 李淳風 (602–670). In *SKQS.*

Muramatsu Shigekiyo 村松茂清 (1608–1695). 1663. *Sanso* 算俎 [Calculation altar]. n.p.

Primary Sources in Western Languages

Boncompagni, Baldassare, ed. 1857. *Scritti di Leonardo Pisano matematico del secolo decimoterzo. I. Il Liber abbaci di Leonardo Pisano.* Roma: Tipografia delle Scienze Matematiche e Fisiche.

Buteo, Johannes (c. 1492–c. 1572?) 1559. *Ioan. Buteonis Logistica: quae & arithmetica vulgò dicitur, in libros quinque digesta quorum index summatim habetur in tergo eiusdem, ad locum Vitruuij corruptum restitutio, qui est de proportione lapidum mittendorum ad balistae foramen, libro decimo.* Lugduni: apud Gulielmum Rouillium.

Nunes, Pedro. 1567. *Libro de algebra en arithmetica y geometria, compuesto por el Doctor Pedro Nunez, Cosmographo Mayor del Rey de Portugal, y Cathedratico Iubilado en la Cathedra de Mathematicas en la Vniuersidad ...* Antwerp: en casa de los herederos d'Arnoldo Birckman a la Gallina gorda.

Rudolf, Christoff. 1525. *Behend vnnd Hubsch Rechnung durch die kunstreichen regeln Algebre so gemeincklich die Coss genent werden. Darinnen alles so treulich an tag gegeben, das auch allein auss vleissigem lesen on allen mundtlich evnterricht mag begriffen werden, etc.* Argentorati: Vuolfius Cephaleus Joanni Jung.

Stifel, Michael. 1544. *Arithmetica integra.* Preface by Philip Melanchthon. Nürnberg: Johann Petreius.

Secondary Sources in East Asian Languages

Satō Ken'ichi 城地茂. 1987. *Sanso: gendai yaku to kaisetsu* 算俎:現代訳と解説 [*Sanso*: Modern Translation and Interpretation]. Tokyo: Kenseisha 研成社.

Secondary Sources in Western Languages

Bala, Arun. 2006. *The Dialogue of Civilizations in the Birth of Modern Science.* New York: Palgrave Macmillan.

Bala, Arun. 2009. Eurocentric Roots of the Clash of Civilizations: A Perspective from History of Science. In *The Challenge of Eurocentrism: Global Perspectives, Policy, and Prospects,* ed. Rajani Kannepalli Kanth, 9–23. New York: Palgrave Macmillan.

Bala, Arun, ed. 2012. *Asia, Europe, and the Emergence of Modern Science: Knowledge Crossing Boundaries.* New York: Palgrave Macmillan.

Bala, Arun, and Prasenjit Duara. 2016. *The Bright Dark Ages: Comparative and Connective Perspectives.* History of Science and Medicine Library 53. Boston: Brill.

Bareiss, Erwin H. 1966. *Multistep Integer-Preserving Gaussian Elimination.* Technical report ANL-7213. Argonne National Laboratory.

Bareiss, Erwin H. 1968. Sylvester's Identity and Multistep Integer-Preserving Gaussian Elimination. *Mathematics of Computation* 22: 565–578.

Brentjes, Sonja. 2010. *Travellers from Europe in the Ottoman and Safavid Empires, 16th–17th Centuries: Seeking, Transforming, Discarding Knowledge.* Variorum Collected Studies Series, CS961. Burlington, VT: Ashgate/Variorum.

Buteo, Johannes. 2008. *Johannes Buteo's* The Shape and Capacity of Noah's Ark: *A Translation of Johannes Bureo's 1554 Edition,* trans. Timothy Griffith and Natali H. Monnette. Vol. 2. Issues in Creation (Center for Origins Research). Eugene, OR: Wipf/Stock Publishers.

Cardano, Girolamo. 1968. *The Great Art: Or, the Rules of Algebra,* trans. T. Richard Witmer. Cambridge, MA: MIT Press.

Chemla, Karine. 1994. Similarities Between Chinese and Arabic Mathematical Writings: (1) Root Extraction. *Arabic Sciences and Philosophy: A Historical Journal* 4: 207–266.

Chemla, Karine. 1997. Reflections on the World-Wide History of the Rule of False Double Position, or: How a Loop Was Closed. *Centaurus: International Magazine of the History of Mathematics, Science, and Technology* 39 (2): 97–120.

Chemla, Karine, and Guo Shuchun. 2004. *Les neuf chapitres: Le classique mathématique de la Chine ancienne et ses commentaires.* Paris: Dunod.

Cifoletti, Giovanna C. 1992. Mathematics and Rhetoric: Jacques Peletier, Guillaume Gosselin and the Making of the French Algebraic Tradition. PhD diss.

Cullen, Charles G. 1990. *Matrices and Linear Transformations*, 2nd ed. Dover Books on Advanced Mathematics. New York: Dover.

Dold-Samplonius, Yvonne, Joseph W. Dauben et al., eds. 2002. *From China to Paris: 2000 Years Transmission of Mathematical Ideas.* Boethius 46. Stuttgart: Franz Steiner Verlag.

Giglioni, Guido. 2021. Girolamo [Geronimo] Cardano. In *The Stanford Encyclopedia of Philosophy*, Winter 2021, ed. Edward N. Zalta. Metaphysics Research Lab, Stanford University.

Gosselin, Guillaume. 2016. *De arte magna libri quatuor =: Traite d'algèbre; suivi de, De ratione discendae docendaeque mathematices repetita praelectio = Leçon sur la mathématique*, trans. Odile Le Guillou-Kouteynikoff. Sciences et savoirs. Paris: Les Belles Lettres.

Grafton, Anthony. 1999. *Cardano's Cosmos: The Worlds and Works of a Renaissance Astrologer*. Cambridge, MA: Harvard University Press.

Hannah, John. 2007. False Position in Leonardo of Pisa's *Liber Abbaci. Historia Mathematica* 34 (3): 306–332.

Hannah, John. 2010. Conventions for Recreational Problems in Fibonacci's *Liber Abbaci. Archive for History of Exact Sciences* 65 (2): 155–180.

Hart, Roger. 2011. *The Chinese Roots of Linear Algebra.* Baltimore: Johns Hopkins University Press.

Hart, Roger. 2012. Tracing Practices Purloined by the 'Three Pillars.' *The Korean Journal for the History of Science* 34 (2): 287–358.

Hart, Roger. 2013. *Imagined Civilizations: China, the West, and Their First Encounter*. Baltimore: Johns Hopkins University Press.

Heng, Geraldine. 2018. *The Invention of Race in the European Middle Ages.* New York: Cambridge University Press.

Heng, Geraldine. 2022. *Teaching the Global Middle Ages.* Options for Teaching 54. New York: Modern Language Association of America.

Hoffman, Kenneth, and Ray Alden Kunze. 1971. *Linear Algebra*, 2nd ed. Englewood Cliffs, NJ: Prentice-Hall.

Horiuchi, Annick. 2010. *Japanese Mathematics in the Edo Period (1600–1868): A Study of the Works of Seki Takakazu (?–1708) and Takebe Katahiro (1664–1739)*, trans. Silke Wimmer-Zagier. Science Networks Historical Studies 40. Basel: Birkhauser.

Horn, Roger A., and Charles R. Johnson. 1985. *Matrix Analysis*. New York: Cambridge University Press.

Horn, Roger A., and Charles R. Johnson. 1991. *Topics in Matrix Analysis*. New York: Cambridge University Press.

Ji, Zhigang. 2015. Needham's 19 (j) and Fibonacci's *Liber Abaci*: A New Approach to the Communication and Transmission of Mathematical Knowledge from China to the West. In Rowe and Horng 2015.

Katz, Victor J., and Annette Imhausen, eds. 2007. *The Mathematics of Egypt, Mesopotamia, China, India, and Islam: A Sourcebook*. Princeton: Princeton University Press.

Knobloch, Eberhard, Hikosaburo Komatsu et al., eds. 2013. *Seki, Founder of Modern Mathematics in Japan: A Commemoration on His Tercentenary*. Springer Proceedings in Mathematics & Statistics 39. New York: Springer.

Lau, Raymond W. K. Forthcoming. *The Needham Question: A Non-Eurocentric Approach Transcending Dialogism.*

Muir, Thomas. 1906–1923. *The Theory of Determinants in the Historical Order of Development.* 4 vols. London: Macmillan. Reprint, 4 vols. bound as 2, New York: Dover, 1960.

Ogawa, Tsukane, Mitsuo Morimoto et al., eds. 2018. *Mathematics of Takebe Katahiro and History of Mathematics in East Asia*. Advanced Studies in Pure Mathematics 79. Tokyo: Mathematical Society of Japan.

Rowe, David E., and Wann-Sheng Horng, eds. 2015. *A Delicate Balance: Global Perspectives on Innovation and Tradition in the History of Mathematics: A Festschrift in Honor of Joseph W. Dauben*. Trends in the History of Science. Cham: Springer.

Serre, Denis. 2002. *Matrices: Theory and Applications*. Graduate Texts in Mathematics 216. New York: Springer.

Sesiano, Jacques. 1985. The Appearance of Negative Solutions in Mediaeval Mathematics. *Archive for History of Exact Sciences* 32 (2): 105–150.

Shen Kangshen, Anthony W.-C. Lun et al. 1999. *The Nine Chapters on the Mathematical Art: Companion and Commentary.* New York: Oxford University Press.

Shigeru, Jochi. 2000. The Dawn of Wasan (Japanese Mathematics). In *Mathematics Across Cultures: The History of Non-Western Mathematics*, 423–454. Science Across Cultures. Dordrecht: Kluwer Academic.

Sigler, Laurence E. 2002. *Fibonacci's Liber Abaci: A Translation into Modern English of Leonardo Pisano's Book of Calculation*. Sources and Studies in the History of Mathematics and Physical Sciences. New York: Springer.

Spiesser, Maryvonne. 2000. Problèmes linéaires dans le Compendy de la praticque des nombres de Barthélemy de Romans et Mathieu Préhoude (1471):

Une approche nouvelle basée sur des sources proches du Liber abbaci de Léonard de Pise. *Historia Mathematica* 27 (4): 362–83.
Stewart, Gilbert W. 1973. *Introduction to Matrix Computations*. New York: Academic Press.
Stewart, Gilbert W. 1998. *Matrix Algorithms: Volume I: Basic Decompositions*. Philadelphia: Society for Industrial/Applied Mathematics.
Strang, Gilbert. 1993. *Introduction to Linear Algebra*. Wellesley, MA: Wellesley-Cambridge Press.
Strang, Gilbert. 2006. *Linear Algebra and Its Applications*, 4th ed. Belmont, CA: Thomson, Brooks/Cole.
Tropfke, Johannes, Kurt Vogel, et al. 1980. *Geschichte der Elementarmathematik, Band 1, Arithmetik und Algebra*, 4th ed. Berlin: Walter de Gruyter.
Vogel, Kurt. 1983. Ein Vermessungsproblem reist von China nach Paris. *Historia Mathematica* 10 (3): 360–367.
Zhang, Fuzhen. 1999. *Matrix Theory: Basic Results and Techniques*. Universitext. New York: Springer.

10

Ibn al-Haytham's *Optics* and the European *Perspectiva* Legacies in Science and Art

Nader El-Bizri

1 Prologue

The European contributions to the unfolding of science, philosophy, and the arts benefited from the transmitted intellectual legacies of Andalusia, Egypt, the Levant, Mesopotamia, Persia, India, and China. These historical phenomena are relatively well documented in various forms of scholarship, and it encompasses the original insights of Joseph Needham in demonstrating the dialogical exchanges, multicivilizational confluences, intercultural synergies, and interdisciplinary trajectories in scholarly research, which characterized the history of science from its ancient classical origins to its early-modern manifestations. Such a line of inquiry finds one of its evidenced expressions in the impact of the Arabic sciences and philosophies on the European intellectual history, and in how this rested mainly on the adaptive expansion of the classical

N. El-Bizri (✉)
College of Arts, Humanities, and Social Sciences, University of Sharjah, Sharjah, UAE
e-mail: nel-bizri@sharjah.ac.ae

A. Bala et al. (eds.), *Multicivilizational Exchanges in the Making of Modern Science*, https://doi.org/10.1007/978-981-97-3541-9_10

Greek scientific and philosophical corpus within the Islamicate civilization, while being influenced as well by diverse traditions in knowledge from the ancient world with provenances from Egypt, the Near East, Persia, India, and China.

To give a specific example of one legacy from the Arabic sciences that had a direct longstanding impact on the European intellectual history, this chapter focuses on the science of optics of the eleventh-century Arab polymath al-Ḥasan Ibn al-Haytham (known in Latin as 'Alhazen', or 'Alhasen'; born in Abbasid/Buyid Basra ca. 965 CE and died in Fatimid Cairo ca. 1041 CE). The consideration of his legacy will be contextualized in this regard by way of addressing the adaptive assimilation of his theories of vision and light within the European *perspectiva* traditions in science, and subsequently in the architectural visual arts, as stretching from the High Middle Ages to the early-modern period. Ibn al-Haytham's monumental *Optics* (*Kitāb al-Manāẓir*; *Book of Optics*)[1] stands in this context as an evidenced locus of dialogical multicivilizational confluences in science. This is the case given how he was critically engaged with the ancient Greco-Roman optical oeuvres, by revaluating and transcending the explication of the principles of visual perception of figures such as Plato, Aristotle, Ptolemy, Euclid, Galen, and in how his legacy influenced eventually in its own right the European *perspectiva* theories in science and art, from the mediaeval period to the seventeenth century, and passing by the Italian Renaissance.

The reception of Ibn al-Haytham's *Optics* in the European milieu passes in the High Middle Ages via Gerard of Cremona's Toledo circle in

---

[1] Ibn al-Haytham, *Kitāb al-manāẓir*, ed. A. I. Sabra, Books I–III (Kuwait: National Council for Culture, Arts, and Letters, 1983) and Books IV–V (Kuwait: National Council for Culture, Arts, and Letters, 2002); Ibn al-Haytham, *The Optics*, trans. A. I. Sabra, Books I–III (London: Warburg Institute, 1989) and Books IV–V (London: Warburg Institute, 2023); Alhazen, *Opticae Thesaurus*, ed. Friedrich Risner (Basel: Episcopius, 1572), with the Latin critical edition and annotated English translation by A. M. Smith of Ibn al-Haytham's (Alhacen) *De aspectibus* (Philadelphia: American Philosophical Society, 2001–2010). The Arabic title, *Kitāb al-manāẓir*, is usually translated in English as the 'Book of Optics', however, it can also literally mean the 'Book of Perspectives', the 'Book of Visible Aspects', or the 'Book of Sceneries'. This monumental treatise is divided into seven books *qua* volumes as follows: Books I–II dealt with direct vision, light, and colour; Book III examined the errors in direct vision; Books IV–VI focused on catoptrics by studying the reflection of light and its instruments as mirrors; Book VII treated dioptrics by investigating the refraction of light and its instruments as lenses.

terms of its Latinate translation, and then through the Franciscan lineage of Roger Bacon's *Opus maius*, John Pecham's *Perspectiva communis*, Erazmus Ciolek Witelo's *Opticae*, and the Dominican Theodoric of Freiberg's *De iride*. In the Renaissance, the *Optics* of Ibn al-Haytham impacts Biagio Pelacani da Parma's *Questiones super perspectiva communi*, Leon Battista Alberti's *De pictura*, Lorenzo Ghiberti's *Commentarii*, and it culminates in its first printed Latin version in the publication in 1572 CE of Friedrich Risner's *Opticae Thesaurus Alhazeni Arabis*. Then, in the seventeenth century, Ibn al-Haytham's *Optics* influences *François d'Aguilon*'s *Opticorum* within the Antwerp Jesuit mathematical school, and it has a direct impact on Johannes Hevelius' *Selenographia*, and it was also consulted by Girard Desargues in his *Manière universelle pour pratiquer la perspective*, and by René Descartes in his *Dioptrique*, in addition to Johannes Kepler in his *Astronomiae Pars Optica*. This broad survey of the adaptive assimilation of Ibn al-Haytham's *Optics* within the European history of science and the architectural arts will be furthermore supplemented in this chapter with epistemic reflections on the merits of Ibn al-Haytham's innovative geometrizing of physics, along with his development of the rudiments of the scientific experimental procedures of controlled testing, and the articulation of the principal elements of his philosophy of mind in analysing the physiological, neurological, and psychological underpinnings of visual perception; all as supported as well by a phenomenology of embodied ocular-kinaesthetic corporeal experiences. To account for these aspects, the present chapter is divided accordingly into two main sections: the first gives an overview of Ibn al-Haytham's theory of direct vision, in terms of contextualizing the principal *perspectiva* elements in his optical inquiries, as segregated from the technical focus on catoptrics and dioptrics, while the second section of the present chapter traces the lines of influence of his *Optics* in science and art in Europe.

2 Ibn al-Haytham's Theory of Direct Vision

To better appreciate and contextualize the complex nature of the unfolding of Ibn al-Haytham's legacy in optics, and of its subsequent impact on the European *perspectiva* legacies in science and art, this section offers some highlights concerning his theory of direct vision as set in Books-I-III of his *Optics* (*Kitāb al-Manāẓir*).

Ibn al-Haytham's theory of visual perception rested on his investigation of the nature and propagation of physical light and the conditions of the intromission of its rays within the eyes. He explicated this via a series of geometric models that guided his experimental procedures and installations. His theory of direct vision passed as well via an examination of the anatomy of the eyes, the physiology of the optic nerves, and their connections to the anterior part of the brain, while also giving a psychological analysis of the cognitive processes that underpinned vision. He also placed an emphasis on the situated psychosomatic experiences of the observer as an embodied cognitive agent who is always immersed within the ambient scene by way of ocular-motor kinaesthetic bodily acts.[2]

---

[2] My present chapter is partly grounded on my earlier publications on Ibn al-Haytham; refer for example to the following sources: Nader El-Bizri, 'The Confluences of Science, Philosophy, and Art: Ibn al-Haytham's *Optics* and the European *Perspectiva* Traditions', in *From Wisdom to Data: Philosophical Atlas on Visual Representations of Knowledge*, ed. José Higuera Rubio et al. (Porto: University of Porto Press, 2022), pp. 65–68, 112–117; Nader El-Bizri, 'Alhazen's Phenomenology of Visual Perception', in *The Book of Sceneries*, eds. Ala Younis and Tarek ElFetouh (Berlin: Hatje Cantz Verlag, 2022), pp. 202–218; Nader El-Bizri, 'Ibn al-Haytham and Hevelius' *Selenographia*', in *The Mantis Shrimp: A Simon Schaffer Festschrift*, eds. Dániel Margócsy and Richard Staley (Cambridge: The Cambridge History and Philosophy of Science Collective, 2022), pp. 55–59; Nader El-Bizri, 'Alhazen's Theory of Vision', *Micrologus* 29 (2021), 21–33; Nader El-Bizri, 'Arabic Classical Traditions in the History of the Exact Sciences: The Case of Ibn al-Haytham', *The European Physical Journal Plus*, 133 (2018), 271–277; Nader El-Bizri, 'Desargues' oeuvres: On Perspective, Optics and Conics', in *Visioning Technologies in Architecture*, ed. Graham Cairns (Aldershot: Ashgate, 2016), 36–51; Nader El-Bizri, 'Seeing Reality in Perspective', in *the Art of Science: From Perspective Drawing to Quantum Randomness*, eds. Rossella Lupacchini and Annarita Angelini (Dordrecht: Springer, 2014), 25–47; Nader El-Bizri, 'By Way of an Overture: Classical Optics and Renaissance Pictorial Arts', in *Recto-Verso: Redefining the Sketchbook*, eds. Nader El-Bizri with Angela Bartram and Douglas Gittens (Aldershot: Ashgate, 2014), 11–26; Nader El-Bizri, 'Classical Optics and the *Perspectiva* Traditions Leading to the Renaissance', in *Renaissance Theories of Vision*, eds. Charles Carman and John Hendrix (Aldershot: Ashgate, 2010), 11–30; Nader El-Bizri, 'Ibn al-Haytham et le problème de la couleur', *Oriens-Occidens*, 7 (2009), 201–226; Nader El-Bizri, 'In Defence of the Sovereignty of Philosophy: al-Baghdādī's Critique of Ibn al-Haytham's Geometrisation of Place', *Arabic Sciences and Philosophy* 17 (2007), 57–80; Nader El-Bizri, 'A Philosophical Perspective

Ibn al-Haytham did not follow the extramission (emission) thesis in vision of the ancient Greek mathematicians (*aṣḥāb al-taᶜālīm*) within the Euclidean and Ptolemaic traditions, nor did he uphold the antithesis that confronted them as set via the Aristotelian intromission doctrine of vision of the ancient Greek natural philosophers *qua* physicists (*al-ṭabīᶜ iyyūn*). The ancient mathematicians in the Alexandrian Hellenistic Euclidean and Ptolemaic legacies held that vision occurs by way of the emission of subtle un-consuming rays of light from the eyes, and that they issue forth outwardly in the shape of a cone that has its summit in the centre of each of the eyes and its base intersecting with the lit visible surfaces of the objects of vision within the outer ambient environment. As for the ancient natural philosophers *qua* physicists of the Aristotelian tradition, they held that vision occurs by way of the intromission of the forms of the visible objects into the eyes of the observer as abstracted from matter when the ambient transparent external medium, such as air, which is between the eyes of the observer and the objects of vision, is itself actualized by illumination into being a transmitter of these forms as abstracted from their matter. Unlike these classical legacies, Ibn al-Haytham focused on the geometric modelling of vision as supplied by the ancient Greek mathematicians, without advocating their emission thesis or its physical entailments. He thusly argued that a visual cone exists in the shape of how selected rays of light, which are emitted from the lit visible surfaces of the objects of vision, converge when entering the apertures of each of the eyes of the observer. He therefore posits his own version of an intromission theory of vision, while using the geometric modelling of the mathematicians, and at the same time removing the ambiguities he detected in the doctrine of the ancient Greek natural philosophers *qua* physicists, whereby they did not demonstrate how the large visible forms of the objects of vision enter the small apertures in the eyes of the observer. Ibn al-Haytham's accounts constituted a novel theory of vision that transcended the shortcomings of the classical theories of his predecessors, whether in the Euclidean and Ptolemaic emission thesis of the mathematicians, or the antithesis that faced them through

on Alhazen's *Optics*', *Arabic Sciences and Philosophy*, 15 (2005), 189–218; Nader El-Bizri, 'La perception de la profondeur: Alhazen, Berkeley et Merleau-Ponty', *Oriens-Occidens*, 5 (2004), 171–184.

the Aristotelian intromission doctrine of the natural philosophers *qua* physicists.

Ibn al-Haytham studied the comportment of the physical rays of light (*ḍaw'*) in the manner they irradiate spherically from every point on the lit surfaces of the objects of vision, and how they propagate rectilinearly within the intermediate isotropic, continuous, and homogeneous transparent medium (*shafīf*), such as air, which exists between the eyes of the observer and the lit visible things that face them. These physical phenomena were studied by him through an isomorphic mathematizing of natural philosophy *qua* physics, as mediated via hypothetical-deductive observational experimentations (*iᶜtibār*) that were undertaken by him via the procedures of controlled testing and repeated verifications. The geometrizing of physics and the testing of its principles experimentally allowed him to posit his novel theory of vision.

Ibn al-Haytham analysed how the physical light-rays that fell orthogonally on the eye would pass rectilinearly across the pupil, cornea, and uvea transparent membranes until they refract at the outer surface of the crystalline-lens (*jalīdiyya*) and then bend again as they pass into the vitreous humour (*zujājiyya*), which itself touches the filaments of the optic nerve (*'asb al- 'ayn*). The crystalline-lens and the vitreous humour are not simply refractive components of the eye, but they are rather sensing organs as well. The light-rays cause physical sensations on the crystalline in an ordered configuration that corresponds point-by-point with the contours and details of the visible lit surfaces from which these rays irradiate rectilinearly and enter the eye in the shape of a cone of light. These ocular sensoria differ in their sensed intensities in proportion to the degrees of illumination from each emission point on the visible lit surface of the object of vision. The seen image has as such a tactile aspect to it in the sense that it is a sensory arrangement of how the physical light-rays touch the crystalline with varying subtle fiery intensities. The array of sensations in the crystalline is then transmitted via the vitreous humour to the filaments of the optic nerves of the left and right eyes, and these neurological sensations get united in the common optical nerve (*ᶜaṣaba mushtaraka*), which itself splits again into two nerve-conduits that reunite in a single sensing organ in the frontal cerebral cortex (*muqaddam al-dimāgh*) that is the last sentient organ (*al-ḥāss al-akhīr*). Such detailed analysis was itself supported by the diagrammatic descriptions of the anatomy of the eyes and the optical nervous system that connects to the front of the brain, as shown in Fig. 1.

Having considered the geometrized physics of light-rays, and the conditions of their intromission into the eyes, and how this is explicated via anatomical studies of the physiology of the eyes and the optical

Fig. 1 Ibn al-Haytham, *Kitāb al-Manāẓir*, 11th century, 'A diagram of the eye' (MS Fatih 3212, vol. 1, fol. 81b, Süleymaniye Mosque Library, Istanbul)

nervous system that connects to the last sensing organ in the front of the brain, Ibn al-Haytham passes to a psychological analysis of the cognitive processes in visual perception and their situated embodied experiences. He offered a *Gestalt* theory of vision in his analysis of the workings of the mind in discernment (*tamyīz*), syllogistic comparative measure (*qiyās*), and inferential judgement (*ḥukm*), which amount to the recognition (*maʿrifa*) of what is seen, while also evoking in this context the interplay between memory (*dhikr*; *ḥāfiẓa*) and imagination (*takhayyul*; *khayāl*). This relates to the visual perception, via inferential judgement and recognition, of twenty-two intentional visible properties (*maʿānī mubṣara*) that cover the cognizance of the light and colours of the objects of vision, their distance, position, solidity, shape, size, separation from other things or continuity with them, the numbering of what is seen, its motion or rest, roughness or smoothness, transparency or opacity, shadows and darkness, beauty or ugliness, similarities or dissimilarities. According to him, the errors in visual perception occur due to the cognitive erring in inferences that are made by the observer when the principal visible properties of distance, position, illumination, size, opacity, transparency, duration of perception, and the natural conditions of the eyes fall outside their respective moderate range and cause as such a specific error in sight. These visible intentional properties show that direct vision or its errors occur due to cognition, and not simply in terms of ocular sensations in physiological and neurological terms. Ibn al-Haytham's theory of vision is therefore connected to a philosophy of mind and cognitive psychology that are grounded on a phenomenology of situated lived ocular-kinaesthetic experiences in embodiment. This was itself demonstrated by him in terms of an experimental installation that he designed and tested, and that evokes *trompe l'oeil* phenomena that are akin to the twentieth-century 'Ames Room' optical illusions.[3]

[3] The 'Ames Room' refers to an experimental peephole space that is architecturally manipulated to give an optical illusion in terms of distance, size, and location, while supporting the investigation of the multiple situational clues that are needed in the visual perception of spatial depth and its estimation. The optical illusion arises when most of the details of the room are not visible to the observer who looks at it through a controlled peephole aperture. The room appears as ordinary but in its actual form it is distorted to give *trompe l'oeil* visual illusions concerning the distance, location, and size of the visible entities inside it. This experimental installation was used in 1946 by the specialist in physiological optics, Adelbert Ames Jr., as

3 Ibn al-Haytham's *Optics* and the European *Perspectiva*

Based on what has been evoked in the earlier section regarding Ibn al-Haytham's theory of direct vision, this part of the chapter focuses on the transmission and adaptive assimilation of his *Optics* (*Kitāb al-manāẓir*) within the European milieu, in view of highlighting how his legacy grants us a concrete example in the history of ideas of fruitful transferences across geographical, cultural, ethnic, religious, disciplinary, civilizational, and historical boundaries. Transcending the legacies in optics of the ancient Greek mathematicians and physicists alike, while reshaping their premises and setting new conclusions, experimentally and by way of mathematizing physics, Ibn al-Haytham's own legacy in optics, as enshrined in his *Kitāb al-manāẓir*, was itself destined to underpin subsequent *perspectiva* traditions in science and art in Europe up till the early-modern era.

The Arabic text of Ibn al-Haytham's *Optics* (his *Kitāb al-manāẓir*) was translated into Latin in the twelfth century in Toledo, possibly within the circle of the Italian scholar Gerard of Cremona (Gerardus Cremonensis; 1114–1187 CE). The manuscripts of the Latinate renditions of this opus were disseminated widely from Andalusia and Sicily across central and western Europe under the titles: '*Perspectiva*', '*De aspectibus*', and '*Optica*'. This adaptive transmission had an impact on the opticians at the Franciscan college in Oxford in the thirteenth century, as exemplified in the *Opus maius* (*Major Opus*) of the Franciscan natural philosopher Roger Bacon (1219–1292 CE), especially in its 'Part V' that contained the *perspectiva* discussion of the physiology of sight, the anatomy of the eyes, the optic nerves, and the brain, along with the physics of light-rays, and their reflection and refraction. Ibn al-Haytham also impacted John Pecham's *Perspectiva communis*. A detailed attentive commentary on Ibn al-Haytham's treatise was furthermore composed by the Silesian Franciscan natural philosopher Vitello (Erazmus Ciolek Witelo; 1230–1280

partly impacted in his inquiries by the optical investigations of Hermann von Helmholtz on the visual perception of space. I investigated this phenomenon of the visibility of space in various published studies, and in particular: El-Bizri, 'La perception de la profondeur: Alhazen, Berkeley et Merleau-Ponty', *art.cit.*

CE). The Latin version of the *Optics* had moreover a direct impact in the fourteenth century on the German Dominican natural philosopher Theodoric of Freiberg (1250–1311 CE), particularly in his study on the rainbow. It is rather worthy noting in this context that the Arabic *Kitāb al-manāẓir* received around that same period in the early fourteenth century an influential recension by Kamāl al-Dīn al-Fārisī (1267–1320 CE) in Iran, which carried the title: *Tanqīḥ al-manāẓir* ('Recension of the Optics').[4] Al-Fārisī's analytics were like those of Theodoric, without any evidence of an existing line of transmission between these two opticians other than their keen commentaries in parallel on Ibn al-Haytham's *Optics* (al-Fārisī on an Arabic copy of this opus in Iran; Theodoric on a Latin translation of it in Germany), along with their critical revisions of the theory of the rainbow.[5]

It must be noted that Ibn al-Haytham's sensibilities as an architect (*muhanddis*) were also noticeable in his oeuvre, besides him being a polymathic optician, astronomer, and geometer. The architectural dimensions in his *Optics* as they related to spatiality, geometry, and visual perception, all resonated as such with the *perspectiva artificialis* of the Florentine Renaissance, and how this was mediated via the reception of Ibn al-Haytham's legacy in Biagio Pelacani da Parma's *Questiones super perspectiva communi*. This channel of transmission is mainly embodied in the impact of Ibn al-Haytham's *Optics* on the *Della Pittura* (*De Pictura*; 'On Painting') of Leon Battista Alberti (1404–1472 CE), and *a fortiori* in the direct influence of Ibn al-Haytham on the '*Commentario terzo*' ('Third Commentary') of Lorenzo Ghiberti (1378–1455 CE). The geometry of the cone of vision in the *Optics* of Ibn al-Haytham was turned into a method of perspectival projection in the architectural representation and design of Alberti. As for the cognitive determinants of Ibn al-Haytham's theory of direct vision, they underpinned Ghiberti's

[4] Kamāl al-Dīn al-Fārisī, *Tanqīḥ al-manāẓir*, MS Istanbul: Ahmet III 3340 (copied at Nīšāpūr, 15 Šaʿbān 716/1316), Istanbul: Topkapı Kütüphanesi; printed in 2 volumes in Hyderabad Deccan, 1928–1930. Refer also to Roshdi Rashed (ed. trans.), *Geometry and Dioptrics in Classical Islam* (London: Al-Furqān Islamic Heritage Foundation, 2005).

[5] I investigated this matter in various published studies, and in particular: El-Bizri, 'Ibn al-Haytham et le problème de la couleur', *art. cit.*

analysis of the functions of sight in shaping the architectural visual and plastic arts.

The Latin translation of Ibn al-Haytham's *Optics* was eventually printed in 1572 CE in Basel under the editorship of the German mathematician at the Collège Royale de France, Friedrich Risner (Fridericus Risnerus; 1533–1580 CE). This edition carried the title: *Opticae thesaurus. Alhazeni Arabis libri septem, nunc primùm editi, Eiusdem liber De Crepusculis et nubium ascensionibus, Item Vitellonis Thuringopoloni libri X*.[6] This printed volume was widely circulated and consulted by René Descartes, Christiaan Huygens, Johannes Kepler, and Girard Desargues. For example, the illustration in Fig. 2 refers to the 'Frontispiece' from the Risner edition, which illustrates the optical phenomena of perspective in direct vision, the rainbow, reflections in mirrors, and refractions in a pool of water (*Triplicis visus, directi, reflexi & refracti, de quo optica disputant, argumenta*). Moreover, this image conveys a visual historicized cross-civilizational classical narrative around the Roman siege of the Hellenistic Syracuse (213–212 BCE), and the use of Archimedean burning conical mirrors to incinerate the ships of the invaders.

Furthermore, Ibn al-Haytham's legacy in the science of optics, with its geometrized physics and experimentations, inspired the seventeenth-century Belgian architect, optician, and geometer François d'Aguilón (1567–1617 CE) at the Jesuit mathematical school in Antwerp, who in his turn influenced the projective geometry of the French architect Girard Desargues (1591–1661 CE). The impress of the Latin translation of Ibn al-Haytham's *Optics* on d'Aguilón was set in the latter's treatise that was published in print in Antwerp in 1613 CE under the title: *Opticorum libri sex: philosophiis juxtà ac mathematicis utiles* ('Six Books of Optics, useful for philosophers and mathematicians alike').[7] This text was illustrated by the Baroque Flemish painter Sir Peter Paul Rubens in a series of vignettes that were engraved by Théodore Galle.

[6] Friedrich Risner (ed.), *Opticae thesaurus*, Basel, 1572; reprinted with an introduction by David Lindberg, New York: Johnson Reprint, 1972. Basileae: Episcopios, 1572; Federici Risneri in Alhazeni Arabis *Opticae thesaurus. Alhazeni Arabis libri septem, nunc primùm editi, Eiusdem liber De Crepusculis et nubium ascensionibus, Item Vitellonis Thuringopoloni libri X*.

[7] François d'Aguilón (Francisci Aguilonii e societate Iesu), *Opticorum libri sex: philosophiis juxtà ac mathematicis utiles* (Antverpiæ: Ex officina Plantiniana, apud Viduam et filios J. Moreti, 1613 / Antwerp: Jan Moretus Widow & Sons, 1613).

Fig. 2 Frontispiece of the *Opticae Thesaurus* edition of Friedrich Risner of the first printed Latin translation of Ibn al-Haytham's *Kitāb al-Manāẓir* (Basel, 1572): https://en.wikipedia.org/wiki/Ibn_al-Haytham#/media/File:Thesaurus_opticus_Titelblatt.jpg

Ruben's vignette, which is shown in Fig. 3, appears on folio 356 of the *Opticorum* (*Liber quintus*: *De luminoso et opaco*). The image reconstructs one of the many experiments that were described in detail by Ibn al-Haytham. In this case, it shows how the rays of light propagate rectilinearly across a homogeneous isotropic transparent medium, such as air, and that they do not change the directional straight trajectory of their irradiation, nor intermix when crossing each other. The fires in the two lamps irradiate spherically in every direction, and the light-rays that are emitted from them pass as such via the circular hole in the surface facing them, and they leave a circular corresponding illuminated area within the obscured small room behind the hole. Each of the lit circles corresponds to straight-line isomorphic mappings of the connections to its correlative lamp via the contours of a single circular hole. The illustration displays a bearded man who represents the experimenter as he is inspired by angelic muses. Similar depictions of the experimenter appear in all the vignettes of Rubens in d'Aguilón's *Opticorum* by way of illustrating the scientific experimental method through artistic visual renderings, including anatomical illustrations.

The *Opticorum* itself belonged to a longstanding tradition in architectural thinking and praxis, whereby notable architects composed treatises that related to the liberal arts, and that made contributions as well to optics and geometry. This goes back, for example, in antiquity to the *De architectura* of Marcus Vitrivius Polio (80–15 BCE), and it is evident in the oeuvres of Ibn al-Haytham who started his career in Abbasid/ Buyid Basra as an architect (*muhanddis*), and then travelled to Fatimid Egypt on the account that he will design and construct a water-barrier

Fig. 3 Peter Paul Rubens, Vignette for the *Opticorum libri sex* of François d'Aguilon (Antwerp, 1613): https://rkd.nl/nl/explore/images/262882

through architectural engineering techniques in order to control the ebb and flow of the Nile. Realising after his field surveys that such project was unfeasible in his time, and due to the lack of building technologies to realize it, Ibn al-Haytham became concerned over his own safety from the punishment that can fall upon him from the dissatisfied Fatimid Caliph, al-Ḥākim bi-Amr Allāh (985–1021 CE). So, the vexed architect Ibn al-Haytham feigned madness and stayed confined to his apartment in the vicinity of the Azhar Mosque until the death of the Caliph in 1021 CE. It is perhaps in this period that the experimentation and composition of his *Optics* was underway and that his genius as an optician and a mathematizing physicist shined through. We witness again in this case an architect who is polymathic as a geometer, optician, and astronomer, with an impact of his legacy on like-minded architects such as Alberti, Ghiberti, d'Aguilón, and Desargues. This also influences the Roman Baroque architecture, especially in terms of the optical illusion of the *quadratura* (quadrature) technique in the visual representation of the semblance of spatial depth in the painting of the frescoed ceilings, copulas, and barrel-vaults of Catholic churches.

The influence of the *Optics* of Ibn al-Haytham and of his studies in geometry and astronomy is also visible in the oeuvre of the Polish-German astronomer Johannes Hevelius (Johannis Hevelii; 1611–1687 CE). This is particularly the case with the latter's treatise: *Selenographia, sive Lunae descriptio* ('Selenography, or a Description of The Moon'), which was published in 1647 CE in Gdańsk in the Kingdom of Poland.[8] This text constituted a milestone on the study of the moon and its topographic mapping. Its 'Frontispiece', as shown in Fig. 4, depicts two foundational figures of this discipline, who are standing on pedestals. To the right-side, we see the astronomer and physicist Galileo di Vincenzo Bonaiuti de' Galilei (1564–1642 CE) standing on the pedestal of 'sensu' ('sensing') and holding a telescope in highlighting his contributions to selenography by way of visual observation as aided by an optical instrument. As for the left side we see an artist's depiction of Ibn al-Haytham (named in a variation of the Latinate rendering of his name as 'Alhasen'

[8] Johannes Hevelius, *Selenographia, sive, Lunae descriptio* (Gedani: Autoris sumtibus, typis Hünefeldianis, 1647).

['Alhazen']) and standing on the pedestal of '*ratione*' ('reasoning') while holding a compass and a geometric diagram in highlighting his mathematical grounding of selenography. The significance of this frontispiece is engaging from the perspective of the history of science in the seventeenth century, since it credits not only a contemporaneous figure from the European culture and its Christian milieu, such as Galileo, but it highlights more tellingly the indebtedness of selenography to the premodern contributions of Ibn al-Haytham, not only more than 600 years before in time, but by also crossing the cultural, ethnic, religious, civilizational, geographic, and historical boundaries in recognising the impact of the transmitted legacy of a Muslim Arab polymath within Europe, from the mediaeval epoch and up until the early-modern era.[9]

4 Epilogue

The legacy of Ibn al-Haytham offers us one example amongst many in which the polymaths of the Arabic sciences and philosophies of the Islamicate civilizations influenced the history of ideas in Europe from the high mediaeval period and up to the early-modern era while passing by the Italian Renaissance. Many other inquiries have been also conducted on the European adaptive assimilation of the oeuvres of Avicenna, Averroes, Alkindi, Geber, and many others in this regard. Each of these legacies gives us perspectives on the dialogical exchanges, multi-civilizational confluences, intercultural synergies, and interdisciplinary trajectories in scholarly research that characterized the history of ideas from its ancient origins to its early-modern expressions. In the spirit of

[9] Besides the studies of A. I. Sabra and Roshdi Rashed that have been conducted in tracing the reception of the oeuvres of Ibn al-Haytham in the Arabic context and the European milieu, the influence of Ibn al-Haytham's *Optics* in the history of science in Europe is investigated amongst others by David C. Lindberg in his *Beginnings of Western Science: The European Scientific Tradition in Philosophical, Religious, and Institutional Context, Prehistory to A.D. 1450* (Chicago: University of Chicago Press, 2010), 2nd ed., while the impact of Ibn al-Haytham on art in Europe is studied *inter alia* by Hans Belting in his *Florence and Baghdad: Renaissance Art and Arab Science* (Cambridge, Mass.: Belknap Press of Harvard University Press, 2011). I also investigated these aspects in science and art in the various publications that I highlighted in Footnote 2 above in this present study.

Fig. 4 Johannes Hevelius, *Selenographia*, Frontispiece (1647): https://commons.wikimedia.org/wiki/File:Hevelius_Selenographia_frontispiece.png

Needham's scholarship, this enriches the growing awareness and realization that modern science and the historical European contributions to its unfolding have been underpinned and inspired by encounters with the transmitted intellectual legacies from Andalusia, Egypt, the Levant,

Mesopotamia, Persia, India, and China; all being elements that are recognized and showcases in the chapters of this present edited volume of proceedings.[10]

References

Alhazen (Ibn al-Haytham). 1572. *Opticae Thesaurus*, Friedrich Risner (ed.). Basel: Episcopius.

———. 1972. *Opticae Thesaurus*, Friedrich Risner (ed.). New York: Johnson Reprint.

———. 1983. *Kitāb al-manaẓır I-III*, A. I. Sabra (ed.). Kuwait: National Council for Culture, Arts, and Letters.

———. 1989. *Book of Optics I-III*, A. I. Sabra (trans.). London: Warburg Institute.

———. 2001–2010. *De aspectibus*, A. M. Smith (ed. trans.). Philadelphia: American Philosophical Society.

———. 2002. *Kitāb al-manāẓir IV–V*, A. I. Sabra (ed.). Kuwait: National Council for Culture, Arts, and Letters.

———. 2023. *Book of Optics IV–V*, A. I. Sabra (trans.). London: Warburg Institute.

Aguilón, François. 1613. *Opticorum.* Antwerp: Jan Moretus Widow & Sons.

Belting, Hans. 2011. *Florence and Baghdad: Renaissance Art and Arab Science.* Cambridge, Mass.: Belknap Press of Harvard University Press.

Fārisī, Kamāl al-Dīn. 1316. *Tanqīḥ al-manāẓir*, MS Istanbul: Ahmet III 3340. Istanbul: Topkapı Kütüphanesi.

El-Bizri, Nader. 2004. La perception de la profondeur: Alhazen, Berkeley et Merleau-Ponty. *Oriens-Occidens* 5: 171–184.

El-Bizri, Nader. 2005. A Philosophical Perspective on Alhazen's *Optics*. *Arabic Sciences and Philosophy* 15: 189–218.

El-Bizri, Nader. 2007. In Defence of the Sovereignty of Philosophy: Al-Baghdādī's Critique of Ibn al-Haytham's Geometrisation of Place. *Arabic Sciences and Philosophy* 17: 57–80.

[10] I dedicate this paper *in memoriam* to my aunt Fatima Yahya and to my uncle AbdulRahman Kotob who both passed away in 2023.

El-Bizri, Nader. 2009. Ibn al-Haytham et le problème de la couleur. *Oriens-Occidens* 7: 201–226.
El-Bizri, Nader. 2010. Classical Optics and the *Perspectiva* Traditions Leading to the Renaissance. In *Renaissance Theories of Vision*, ed. Charles Carman, et al., 11–30. Aldershot: Ashgate.
El-Bizri, Nader. 2014a. Seeing Reality in Perspective. In *The Art of Science: From Perspective Drawing to Quantum Randomness*, ed. Rossella Lupacchini, et al., 25–47. Dordrecht: Springer.
El-Bizri, Nader. 2014b. By Way of an Overture: Classical Optics and Renaissance Pictorial Arts. In *Recto-Verso: Redefining the Sketchbook*, ed. Nader El-Bizri, et al., 11–26. Aldershot: Ashgate.
El-Bizri, Nader. 2016. Desargues' oeuvres: On Perspective, Optics and Conics. In *Visioning Technologies in Architecture*, ed. Graham Cairns, 36–51. Aldershot: Ashgate.
El-Bizri, Nader. 2018. Arabic Classical Traditions in the History of the Exact Sciences: The Case of Ibn al-Haytham. *The European Physical Journal plus* 133: 271–277.
El-Bizri, Nader. 2021. Alhazen's Theory of Vision. *Micrologus* 29 (2021): 21–33.
El-Bizri, Nader. 2022a. Alhazen's Phenomenology of Visual Perception. In *The Book of Sceneries*, ed. Ala Younis, et al., 202–218. Berlin: Hatje Cantz Verlag.
El-Bizri, Nader. 2022b. The Confluences of Science, Philosophy, and Art: Ibn al-Haytham's *Optics* and the European *Perspectiva* Traditions. pp. 65–68. In *From Wisdom to Data: Philosophical Atlas on Visual Representations of Knowledge*, ed. José Higuera. Rubio, et al., 112–117. Porto: University of Porto Press.
El-Bizri, Nader, Ibn al-Haytham, and Hevelius. Selenographia. In *The Mantis Shrimp: A Simon Schaffer Festschrift*, eds. Dániel Margócsy and Richard Staley, 55–59. Cambridge: The Cambridge History and Philosophy of Science Collective.
Fārisī, Kamāl al-Dīn. 1928–1930. *Tanqīḥ al-manāẓir*, 2 volumes. Hyderabad Deccan.
Hevelius, Johannes. 1647. *Selenographia, sive, Lunae description.* Gedani: Autoris sumtibus.
Lindberg, David C. 2010. *Beginnings of Western Science: The European Scientific Tradition in Philosophical, Religious, and Institutional Context, Prehistory to A.D. 1450* (2nd ed). Chicago: University of Chicago Press.
Rashed, Roshdi. 2005. *Geometry and Dioptrics in Classical Islam.* (ed. trans.). London: Al-Furqān Islamic Heritage Foundation.

11

The Survival of Old Book Forms on the Periphery: Chinese Book Forms in Dunhuang and Beyond

Imre Galambos

Until the manuscript discoveries in the early part of the twentieth century, the history of the Chinese book had been envisaged according to traditional accounts from transmitted sources. The newly excavated materials, however, provide first-hand evidence for the study of Chinese manuscripts from the pre-modern period and force us to reconsider some long-held assumptions. Intriguingly, most of the discoveries happened along the peripheries of Sinitic sphere at sites along the Silk Road (e.g. Dunhuang, Turfan and Khara-khoto) and in places such as Japan and Korea, as well as among the non-Han minorities of China. This paper focuses on the physical form of manuscripts, arguing that while some book forms have disappeared without a trace from the central areas of Sinitic culture, they continued to be used on the peripheries up to modern times. The survival of these book forms on the peripheries corroborates patterns of transmission documented in other areas of

I. Galambos (✉)
School of Literature, Zhejiang University, Hangzhou, China
e-mail: iig21@cam.ac.uk

A. Bala et al. (eds.), *Multicivilizational Exchanges in the Making of Modern Science*,
https://doi.org/10.1007/978-981-97-3541-9_11

research, including linguistics and the history of science and technology.[1] In this paper, I would like to look at two book forms, namely, the codex and the multi-layered roll (commonly known as 'whirlwind binding'), both of which are attested among manuscripts found at Silk Road sites but not in central China.

What I aim to document here is a phenomenon similar to what the late palaeographer J.-P. Gumbert has observed with regard to Ethiopian manuscripts, which often preserve codicological features that had long disappeared from European and Mediterranean manuscript cultures. He noted that Ethiopia is 'a sort of Jurassic Park of codicology' (Gumbert 2011, p. 300). I follow his approach and examine the physical characteristics of manuscripts along the margins of the Sinitic sphere to look for features that have since then been discontinued.[2] Being more volatile and likely to embrace innovations, the centre was also subject to periodic rounds of standardisation, while groups on the outside could preserve aspects of earlier culture.

1 Manuscripts on the Periphery

The late nineteenth and early twentieth centuries saw the discovery of large groups of medieval paper manuscripts in Chinese and other languages. Interestingly, most of these finds happened along the peripheries of the Sinitic world, rather than in central areas, such as the Central Plains or the region south of the Yangtze River. The largest sites yielding manuscripts were at Dunhuang, Turfan and Khara-khoto, all located in the north-western part of modern China, along the region known today as the Silk Road. For the first three decades of the twentieth century, that is, until the late 1920s, a number of foreign expeditions conducted

[1] A relevant example is the survival of a considerable body of early Buddhist scriptures in Chinese translations, the originals of many of which, written in Indic languages, are no longer extant.

[2] The concepts of centre and periphery are somewhat arbitrary and depend on the focus of examination. As this study is concerned with Chinese manuscripts, it seems reasonable to treat central China as the cultural core and the regions around it as the peripheries. This does not automatically assign a central position to China, and should we be focusing, say, on Tibetan manuscripts, our centre would naturally shift to the Tibetan Plateau.

excavations at these sites, being interested, for a variety of reasons, in the borderlands.[3] Chinese scholars such as Yang Shoujing 楊守敬 (1839–1915), in turn, focused on tracking down lost Chinese manuscripts, texts and editions in Japan.[4] A common trait of these places was that they lay beyond the core of the Sinitic world, in areas that could be seen as peripheral or marginal. Even sites that are within the territory of China today were part of non-central regions traditionally inhabited by peoples forming multicultural societies. These are also the regions which had historically for longer periods been under non-Chinese rule.

Dunhuang, for example, was part of the Tang Empire only until the 770s, after which it fell under Tibetan control, marking the beginning of several centuries of non-Chinese rule. It was only in the 1720s that the region, once again, became part of China, when the Kangxi Emperor's 康熙 (r. 1661–1722) troops brought it under Qing control.[5] The Dunhuang manuscripts had been sealed inside a cave around 1006, during a period when the region operated as an independent kingdom, and was not part of the newly formed Song state. The manuscripts, therefore, had for the most part been produced by a mixed population at a time when, from the perspective of central China, the region was just another Central Asian polity. Naturally, this was true not only of manuscripts in Tibetan and a variety of Central Asian languages but also of those in Chinese.[6]

The newly discovered manuscripts were unprecedented in terms of their quantity and the diversity of new sources they offered. Some of them were the earliest extant versions of canonical scriptures, others were long-lost works only known by title or in quotations, yet others

---

[3] During the period in question, these regions were often the grey areas where China's political control was weaker, prompting colonial powers such as Britain and Russia to compete for dominance there. Despite their scientific pretext, the foreign exploration of Western China was directly linked with colonial designs. On the discovery of manuscripts and artefacts at these sites, as well as their removal from the country, see Jacobs (2020).

[4] On Yang Shoujing's antiquarian pursuits in Japan, see Brown (2012) and Mayanagi et al. (2015).

[5] For a brief overview of Dunhuang's history from the Tibetan period through the Qing dynasty, see Rong (2013, pp. 38–49).

[6] Considering the quantity of Dunhuang manuscripts, it was inevitable that some manuscripts came into collection from other parts of China and Central Asia. These non-local additions demonstrate the interconnected nature of Central and East Asia; see Hao and Wu (2023).

were texts entirely unknown to modern scholarship. The material also contained texts related to other religions, as well as a substantial body of administrative and economic documents related to the daily life of local communities. Because Chinese historical records do not go into much detail regarding the affairs of this region in periods when it was cut off from China, the new documents proved crucial for reconstructing the local history of the region, including its economic and religious dimensions.

The manuscripts were also highly valuable from the point of view of the history of the book in China, and in East and Central Asia in general. They offered tangible evidence for diverse book forms, sewing and binding methods, paper production techniques and scribal practices, many unmentioned in transmitted texts. Even though there have been some manuscripts preserved in central China, they have typically been part of monastic, imperial, or private collections, which also meant that they had been subject to periodic interventions, such as conservation, rebinding, or the addition of marginalia and collectors' notes. This was also true of the manuscripts preserved in Japanese collections, which have been curated by generations of scholars or abbots. By contrast, the manuscripts excavated from Silk Road sites had been deposited in their resting place many centuries ago, with no subsequent alterations. Not being subject to the intervention of well-meaning collectors and conservators in later centuries, they preserve the physical form of pre-modern books in a much more authentic way.

The first wave of research on the new material was primarily concerned with their content, that is, the texts. As manuscripts first and foremost are careers of textual material, it was scholars in textual studies and history who made use of them, approaching them from a variety of angles, including literary history, philosophy, religious studies and linguistics. In a sense, the textual focus was also the result of such manuscripts not having been available earlier in sufficient quantities, which prevented the development of a field of research that would have taken into consideration their material form. Although some existing fields of research could have provided a much-needed methodological framework, these dealt with materials that were largely unrelated to medieval Buddhist manuscripts from north-western China.

One such field was the well-developed and sophisticated science of Chinese bibliographic scholarship known as *banben xue* 版本學 ('bibliography; study of editions'), devoted to the analysis of printed books and editions. Despite the obviously overlap with manuscript studies, this field of research found only limited application in the study of newly excavated manuscripts. Another relevant field was *komonjogaku* 古文書學 (lit. 'study of ancient documents') in Japan, which focused on the diplomatic analysis of old historical documents, principally with the aim of judging their authenticity and value. This field had developed partly under Western influence during the Meiji 明治 period (1868–1912) and began gaining momentum just about the time when the first manuscript discoveries along the historical Silk Road were happening.[7] As a field, *komonjogaku* was concerned with old Japanese documents, which is probably why it was not systematically applied to the study of Dunhuang and Turfan manuscripts, even though its technical apparatus would have been highly relevant. Both of these methodologies engaged with a specific type of material and, perhaps because of that, were not directly utilised in the study of medieval manuscripts.

That the physical form of manuscripts was of import for the history of the book was clear from the start. Both M. Aurel Stein (1862–1943) and Paul Pelliot (1878–1945), the two European scholars who first had the chance to examine the contents of the Dunhuang library cave, ommented on the form and binding of manuscripts, drawing attention to their significance for contacts between China and other cultures. Later scholarship, however, seems not to have shared their fascination with this aspect of the material and instead tried to situate the finds within received Chinese narratives of the history of the book (e.g. Ma 1926; Hummel 1941). Unsurprisingly, this approach led to seeing developments as having transpired exclusively within the confines of the Sinitic world, explaining changes as local inventions arising in response to the needs of a changing society. Occasionally, scholars of non-sinological

[7] On the origins of the field of *komonjogaku* in Japan, see Mehl (2013).

training tried to look at China as part of a wider world and integrate the new data into broad transcultural processes.[8]

Looking at it from a modern perspective, it is undeniable that the vast body of material discovered from Silk Road sites attests to the existence of diverse book forms, including ones that have since disappeared in China. Although the vast majority of the manuscripts were in the form of scrolls and smaller sheets of paper, there were also other, less common book forms. Discounting some idiosyncratic and genuinely rare examples, the main 'other' book forms were the booklet, pothi, concertina and whirlwind binding.[9] Although the pothi and the concertina were later also in use throughout East Asia, the ones found at Dunhuang were by far the earliest exemplars. The booklet appeared in Dunhuang in two major variations, both of which were similar in shape to the thread-bound book of later print culture, but different from those structurally. The sewn variant, also seen in Khara-khoto and Turfan, was morphologically similar to the codex of Western scribal cultures, whereas the glued one seemed to have had very few parallels elsewhere. Similarly, whirlwind binding was a form that did not achieve wider distribution.

2 The Codex

First, let us turn our attention to the sewn booklet. Structurally, this form matches the Christian codex, which began to spread among early Christians in the second and third century, but was later also adopted into Jewish and Islamic scribal cultures.[10] Considering the morphological similarities of the Dunhuang copies to the Western examples, it is reasonable to apply the term codex for them. By doing this, we are explicitly connecting them with the same book form used by Christian and Islamic communities across West Asia and Europe. Discussing the structural

---

[8] See, for example, Diringer (1953, pp. 382–425). An identical but more readily accessible edition of this book, bearing a new title, is Diringer (2017).

[9] On the alternate book forms in Dunhuang, see Galambos (2020, pp. 24–36). Fujieda Akira called these 'irregular forms,' highlighting their difference from the scroll which he considered the 'standard form'; Fujieda (1966, pp. 24–32).

[10] On the spread of the codex, see Harnett (2017).

similarities with the Western codex, Jean-Pierre Drège has cautiously raised the possibility that this book form may have been introduced to Dunhuang by Nestorian Christian or Manichaean communities (Drège 2014b, p. 376; Drège 2018, pp. 27–28). In either case, the Dunhuang codex provides tangible evidence of a long-distance transmission of material culture across the entire Eurasian continent. Dunhuang's location along the Silk Road and its distance from sea routes in general suggest that this form spread via an overland route directly through Central Asia.[11]

In this place, however, I am concerned not with the introduction of the codex to East Asia but with its subsequent use in China and among its neighbours. Sometime after the adoption of this Western book form in China, the localised Chinese form also began to spread to adjacent regions, including Central Asia and Japan.[12] Although the available material is limited, this secondary transmission was most likely triggered by the adoption of Chinese Buddhist texts in cultures which already had a tradition of using Sinitic writings. This was an altogether different process, as in this case it was specifically the Chinese version of this book form that spread to China's neighbours.

Structurally, the codex is formed of one or more quires or gatherings, which in turn consist of folded bifolios stacked and sewn together along the centrefold. In turn, the quires themselves are also sewn together into a complete booklet, much like our modern books. For example, manuscript Pelliot chinois 3823 is a small codex composed of three quires, each containing three bifolia. Figure 1 shows the centrefold of the first quire, where the thread that binds the three bifolios together is clearly visible. In terms of its content, the manuscript contains an array of textual fragments, including text titles and lists of difficult characters extracted from Buddhist scriptures. The disorganised content and the scruffy handwriting suggest that this may have been a

[11] In this respect, I would rule out the possibility of the role of Chinese contacts with the Islamic world, expertly discussed by Hyunhee Park (2012), as there is no evidence of interaction with the Islamic world in the early tenth century in Dunhuang. The contacts with Islam must have begun towards the end of the tenth century, when the Khara-khanids occupied Khotan, prompting a large wave of refugees to flee to Dunhuang.

[12] Once again, this pattern of transmission parallels the adoption of Buddhist teachings in China and then the spread of Chinese type of Buddhism to areas around China.

manuscript belonging to a student. This codex, like most of the ones from Dunhuang, has no hard cover and thus corresponds to a Western form that Janos Szirmai described as having 'limp binding' (Szirmai 1999, pp. 285–319).[13]

Although codices were also found at Dunhuang, Turfan and Khara-khoto, this form did not seem to have continued into the Yuan and later periods in China. It simply disappeared from the scene.[14] Instead, it was superseded by the traditional Chinese thread-bound book (*xianzhuang shu* 綫裝書), in which the folded folios were stab-stitched at their open

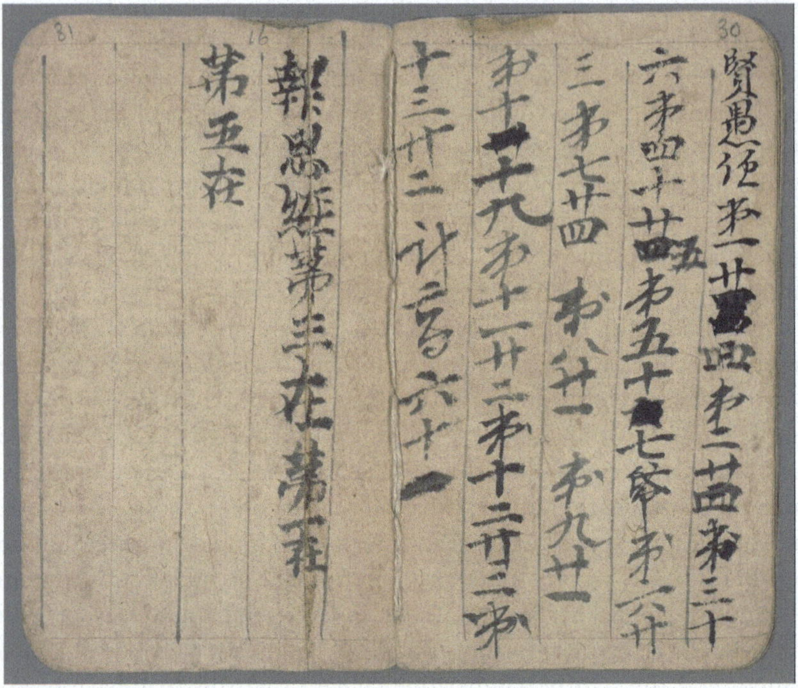

Fig. 1 A tenth-century codex from Dunhuang. Pelliot chinois 3823; courtesy of the Bibliothèque nationale de France

[13] I am grateful to my colleague Teresa Webber for pointing this out to me.

[14] This is true up to the nineteenth century, when the modern Western book, which is essentially also of the same form, was introduced once again.

end, and it was the stacked centrefolds that formed the fore-edge. It is tempting to conclude that in China the codex was a dead-end development, limited to regions along the Northwest, as no copies have been excavated from more central sites. This assumption, however, would not be accurate for two important reasons. The first is that the thread-bound book seems to have been a development of the codex, which is supported by some in-between examples from Dunhuang and Khara-khoto. In other words, the codex did not disappear but evolved and the Chinese books of later periods are its descendants.

The second reason is that manuscripts with the same structure were also transmitted to Japan and survived into modern times. Modern Japanese scholarship coined the term *tetsuyōsō* 綴葉装 (i.e. 'stitched leaves binding') to describe them, and for some time considered them to have been a Japanese development with no continental precedents. The Dunhuang examples, however, clearly show that this was not the case, and this is now generally recognised (Sasaki 2016). The earliest manuscripts of this form were allegedly brought to Japan in 806 by the monk Kūkai 空海 (774–835) following his visit to the Tang. Since then, this form became associated with the Shingon School 真言宗, but was also favoured for writing diaries, poetry and other less formal types of content. In either case, the physical form was closely tied to the context, which is probably also why it was not replaced by more common forms and managed to linger on as a relatively marginal phenomenon for more than a millennium. A concrete example is the Japanese manuscript collection of *waka* 和歌 poems from the Cambridge University Library, probably dating to the mid-Edo period (1603–1868). The spine of the volumes is left uncovered (Fig. 2), displaying the individual quires that make up the volume.

The traditional name for such manuscripts in Japan was *sasshi* 册子 (Chinese: *cezi*), which matches how it was called in Dunhuang and Turfan. Some Dunhuang manuscripts expressly used the term *cezi* 册子 (also written as *cezi* 策子) in reference to themselves. In Turfan, the Uyghurs called their own booklets *čagsı*, a term derived from the

Fig. 2 The spine of a Japanese manuscript collection of *waka* poems from the mid-Edo period, Cambridge University Library, FJ.636.17

same Chinese word (*cezi* 册子, LMC *tʂhaijk-tsẓ'*).[15] The Uyghur and Japanese examples not only demonstrate that the Chinese term was widely used throughout East and Central Asia but also implies that those regions adopted the form itself from Chinese manuscript culture. This is the case even if the term *cezi* did not exclusively refer to sewn codices but could also designate other types of booklets.

The tradition that links this form with the Shingon School, and more concretely with the figure of Kūkai, identifies China as the ultimate source. It is unlikely, however, that the introduction of this book form to Japan happened as the result of the actions of a single person, carrying a few exemplars on himself. Although Kūkai may have indeed brought back some manuscripts in this form, many more copies must

[15] Galambos (2020, p. 12). The Turfan examples also include a third type of booklet, which was neither sewn nor glued but folded from larger sheets of paper, typically taken from scrolls with writing on them. This variant is also attested in Dunhuang but only in limited numbers.

have reached Japan around that period, possibly through different channels. We know that Kūkai had travelled along the coast and into central part of the Tang through Luoyang 洛陽 and Chang'an 長安 and had never ventured farther north or west. This indicates that the codex form was also used in central China at that time. Moreover, it must have been common enough to be included in the selection of books he took back to Japan.

It is of significance that the sewn codices in Dunhuang were used not only for Chinese texts but also for those written in Tibetan, Uyghur, and even Old Turkic. In Khara-khoto, they were used for both Chinese and Tangut manuscripts, and in Turfan, for Uyghur ones. The translingual pattern of use was true for not only codices but most book forms, including the scroll and the pothi. This situation demonstrates how fluidly technologies of the book moved across linguistic and state boundaries, which is likewise among the main reasons why the codex was able to spread to East Asia in the first place.

3 The Multi-layered Roll

'Whirlwind binding' is the translation of the Chinese term *xuanfeng zhuang* 旋風裝, a name for manuscripts that consist of several sheets of paper stacked on top of each other and joined on one end with a wooden stick. The stacked sheets are then rolled up from one end as we would sometimes roll up a newspaper or an exercise book.[16] There is some debate in modern scholarship about what exactly 'whirlwind binding' was and whether it designated the same thing as *longlin* 龍鱗 ('dragon scales') and other types of binding, or whether specific manuscripts belonged to this category (e.g. Li 1989; Hou 2012). The term itself is not attested in contemporaneous manuscripts, and whether such manuscripts looked like 'whirlwind' or 'dragon scales' is a highly subjective call. In this paper, I am not primarily concerned with terminology and use the name 'multi-layered roll' as an umbrella term for similar types of manuscripts (Fig. 3). Their shared feature is that in each

[16] For an overview of whirlwind binding, see Drège (2014a).

Fig. 3 A multi-layered roll from Dunhuang. Manuscript Or.8210/S.6349, The British Library

case stacked-up sheets of paper are joined together on one side and then rolled up into a roll.

The main problem with analysing this book form is the lack of data, as there are only a few Chinese examples known from the pre-modern period, and almost all of them come from the Dunhuang library cave. Moreover, the few examples are not entirely consistent in form, which is part of the reason why there is disagreement whether they all belong to this category. The variability is in itself significant because, as Jean-Pierre Drège observed, it signals that this form was not well developed and remained relatively uncommon (Drège 2014a, p. 371). As a Tibetan example, he cited manuscript Pelliot tibétain 1257, which contained a Sino-Tibetan glossary of Buddhist terms and consisted of ten sheets of paper attached to two bamboo sticks.[17] He also pointed out that the Dunhuang library cave contained large Tibetan pothi books similarly rolled up into rolls.

Some scholars stressed that this book form had a great advantage over the scroll when searching for and looking up data quickly. Li Zhizhong, for example, stated that it developed 'as an answer to the difficulties of scanning a text in the scroll format' (Li 1989, p. 108). Indeed,

[17] On this manuscript, see Apple and Apple (2017).

the manuscripts that can be ascribed to this category typically contain database-type texts such as dictionaries and divination texts, which users did not read continuously but used as reference tools. Nonetheless, we should also keep in mind that this form does not seem to have existed prior to the Tang period and similar texts would have been written on scrolls. In addition, the overall rarity of the multi-layered book means that majority of such texts continued to be written on scrolls even after this allegedly superior form was invented. Therefore, it is questionable whether we can attribute the advent of this form to a single act of inspired invention, following a centuries-long stretch of having to put up with a less convenient form.

Despite its scarcity in East Asia, the multi-layered roll has some parallels in other parts of the world. Its structure resembles, for example, the modern newspaper stick (Fr. *baguette de lecture*) used in hotel lobbies for providing daily papers for hotel guests. The large and bulky sheets of newspaper are clamped together on one side, providing physical support and making it possible to leaf through them without taking the newspaper apart. A medieval European parallel is the Exchequer's pipe rolls used in England since the twelfth century.[18] These are extremely large rolls of parchment which contain the financial records of the annual audit process.[19]

Even though the European examples provide interesting parallels, they are unrelated to the Chinese and Tibetan manuscripts from Dunhuang. A more plausible actual link is with a similar book form used among the Yi 彝 (i.e. Nuosu or Lolo) people, who write their manuscripts on large sheets of paper joined on one end, either clamping them together with a wooden stick or folding back the beginning of the paper block and strengthening it with a thread. Usually, the manuscript also has a cloth cover, protecting the thin sheets of paper. For example, manuscript Or. 6911 from the British Library is part of a collection of Yi manuscripts that entered the British Museum's collection in the late nineteenth century. In its unrolled state, the manuscript's bottom layer is a longer

[18] I am grateful to Teresa Webber, Maree Shirota, and Stefan Holz for generously sharing their knowledge about the pipe rolls with me.

[19] On the function of pipe rolls, see Cassidy (2012).

piece of dark canvas, onto which paper sheets of equal size are placed, occupying the middle third of the canvas. The third of the canvas on the left is folded over to cover the paper sheets in the middle third, and then it is stitched at the fold (i.e. the left side) to the stacked sheets. Then the entire manuscript is rolled up starting from the stitched left side, and the third of canvas on the right becomes a protective cover. Although this particular manuscript does not have a wooden stick, the stab-stitched edge of the thick stack of paper creates a strong support at the core of the rolled-up scroll. Finally, a long string is tied around the roll to bind it tightly together.

Many of the other Yi manuscripts at the British Library have a similar binding, showing that we are dealing with a specific book form, rather than unrelated idiosyncratic cases. That many of the Yi manuscripts from other collections have the same type of binding corroborates this assumption. For example, manuscript EAP217/4/1 is a late nineteenth-century Yi manuscript of a divination text from the collection of Fang Zhengrong from Xinping, Yunnan.[20] Figure 4 shows the unrolled manuscript, with the block of stacked sheets atop the darker canvas, stitched together on the left side. At the same time, despite the similarities between manuscripts from different collections, they were not all made exactly the same way. In fact, their physical structure varies quite a bit, attesting to the lack of standardisation. In this, they parallel what we have seen with regard to multi-layered rolls in Dunhuang, where the few extant manuscripts also displayed considerable variation. From the perspective of the current paper, it is important that the overall structure of Yi manuscripts from the nineteenth and twentieth centuries is analogous to that of the Dunhuang manuscripts, suggesting an uninterrupted continuity stretching across linguistic, cultural and geographical boundaries for an entire millennium.

The connection between the Chinese and Tibetan manuscripts from Dunhuang and those of the Yi in Southwestern China may not be immediately obvious. There is, however, evidence of some continuity between

[20] This manuscript has been digitised by Sun Yat-sen University as part of the British Library's Endangered Archives Programme (Project EAP217).

Fig. 4 A Yi manuscript (EAP217/4/1) from the Fang Zhengrong collection in Xinping, Yunnan. Digitised by Sun Yat-sen University as part of the Endangered Archives Programme, British Library

Tibetan manuscripts from Dunhuang and the religious traditions of non-Chinese peoples of Yunnan and Sichuan.[21] Moreover, the multi-layered roll is certainly not the only book form shared between these cultures. But the co-occurrence of more common forms such as the pothi and the scroll in distant regions and periods is less conspicuous than the survival of otherwise extremely rare forms. In addition, the fact that the multi-layered roll is not present in other traditions makes the link between Dunhuang and modern Southwest China all that more apparent. Naturally, the book culture of central China may have also contributed to the transmission of this form, even if no traces of it can be found there.

This is not to say, however, that multi-layered rolls were unknown in other regions of Asia. We have already seen that Drège mentioned the large rolled-up Tibetan pothi books found inside the Dunhuang library cave. Although the pothi leaves were not clamped or stitched together,

[21] See, for example, the studies in Agnieszka Helman-Ważny and Charles Ramble's volume on Bon and Naxi manuscripts; Helman-Ważny and Ramble (2023).

Fig. 5 A Thai manuscript roll. Manuscript ARP.Tai.Nüa.001, Chiang Mai Rajabhat University Library (Helman-Ważny et al. 2021, p. 158)

in their rolled-up form the manuscripts would have technically qualified as multi-layered rolls. Just as interesting is the similarity of some Thai booklets with stab-stitched binding, rolled up into a roll and wrapped in protective clothing (Fig. 5). A long string securely tied the roll together (Helman-Ważny et al. 2021, pp. 156–160). These are oblong booklet-type manuscripts in which the stitched spine is along the shorter edge of the book, making rolling them up a convenient method for protection and storage. In this case, the morphological correspondence with the Yi manuscripts is not particularly surprising, considering the range of cultural and linguistic contacts between Thailand and China's Southwest.

4 Conclusions

This paper examined two book forms attested among Chinese manuscripts from Dunhuang but not present in the Chinese book culture of later periods. Strictly from this perspective, these two forms appear to have been dead-end developments. Yet if we look at cultures along the fringes of the Sinitic world, or even beyond its boundaries, we can see that these forms remained in use down to modern times. The codex (i.e. sewn booklet) survived in Japan and the multi-layered roll (i.e. whirlwind binding), among the Yi people of Southwest China. In each case, they were limited to a specific context, revealing that the transmission of the manuscripts' physical form was inextricably tied to their function. Significantly, the context in which these manuscripts continued to circulate fell outside the state's authority, which is why they were not subject to assimilation and regularisation.

Although on the surface these marginal book forms may appear irregular, the very fact that they have parallels among medieval manuscripts from geographically distant sites evidences the conservative nature of their transmission. While it is inevitable that they would have undergone a multitude of smaller changes, their overall structural consistency suggests a path of transmission in which innovation and change was kept to a minimum. Once again, the reason for this may be that their production and use were linked to a specific context. Whether religious or social, the traditions or domains in which they circulated created a type of environment that did not encourage experimentation and innovation.

The manuscripts discussed above are part of material culture, closely tied to the beliefs and practices of pre-modern societies. The expertise that formed the foundation of their production was a technology that travelled along with the manuscripts across vast distances. In fact, it was the transmission of the technology itself that ensured that artisans in other places could produce new manuscripts of the same format. Today, we can rely on the manuscripts and their codicological characteristics to document the patterns of transmission for this type of technology, which in turn attests to direct contact between peoples and cultures around China. Although there is little concrete evidence for the existence of these two book forms in central China, they undoubtedly existed there before spreading to neighbouring regions. This is true even if there were some direct contacts between the peripheries themselves, as it was likely the case for the multi-layered roll.

The phenomenon of older codicological features surviving in locations far away from where they were used most extensively is not limited to these cases, or to the case of Ethiopian manuscripts mentioned at the beginning of this paper. Rather than being exceptional, this pattern seems to have been relatively common in different cultures. For example, the earliest surviving examples of the pothi form are from the northwest of the subcontinent and the areas around the Tarim Basin in China, far away from South India where it must have originated (Baums 2021, p. 343). Similarly, there are also well-known examples in the domain of texts, such as the survival of early Buddhist scriptures in Chinese translation, instead of their original Indic languages.

Finally, the findings presented here force us to revisit the meaning of concepts of periphery and centre. We began our analysis deliberately from the perspective of Chinese book culture, delegating regions such as Dunhuang, Japan, and Southwestern China to the status of periphery. Nonetheless, if we zoom out and consider the entirety of East and Central Asia, we are able to see developments in a much more organic and interconnected way. We begin to realise that each of these local cultures had its own dynamic and the contacts that brought about changes were not exclusively in relation to the Chinese dynastic centres but also between the neighbours themselves.

References

Apple, James B., and Shinobu A. Apple. 2017. A Re-Evaluation of Pelliot Tibétain 1257: An Early Tibetan-Chinese Glossary from Dunhuang. *Revue d'études tibétaines* 42: 68–180.

Baums, Stefan. 2021. Inventing the Pothi: The Adoption and Spread of a New Manuscript Format in Indian Buddhism. In *Body and Cosmos: Studies in Early Indian Medical and Astral Sciences in Honor of Kenneth G. Zysk*, eds. T.L. Knudsen, J. Schmidt-Madsen and S. Speyer, 343–362. Leiden/Boston: Brill.

Brown, Shana J. 2012. Modern Antiquarianism and Sino-Japanese Rivalry: Yang Shoujing in Meiji Japan. In *Role of Japan in Modern Chinese Art*, ed. J.A. Fogel, 69–83. Berkeley/Los Angeles/London: University of California Press.

Cassidy, Richard. 2012. Recorda Splendissima: The Use of Pipe Rolls in the Thirteenth Century. *Historical Research* 85: 1–12.

Diringer, David. 1953. *The Hand-Produced Book*. London: Hutchinson's Scientific and Technical Publications.

Diringer, David. 2017. *The Book Before Printing: Ancient, Medieval and Oriental*. New York: Dover Publications Inc.

Drège, Jean-Pierre. 2014a. Les livres en tourbillon. In *La fabrique du lisible: la mise en texte des manuscrits de la chine ancienne et médiévale*, eds. J.P. Drège and C. Moretti, 369–372. Paris: Collège de France, Institut des hautes études chinoises.

Drège, Jean-Pierre. 2014b. Les codices. In *La fabrique du lisible: La mise en texte des manuscrits de la chine ancienne et médiévale*, eds. J.P. Drège and C. Moretti, 373–376. Paris: Collège de France, Institut des hautes études chinoises.

Drège, Jean-Pierre. 2018. Dunhuang and Two Revolutions in the History of the Chinese Book. In *Great Journeys across the Pamir Mountains: A Festschrift in Honor of Zhang Guangda on His Eighty-fifth Birthday*, eds. H. Chen and X. Rong, 20–32. Leiden/Boston: Brill.

Fujieda, Akira. 1966. The Tunhuang Manuscripts: A General Description. Part I in the *Zinbun* 9: 1–32.

Gumbert, J.-P. 2011. The Tacketed Quire: An Exercise in Comparative Codicology. *Scriptorium* 65 (2): 299–320.

Galambos, Imre. 2020. *Dunhuang Manuscript Culture: End of the First Millennium*. Berlin: De Gruyter.

Hao, Chunwen & Shaowei Wu. 2023. The Circulation of Texts between Dunhuang and Other Regions as Viewed from the Dunhuang Manuscripts. In *Beyond the Silk and Book Roads Rethinking Networks of Exchange and Material Culture*, eds. M.C. Wang and R.R. Overbey, 151–193. Leiden/Boston: Brill.

Harnett, Benjamin. 2017. The Diffusion of the Codex in the *Classical Antiquity* 36 (2): 183–235.

Helman-Ważny, Agnieszka, and Charles Ramble, eds. 2023. *Bon and Naxi Manuscripts*. Berlin: De Gruyter.

Helman-Ważny, Agnieszka, Direk Injan, Khamvone Boulyaphonh, and Volker Grabowsky. 2021. The Stab-stitched Binding of Tai Manuscripts: A Survey of History, Technique, and Function. *Journal of the Siam Society* 109 (1): 139–168.

Hou, Chong 侯沖. 2012. Cong Fengyi Beitangtian Dali xiejing kan xuanfengzhuang de xingzhi 從鳳儀北湯天大理寫經看旋風裝的形制. *Wenxian* 文獻 1: 40–48.

Hummel, Arthur W. 1941. The Development of the Book in China. *Journal of the American Oriental Society* 61 (2): 71–76.

Jacobs, Justin. 2020. *The Compensations of Plunder: How China Lost its Treasures*. Chicago: University of Chicago Press.

Li, Zhizhong. 1989. Problems in the History of Chinese Bindings. *British Library Journal* 15 (1): 104–119.

Ma, Heng 馬衡. 1926. Zhongguo shuji zhidu bianqian zhi yanjiu 中國書籍制度變遷之研究 in the *Tushuguanxue jikan* 圖書館學季刊 1 (2): 197–213.

Mayanagi, Makoto, Takashi Miura, and Mathias Vigouroux. 2015. Yang Shoujing and the Kojima Family: Collection and Publication of Medical Classics. In *Antiquarianism, Language, and Medical Philology from Early Modern to Modern Sino-Japanese Medical Discourses*, ed. B.A. Elman, 186–213. Leiden/Boston: Brill.

Mehl, Margaret. 2013. The European Model and the Archive in Japan: Inspiration or Legitimation? *History of the Human Sciences* 26 (4): 107–127.

Park, Hyunhee. 2012. *Mapping the Chinese and Islamic Worlds*. Cambridge: Cambridge University Press.

Rong, Xinjiang. 2013. *Eighteen Lectures on Dunhuang*. Leiden/Boston: Brill. Translated by Imre Galambos.

Sasaki, Takahiro 佐々木孝浩. 2016. *Nihon koten shoshigakuron* 日本古典書誌学論. Tokyo: Kasama shoin.

Szirmai, J.A. 1999. *The Archaeology of Medieval Bookbinding*. London and New York: Routledge.

Part IV

Medical Traditions in Dialogue

12

Healing Traditions and Medicinal Products in the Market of Health, Healing, Beauty, and Vigor in the Dutch East Indies

Hans Pols

In his landmark books entitled *Southeast Asia in the Age of Commerce*, Antony Reid (1993, 1998) has called the Indonesian archipelago the 'lands below the winds' as they have been centers of trade, commerce, and cultural exchange since the late Middle Ages. Its harbor cities—including Ambon, Makassar, Batavia, Semarang, and Surabaya—were sites where traders, visitors, migrants, physicians, natural historians, and locals exchanged medicinal substances and shared ideas about health, disease, and injury as well as insights into practices related to healing, managing disease and injury, strengthening one's health, and maintaining and enhancing beauty and vigor. Chinese, Arab, Indian, and, later, European traders (and settlers) acquired herbs and spices while discussing medical lore with local traders, who sold their medicinal products and passed on knowledge about health, disease, and its management. Several healers among the various groups migrating to the Indies established practices offering their goods and services—at times informally among

H. Pols (✉)
University of Sydney, Sydney, NSW, Australia
e-mail: hans.pols@sydney.edu.au

A. Bala et al. (eds.), *Multicivilizational Exchanges in the Making of Modern Science*,
https://doi.org/10.1007/978-981-97-3541-9_12

their own communities, at other times more visibly aiming to attract customers more widely. In the early twentieth century, a wide range of medical traditions, including traditional herbal medicine from various regions in the archipelago, Chinese Traditional Medicine, Ayurveda, homeopathy, electrotherapy, and various forms of herbal medicine from elsewhere were available in the Indies. From the 1920s, western pharmaceutical products became available and were advertised widely. Pharmacies functioned as sites where powders, pills, potions, tonics, and medicinal substances were sold; pharmacists offered both 'regular' and 'patent medicines,' which were, in the eyes of western-trained physicians, quack cures. Advertisements for various cures, medical preparations, and healing practices were common in the vernacular, Sino-Malay, and the Dutch colonial press, which provides an indication of the global nature of this market. Exchanges between healers, which could include dialogue, appropriation, or creative misunderstanding, created an extraordinarily diverse medical market in the Dutch East Indies. Of course, healers would at times emphasize their unique approaches and present them as incompatible with those offered by others to gain a competitive advantage. Whatever approach they might have taken, the Dutch East Indies were a significant node in a global network of healing, health, vigor, beauty, and medicinal products.

In this chapter, I follow the approach of Laurence Monnais (2019) in her book *Colonial Pharmaceuticals*, which focuses on pills, potions, lotions, and tonics rather than on medical institutions, medical practitioners, and medical theories to map the nature of the medicalization of everyday colonial life in colonial Vietnam. I include medical manuals and magazines providing simple prescriptions that can be prepared at home from ingredients obtained from local pharmacies, markets, and gardens. For the great majority of Indonesians, consulting physicians and buying western pharmaceuticals were out of reach because of the expense involved; most relied on herbal medicine and cheaper everyday medicinal products, which were available everywhere. I present several examples that illustrate both the diversity and the global nature of the medical market in the Dutch East Indies and (early) Indonesia without attempting to provide a comprehensive overview. These examples provide insights into the dynamic market of health, healing, beauty, and vigor in the Dutch East Indies. I conduct 'mundane' history or, to use Roy Porter's (1985) famous phrase, history 'from below' or from the 'patient's point of view.' To focus on the way common Indonesians dealt with disease, injury, and ill-health, I deliberately exclude the ideas and practices of leading practitioners of western medicine in the Dutch East Indies, whether they were Dutch, non-Dutch European, Chinese Indonesian, or Indonesian. I do not take the many publications in the *Medical Journal of the East Indies* (*Geneeskundig Tijdschrift voor Nederlandsch-Indië*) and similar journals publishing western medical research into account. Neither will I focus on the classical texts of Chinese medicine (such as the *Yellow Emperor's Classic of Internal Medicine*), or any type of authoritative. Instead, I focus on 'low brow' publications such as magazines, flyers, pamphlets, booklets, and advertisements published by pharmacists, physicians, and healers because they provide a much better insight into how medicine, medications, and various treatments were presented to average Indonesians.

Most historians investigating the medical market in the Dutch East Indies have focused on two medical or healing traditions: western and traditional herbal medicine. At times they have investigated the dynamic interaction between both; more often, they have emphasized their incompatibility (Hesselink 2011). They thereby overlook various other medical traditions that were present. The medical marketplace in the Dutch East Indies and Indonesia offered a wide range of suppliers who sold medications, cures, and approaches from multiple local and global sources. It remains to be investigated to which extent the diversity of healers engaged in dialogue or were inspired by each other's approaches, or whether they saw each other as competitors.

1 Indonesian Traditional Herbal Medicine (*jamu*)

Very little scholarly research has been conducted on indigenous Indonesian herbal medicine or its Javanese variety—also called *jamu*—even though it was (and is) present everywhere in the Dutch East Indies and Indonesia (Jordaan 1988). Unlike Ayurvedic medicine and Traditional

Chinese Medicine, it does not have a set of authoritative writings on which (most) practice is based, and it has never become a viable export product globally (Pols 2016). Traditional herbal medicine in Indonesia is based on oral traditions that are passed from mother to daughter, and prepared every day fresh for sale along the roads or at markets (*pasar*) (Beers 2001). Its prescriptions include herbs and spices that are also used in the preparation of food—in many traditional healing approaches, food, diet, and medication blend. Most Indonesians use some form of *jamu* regularly, while poor Indonesians rely on it exclusively (in addition to aspirin and other medications that are available over the counter in small quantities), as medical care is expensive and community health centers often far away.

A new dynamic element was added to the production and consumption of *jamu* when Chinese-Indonesian entrepreneurs commenced its mass production in the 1920s and set up extensive marketing campaigns for their products (for a history of the first of these businesses see Sumardono and Hanusz [2007]). By packaging *jamu* in pills and sachets, the bitter taste of some of these concoctions became imperceptible. Even today, several *jamu* companies produce tablets, powders, and solutions which are available all over Indonesia. The mass production of Indonesian herbal medicine followed the example set in China during the 1910s, when several companies started to produce medicinal products or patent medicines (Cochran 2006), alongside the increasing availability of western pharmaceutical products such as Aspirin (Illustration 1). Companies producing Chinese medicine founded their own newspapers to advertise their products. The packaging, marketing, and advertisements of these medical products was colorful and appealing, which contrasted with the sober style adopted by western pharmaceutical companies. First produced in Rangoon, Burma (Myanmar) in the 1870s, tiger balm became one of the most successful remedies; in the 1920s, production shifted to Singapore (Hong and Huang 2007).

Western-trained physicians were generally wary of herbal medicine, partly because it was more popular (and thereby constituted competition to their services), and partly because they considered it unscientific and potentially dangerous. When western physicians mentioned herbal medicine in official sources, they posed a sharp opposition between

Illustration 1 A large advertisement for Aspirin (in Malay) in the main shopping street in Malang (East Java). It is placed on the second floor of a shop selling medications and bandages named *The Rising Sun* (the name of the store appears in Dutch). To the left of the entrance, Chinese characters are present. The advertisement reads: 'Use Aspirin tablets for influenza, headaches, colds, gout, fever, etc.' Author's collection

'real' medicine and folklore. Yet in the nineteenth century, several physicians showed interest in herbal medicine, as the tools of their trade (mercury, bloodletting, purgatives, emetics) had been discredited with limited available alternatives (Pols 2009). Before 1800, physicians such as Garcia de Orta, Bontius, and Georg Eberhard Rumphius had actively studied local medical traditions; their work had been eagerly studied in Europe (Grove 1995; Cook 2007; Yoo 2018). The spice trade was partly motivated to obtain medicinal herbs and spices that had not been available in Europe. Despite their public stance on local medical traditions, several European physicians incorporated local medical lore in their practices. Their disdain for *jamu* was not as widespread as suggested by some proclamations in elite medical journals. It is difficult to ascertain what 'average' physicians prescribed in their everyday routine and what they thought of traditional Indonesian herbal medicine. They were most likely aware that most of their patients consumed it and might have recommended it for specific conditions. It is important to keep in mind that, at the time, most people only consulted physicians when they were in severe pain or seriously ill. In the major urban centers, there was no problem locating physicians, hospitals, and clinics, but in all other areas, medical services were difficult to access. Most medical care at the time was self-care based on oral traditions, medical manuals, advice books, flyers, and marketing materials.

Moeria Magazine

Despite the official stance of elite physicians who often disavowed traditional herbal medicine, in practice western and Indonesian medicine existed side by side more or less peacefully. This was the case in particular for Chinese Indonesians and indigenous Indonesians, who could only access western medicine if they were wealthy or through clinics and hospitals established by missionaries or the businesses they worked for. An illustration of how medical traditions at times intermingled among Indo-Europeans, Chinese Indonesians, and Indonesians—the people for whom Malay was the *lingua franca*—can be seen in the journal *Moeria* [*Muria*], which was first published in 1932. I have only been able to locate the trial issue; it is possible that the journal did not attract sufficient subscribers to be viable and that only one issue was ever published.

Muria refers to a mountainous area of volcanic origin located Northeast of Semarang [Central Java], the city where the journal was published. The publisher was a pharmacy [*Chemicaliënhandel* or *Apotheek*] located on Muria Street in Semarang. Most pharmacies in the Dutch East Indies were run by Chinese Indonesians, except for a small number of elite Dutch pharmacy stores such as those of the Rathkamp company, which (surprisingly) placed an advertisement on the front page. Most pharmacies sold a great variety of medications, potions, powders, salves, and contraptions, many of which did not fit narrow definitions of standard western medicine. Pharmacies were also a place to obtain advice about health matters. The sub-title of the magazine is interesting, as it refers to *ilmu thabib*—the science/wisdom of (Islamic) healers—and not to *ilmu dokter*—which refers to western medical knowledge. The full sub-title is 'Monthly [magazine] to provide guidance on the sciences of health and medicine with prescriptions and important insights.'[1]

The magazine's introduction discusses the economic malaise and the fact that physicians are not always available (because of distance and/or cost); it highlights that the magazine will provide prescriptions for medications that can be prepared at home with ingredients that are available in pharmacies, department stores, and one's own garden (Penerbit 1932). The first article discusses how 'non-physicians [*thabib*] can assist sick individuals' (Reporter 1932). The (anonymous) author emphasizes the importance of medical science, which provides the necessary training to correctly diagnose diseases and distinguish between primary and secondary symptoms. The author also mentions that many disorders can be treated at home when the insights presented in the magazine are followed carefully. The remainder of the journal contains articles on diabetes, issues related to teething in children, and first aid. Most striking are the directions on how to make medications at home. The first section lists typical western cures, including quinine, chloral-hydrate, chlorodyne, chloroform(!), iodine, camphor, and more. The second section contains directions on how to prepare *jamu Jawa*, while the last section lists Chinese medications. The journal concludes with small advertisements for physicians based in Semarang with their respective specializations—all of them practicing orthodox western medicine.

With an advertisement of *Rathkamp* pharmacies on the cover and the advertisements of several Semarang-based European physicians on the last page, the journal appears fully legitimate and its contents endorsed by western medicine. At the same time, the content is quite diverse,

[1] '***Maandblad*** *boeat memoentoen ilmoe kesehatan dan obat-obatan dengan berikoet* ***recept-recept*** *dan ilmoe pengetahoean jang penting-penting*.' On the front page, in large font, is printed: 'Ini **Maandblad** disediakan **speciaal** melainkan boeat ilmoe obat-obatan, pengetahoean tentang penjakit, dan **recept-recept** jang berpenting boeat roemah tangga. Silakan lantas minta djadi langganan soepaja bisa dapat tjoekoep dari **nummer** jang pertama' [This monthly magazine provides special information about medicines, knowledge about diseases, and the important prescriptions for the household. Please immediately ask to become a subscriber so you can get more information than that in the first number]. The words in bold are Dutch.

presenting medications from various medical or healing traditions. Most importantly, it appears that there is no conflict between the various medical traditions it presents. *Moeria* also indicates that Indonesians preferred to make their own medications at home; during the Economic Depression of the 1930s, consulting western physicians was too costly for most. Of course, *Moeria* magazine was published to increase sales at *Muria* pharmacy; for businesses, the purity of various healing approaches came second to pragmatic concerns related to sales. Consumers were guided by pragmatic concerns as well, as they hoped to find products that would alleviate their ills and ailments.

Mohamed Saleh's *Advice Book* [*Boekoe Nasehat*]

Mohamad Saleh was born in Ciamis (West Java, Sunda area) and graduated from the Batavia Medical College (then named *Dokter Djawa School*—Javanese Doctors' School) in 1893. He worked for the colonial medical service for fifteen years before he commenced his private practice, first in Sumedang and, in 1919, in Bandung (both in West Java). He was renowned as a physician and had many patients. In 1917, he realized that his work interfered with his health and started working less. In 1924, he passed away (Reporter 1924).[2] From 1917, advertisements appeared with medications claiming to follow Dr. Saleh's prescriptions. One interpretation suggests that he aimed to make his formulas more widely available by selling them through pharmacies. A less positive interpretation is that pharmacists were taking advantage of his fame by monetizing Dr Saleh's reputation. It is also possible that Dr. Saleh wanted to cash in on his fame by selling his name and image to Toko Marie, a large department store in Bandung owned by Yo Bok Seng, a Chinese-Indonesian businessman who also ran a mail-order and publishing business.

In 1920, Saleh published *Boekoe Nasehat* (Advice Book) with Toko Marie (Saleh 1920). He went to great lengths to make the book appear

[2] Dr. Saleh's full name was *Raden* Mohamad Saleh Mangkoepradja. *Raden* is a (modest) aristocratic title. One of his sons, Abdurachman Saleh, taught at the Batavia Medical School during the Japanese occupation. His brother, Gatot Mangkoepradja, was active in the Indonesian nationalist movement.

authentic and authoritative by including his picture and signature, highlighting the latter by using a different ink color (Illustration 2). The book uses western medical terms to describe various disorders and diseases, with illustrations that would not be out of order in a western textbook. The prescriptions in the book combine ingredients commonly used in *jamu* with chemicals and substances used by western physicians added. The terms *jamu* or *jamu Jawa* are not mentioned. A sharp distinction between western and Javanese medicine does not apply to the contents of this book—quite the contrary: Dr. Saleh's prescriptions are unique because he combined elements from both.

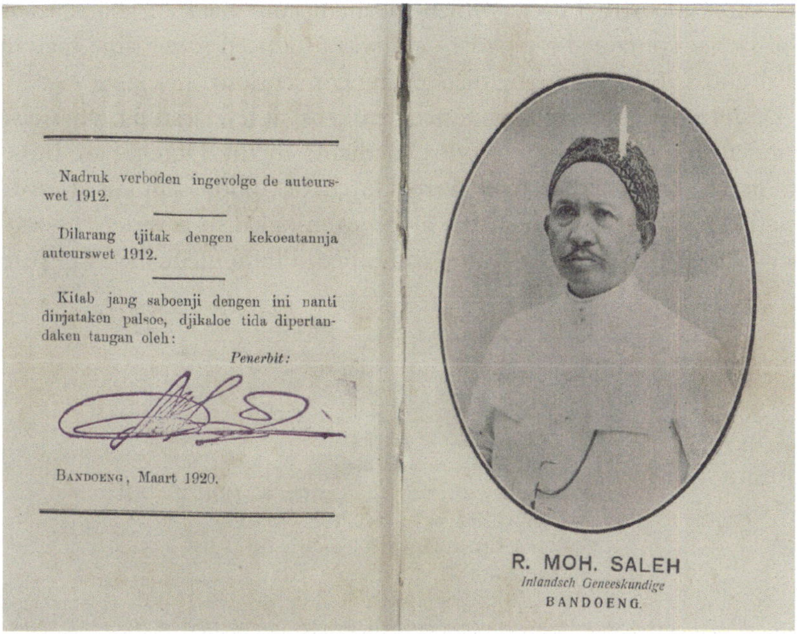

Illustration 2 The publisher of *Nasehat Dokter* clearly attempted to make the book appear authentic and authoritative, with an image of R. Moh Saleh and his signature appearing after the title page, a practice that was not common at the time Source: University of Leiden library

Several newspaper advertisements refer to Dr. Saleh. In 1924, an advertisement for stomachache and dysentery appeared in a Dutch-language newspaper (Illustration 3). It promised a total cure if used for several months.

A second example of a jamu prescription bearing Moh. Saleh's name is *Djamu Kain Kotor* (dirty rags jamu—i.e. jamu for conditions related to menstruation). It was sold in the 1950s under the brand name *Rumah Obat* [pharmacy] *Karuhun, dahulu* [formerly] *Rumah Obat Dokter Saleh*—both of which were established by Yo Bok Seng in the 1930s. On the sachet the phrase '*terlambat*' (delayed) is printed after the prescription's name, indicating that it contained a medication to 'regulate' one's menstrual cycle, which meant it contained an abortifacient. The sachet contained a powder, and was produced some time between 1951 and 1967, judging by the spelling conventions.

Despite the official pronouncements of a few leading physicians representing western or orthodox medicine in the Dutch East Indies, irrespective of whether they were Dutch or Indonesian, in everyday practice both types of medicine at times mingled in a variety of ways: they existed side by side or were combined. Plural medicine has existed

Illustration 3 Powder for stomach aches (dysentery and other stomach illnesses), following the prescription of Dr. Rd. Moh. Saleh, promises, 'after use for several months, total recovery, for both adults and children.' Main agent: Toko Marie, Bandung. The advertisement appeared in the *Preangerbode* on 9 May 1919. Source: delpher.nl

since colonial times. Pharmacies sold both types of medicine and physicians (probably) used both in their practice. Because these prescriptions were offered by pharmacies, most buyers would not need to consult a physician to access them in the first place.

2 Chinese Medicine in the Dutch East Indies

Migrants from various parts of China were among the first arriving in the archipelago; many settled before the VOC commenced operations in the early seventeenth century (Tagliacozzo and Chang 2011; Sai and Hoon 2013). Several brought Chinese medical traditions with them (Pols and Suci 2021). Chinese people (initially mostly men) migrated for various reasons; poverty and attempting to build a better life elsewhere were prominent. Many migrants married local women, learned the local language and, over generations, lost the ability to speak the Chinese language of their region of origin (so-called *peranakan* Chinese). Others converted to Islam. More recent immigrants still spoke one of the various languages spoken in China. Chinese healers (*sinses*) moved to the Indies and offered their services to Chinese Indonesians, although their reputation spread more widely, particularly in relation to treating syphilis (Lambord-Salmon and Sidharta 2007; Schijf and The 1992).[3]

It is unclear whether, and to what extent, *sinses* modified their approaches to healing and the prescriptions they dispensed after their arrival in the Dutch East Indies. It is likely that they were unable to find all necessary ingredients for their prescriptions in the Indies and substituted local ingredients. Further exploration of this topic would be an ambitious undertaking involving a team of researchers with at least one member who is able to read Chinese and is familiar with Chinese medicine. In this chapter, I only provide a single example: a hand-written

[3] For analyses on how medical traditions are transferred and translated in general, see Alberts et al. (2022) and Yoeli-Tlalim (2021). For explorations during colonial times see, for example, Vorderman (1889).

prescription book or formulary used in a store selling Chinese prescriptions in Magelang (Central Java) from before the Pacific war (Illustration 4).

On the pages of this formulary, Malay notes are interspersed with the Chinese text. I assume that the former were added at a later date. It is possible that a Chinese physician or pharmacist settled in Magelang and established a pharmacy to sell Chinese medical preparations. The second generation of storekeepers were unlikely to be as proficient in Chinese as the first one; notes in Malay were therefore added. These notes also provided guidelines for local shop assistants. It is possible that one of the (later) authors became conversant with local medical lore and tried to incorporate it in the original prescription book. This formulary

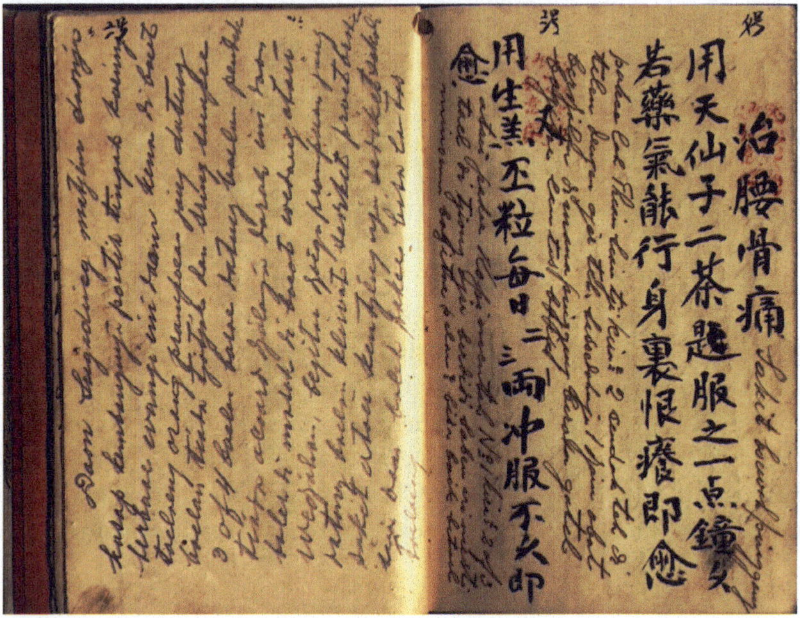

Illustration 4 Two pages from a hand-written formulary or prescription book used in a Chinese-run pharmacy in Magelang, Central Java, before the Pacific War. Text in Malay is written alongside the Chinese text. Author's collection

demonstrates that several authors associated with the original Chinese pharmacy were bi-lingual and 'translated' Chinese medical insights into Malay.

3 Pharmacies

In 2001, the World Bank estimated that two-thirds of healthcare in Vietnam involved purchasing medications from pharmacists or other traders. Only 7% of these transactions were based on prescriptions by physicians. It is likely that similar figures obtain in other Asian countries (cited by Monnais [2019, pp. 15–16]). In colonial times, the percentage of physician-prescribed medications was probably much lower. The role of pharmacies in individual and family-based practices of healing and health management in the Dutch East Indies and Indonesia has rarely been investigated from both historical and anthropological perspectives. The contents of flyers, newsletters, and advertisements indicate that pharmacies sold a large variety of medications in addition to medicinal substances, patent medicines, alternative formulations, potions, elixirs, tonics, and other compounds. In the Dutch East Indies, most pharmacies were run by Chinese Indonesians. Their offerings demonstrated the presence of global networks for the production, transport, marketing, and sales of a wide range of medical products.

In the 1920s, the marketing of pharmaceutical products expanded. Newspapers, both vernacular and Dutch, featured advertisements for Aspirin and other everyday remedies daily; billboards, posters, and banners advertising modern, mass-produced medications appeared everywhere. Yet western medications were expensive and therefore unavailable to almost all Indonesians, who continued to resort to various forms of herbal medicine, whether these were prepared fresh by jamu sellers, or mass-produced and available in affordable small packages. Most people looked for 'do-it-yourself' everyday remedies that did not require a consultation with a healer or a physician. The marketing of pharmaceuticals and mass-produced local remedies indicated how their products could be used, assisting clients. Although visits to physicians and hospitals were rare during the colonial period as far as the majority of the Indonesian population was concerned, their consumption of medicinal products was continuous and copious.

For *sinses* practicing Chinese medicine in the Indies, prescriptions were individually based, because they viewed diseases as the consequence of an imbalance in various bodily fluids or interruptions of *qi*; disease states were different for everybody. Prescriptions that applied to common conditions or disease categories were incompatible with the tenets of Chinese medicine. Yet Chinese-Indonesian pharmacies sold exactly those, and thereby practiced an interesting variation of 'orthodox' Chinese medicine. To sell medications to people with average incomes, a pragmatic approach was taken.

Kasehatan Magazine

In the Dutch East Indies, most pharmacies and traders in herbs and spices were Chinese Indonesian. Pharmacists offered an extensive range of products to maintain health, beauty, and vigor, to alleviate the effects of injury, illness, and disease, and to address common ailments. The magazine *Moeria*, discussed above, was published by a pharmacy. A similar magazine, *Kasehatan* [Health] (Illustration 5) was published by pharmacist (and publisher) Probitas, located on the Molenvliet (today: Jalan Gadjah Mada), the main road connecting old Batavia to the new neighborhood of Weltevreden (today: Central Jakarta), and very close to Chinatown. Probitas was part of a Chinese-Indonesian conglomerate, which owned a pharmacy, a printing press, and, possibly, several other businesses. Only three issues have been preserved: July and August 1930, the first and second issues published, and June 1931; all with recorded print-run of 10,000. The language of the magazine is a mixture of Bahasa Betawi, the local Batavia dialect, and a local street version of Bahasa Melayu. Dutch phrases were used as well, often for technical terms. This magazine was similar to the publications sold elsewhere in Asia that advertised patent medicines in that they contained both articles and advertisements (Cochran 2006). *Kasehatan* advertised several elixirs

Illustration 5 The masthead of magazine Kasehatan [Health], published in 1931 by Pharmacy Probitas, Batavia. Author's collection

(*anggur*, a fermented drink) that claimed to strengthen the constitution and cured *kurang darah* ('less blood'—anemia).

The magazine featured many advertisements. On the cover of the first issue a full-page advertisement for *Buffalo Head Brand Elixir* (Anggoer Tjap Kapala Banteng) appeared. It featured, not surprisingly, a large buffalo head, lowered and snorting, followed by indications for its use: it cured 'less blood' (anemia), lethargy, slimness, weak bones, rheumatism, lack of appetite, inability to get pregnant, trouble during pregnancy, and irregular menstruation. The taste is described as 'sweet, fragrant, not bitter' the elixer could be used by 'adults and children, men and women, poor and rich.' Inside, advertisements for Dragon Elixir and After-Birth Elixir can be found (Illustration 6), followed by an advertisement to enhance male vigor (Illustration 7).

A third advertisement is for Mei Ling Fan face-whitening powder [*bedak*], which is 'praised all over the world.' In the same issue, a 3-page biography of well-known Beijing-based actor Mei Lan Fan appeared. Mei Lan Fan was a well-known Chinese singer mainly performing in Chinese operas, playing almost exclusively female roles. While touring the United States, he received an honorary degree from Pomona College in California. The advertisement and article are a form of celebrity

Illustration 6 Advertisement for Dragon Elixir (Anggoer Naga) and After-Birth Elixir (Anggoer Branak). The text reads: 'Be careful! These days, there are various dangerous elixirs. This is Dragon Elixir and After-Birth Elixir by Dr. Lie Tjwan Kiat, which is the best and efficacious. We have already received thousands of letters of praise. Hurry and buy [it]!!'. Author's collection

endorsement, but it is unlikely that the actor and singer had any knowledge of this. The advertisement states: 'This powder, which is made from mixed products, does not contain poison. Be careful not to use other powders that are known to have toxic reactions.'

Most facial powders sold in the Dutch East Indies contained lead, despite several regulations prohibiting its use. Consequently, several Chinese-Indonesian women, who were the main consumers of face-whitening powders, developed health problems. In 1923, several Chinese-Indonesian women living in Surabaya showed symptoms of paralysis in the hands, stomach cramps, loss of energy, lethargy, anemia, severe headaches, and, in a few cases, near-blindness. A physician who investigated these women concluded that their symptoms were related to lead poisoning following the use of a specific brand of facial powder that contained large quantities of lead. This facial powder had been acquired

Illustration 7 A medication to revive masculine power. Hercules Power Pills (*krachtpillen*, (Dutch). The text reads: 'Power pills, the efficacy of which can be guaranteed. This pill is so extraordinary that it is hard to find another one that works so quickly, smoothly, and without any poor side effects for individuals of little strength.' Author's collection

in Chinese-Indonesian pharmacies and was branded as Chinese.[4] The advertisements of Mei Lan Fan facial powder claimed that it did not contain lead.

Pharmacies, other than those that were run by Dutch companies, sold a great variety of health products and used clever advertising strategies to sell them. Advertisements often referred to the Chinese origin of their products, but these do not appear to be closely related to traditional Chinese medicine. These advertisements appeared in the vernacular press, indicating that their market primarily consisted of Indonesians, Chinese Indonesians, and less wealthy Indo-Europeans.

[4] See Reporter (1927) and Reporter (1934). For the medical report on lead poisoning among Chinese-Indonesian children, see De Langen (1934).

4 Patent Medicines, Alternative Healers, and Marketable Cures

Little research has been conducted on the various forms of alternative medicine exported from Europe and North America to the Dutch East Indies, despite their ubiquitous presence. Homeopathy, electrotherapy, and various natural healing techniques were advertised and provided. At times, they appeared in the Dutch colonial press, but were advertised more frequently in the vernacular and Sino-Malay press, targeting Chinese Indonesians and the various ethnic groups living in the Dutch East Indies.

Electrotherapy

The following brochure on electrotherapy (*stroom kesehatan*) contains extensive information on a specific electrotherapy device, the ailments that can be alleviated through its use, and brief explanations of the nature of disease. Customers were able to buy the electricity generator and numerous additional parts separately, each of which had a specific purpose (treating conditions in the ear, nose, etc.) (Illustration 8). The text is written in standard or 'civilized' Malay, which was used in publications by the colonial administration. Printing brochures like these was expensive. The head distributor was Dutch, but there is no record of a single advertisement for this device in the Dutch colonial newspapers. The marketing for this device therefore solely targeted individuals for whom Malay was the first language. Given the expense involved, it is likely that it was primarily marketed to Chinese Indonesians. There are other indications that this was the case.

The list of ailments that could be alleviated by using this device does not represent the average 'western' disease universe, nor does it refer to Indonesian conceptions of disease such as 'masuk angin' [enter wind—the most common affliction among Indonesians, which includes having a cold, feeling cold, bubbles in one's stomach, etc.]. Apart from some common conditions (including paralysis, boils, coughs, a stiff neck,

Illustration 8 Stroom Kesehatan or electrotherapy. 'Recover – Health – [Be] Fresh, with electrotherapy "Primus", the leader in treating sick and healthy people with electrotherapy. Arranged with the advice of physicians' and 'Health, easy, and happy, with the tools of electrotherapy.'

dizziness, and convulsions), there were several afflictions that resembled neurasthenia or depression, including the 'disease of thinking about trouble, with lethargic body (*penjakit pikiran soesah badan lesoe*) and nerves (forgetfulness, troubled thoughts)'; 'easily getting startled or angry (*lekas kaget dan gampang marah*)'; and 'feeling lonely, no joy, and much fear (*berasa kesipian tida ada kegoembiraan merasa banjak takoet*).' There were yet other conditions that sounded different: 'blood that runs too fast,' 'phlegm disease,' '"less blood" (anemia),' and 'pounding heart' (the list of conditions appears in Hooge (1932, pp. 3–4)). These ailments were related to Chinese conceptions of health and disease.

B. M. Chee [Beda Maximillianus Chee]

B. M. Chee was an extraordinary healer, naturopath, electrotherapist, and herbalist from Magelang, Central Java (Illustration 9). He was born, in 1899, as Djie Kiem Tjhiang, but later changed his name to Beda Maximillianus Chee Schramm. He had been a sickly child. A (Chinese Indonesian) neighbor adopted him and treated him with herbal medicine, through which he recovered his health. This inspired him to pursue a career in herbal medicine. According to a short biography published in 1987 that might contain inflated claims, he worked for some time at the Singapore General Hospital and studied medicine in the evenings (Schramm and Chee 1987). He also studied pharmacy in Shanghai, Hong Kong, and Manila, before moving to California where he trained with leading naturopaths Arthur Schramm, Benedict Lust, and Carl Schultz. At this time, he became a member of a Rosicrucian association and the Freemasons of the 33°. When he returned to the Dutch East Indies in the late 1930s, he established a practice in Tasikmalaya (West Java) and founded the Brayuverdic Medical Society. During the Japanese occupation, his practice was ransacked, all his books burned, and all his medical equipment destroyed. After World War II, Chee set up a practice and a jamu factory in Malang (West Java), and began the most active part of his career.

Beda M. Chee had an expansive vision: he ran a clinic in Malang and sold herbal prescriptions under the brand name *Alam Subur* (prodigious nature). In his practice, he incorporated naturopathy, water cure, osteopathy, massage, electrotherapy, light therapy, and probably several other allopathic and alternative traditions. In his 1952 catalogue, Chee advertised a wide range of medicinal products, which included elements commonly used in Javanese herbal medicine as well as western ingredients. In 1967, when the Suharto regime required Chinese Indonesians to adopt Indonesian-sounding names, he wrote to the Schramm family to ask permission to adopt their family name, which was granted. From that time on, he was known as Dr. nat. Beda M. C. Schramm.

As the examples of electrotherapy and Beda Schramm's naturopathic medical empire make clear, both were global in nature. Electrotherapy originated in the United States and was popular in both North America and Europe in the late nineteenth century as a cure for neurasthenia and other conditions. In the Indies, it was incorporated in a 'medical universe' that emphasized blood and wounds as central features, sidestepping the

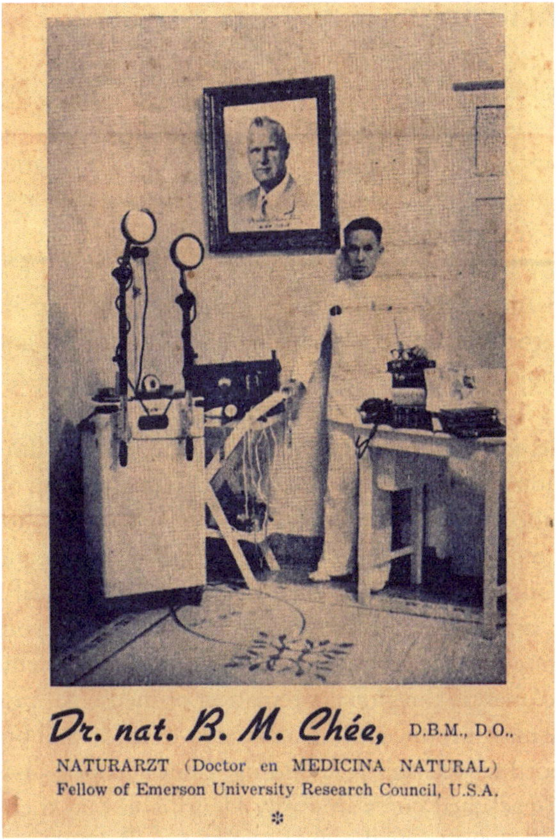

Illustration 9 Beda Maximillianus Chee in his consultation room. Equipment for UV therapy and electrotherapy are clearly visible, as is the portrait of Arthur Schramm. Author's collection

nervous system almost entirely ('neurasthenia' was not popular as a diagnostic term in the Dutch East Indies, unlike elsewhere in Southeast Asia). Schramm incorporated various healing methods in his own approach, most of which he acquired in the United States where he studied with German migrants. Today, Beda M. Chee Schramm is remembered as the author of a book on natural therapeutic methods—mostly jamu (Schramm and Chee 1987).

5 Conclusions

This eclectic overview can serve as a reminder that, in the Dutch East Indies, 'medications' and 'cures' beyond more or less 'official' or 'orthodox' western medicine were very common. The market for patent medicines, quack cures, and healing practices has always been more extensive than medications produced by western pharmaceutical companies that can be obtained by prescription or over the counter. To gain a comprehensive perspective on health and healing in the Dutch East Indies, the full range of healing practices, cures, and medicinal products should be taken into consideration. Focusing on influential texts, irrespective of whether they stem from western medical traditions or Ayurvedic, Chinese, or Arab medicine, unduly limits the focus of investigation. In addition, it is important to focus on the 'material' elements of the medical market: pills, potions, powders, tonics, elixirs, and substances. Medications, concoctions, and medical equipment tend to circulate unchanged both locally and globally. Medical theories, on the contrary, are easily transformed once they travel beyond their area of origin, as are healing practices. A focus on material elements enables researchers to analyze the global circulation of medicinal products. After these became mass-produced, the materiality of medicinal products came to include bottles, packaging, branding, brand names, and marketing. Branded and packaged cures are easily identifiable and thereby acquired a significant 'constancy' in the medical market. Both the pharmacies of the Dutch East Indies, and the work of 'quacks' and sellers of patent cures, demonstrate the global nature of trade and exchange in curative substances.

In *The Colonial Life of Pharmaceuticals: Medicines and Modernity in Vietnam*, Laurence Monnais (2019) analyzes the marketing and sales of various western pharmaceutical products in colonial Vietnam. She expresses surprise about how little attention historians of medicine have given to pharmaceutical products listed in medical publications and health reports. She has been able to locate several sources to analyze their distribution. In her discussion, Monnais includes the large market in 'unofficial' medications and pharmaceuticals (Monnais 2019, pp. 233–253). In *Chinese Medicine Men*, Sherman Cochran focuses exclusively on these alternative medications, how they are branded, advertised, and marketed (Cochran 2006). A combination of both approaches is promising to investigate the medicalization of Indonesian society. A focus on medicinal substances (powders, pills, and liquids) is not sufficient; instruments (UV and infrared lamps, electrotherapy machines, acupuncture needles) should to be included as well. The resulting analysis would reveal the globalization of approaches to healing in novel and insightful ways.

As is clear in this overview, Chinese-Indonesian healers and Chinese-Indonesian pharmacists were important intermediaries in producing, selling, and marketing medicinal products, potions, lotions, powders, concoctions, including traditional Indonesian herbal medicine (*jamu*). As others have argued, the marketplace was important for the transfer and circulation of medical ideas along with medical herbs, spices, substances, and potions (Cook 2005, 2007). The market of medications, health, beauty, and vigor has always been global, and is the first one accessed by individuals plagued by afflictions and adverse conditions. Its inclusion in the history of health and medicine is critical.

References

Alberts, Tara, Sietske Fransen, and Elaine Leong. 2022. Translating Medicine, ca. 800–1900: Articulations and Disarticulations. *Osiris* 37: 1–21.

Beers, Susan-Jane. 2001. *Jamu: The Ancient Indonesian Art of Herbal Healing*. Singapore: Periplus.

Cochran, Sherman. 2006. *Chinese Medicine Men: Consumer Culture in China and Southeast Asia*. Cambridge: Harvard University Press.

Cook, Harold J. 2005. Global Economies and Local Knowledge in the East Indies: Jacobus Bontius Learns the Facts of Nature. In *Colonial Botany: Science, Commerce, and Politics in the Early Modern World*, ed. Claudia Swan and Londa Schiebinger, 100–118. Philadelphia, PA: University of Pennsylvania Press.

Cook, Harold J. 2007. *Matters of Exchange: Commerce, Medicine, and Science in the Dutch Golden Age*. New Haven, CT: Yale University Press.

De Langen, C.D. 1934. Loodencephalitis en -Meningitis bij Kinderen op Java. *Geneeskundig Tijdschrift Voor Nederlands Indië* 74 (3): 151–158.

Grove, Richard H. 1995. *Green Imperialism: Colonial Expansion, Tropical Island Edens and the Origins of Environmentalism, 1600–1860*. Cambridge: Cambridge University Press.

Hesselink, Liesbeth. 2011. *Healers on the Colonial Market: Native Doctors and Midwives in the Dutch East Indies*. Leiden: KITLV Press.

Hong, Lysa, and Jianli Huang. 2007. Chinese Diasporic Culture and National Identity: The Taming of the Tiger Balm Gardens in Singapore. *Modern Asian Studies* 41 (1): 41–76.

Hooge, H. 1932. *Semboeh, Sehat, Segar oleh Stroom Kesehatan 'Primus'*. Bandoeng: H. Hooge.

Jordaan, Roy E. 1988. On Traditional and Modern *Jamu* in Indonesia and Malaysia: A Review Article. *RIMA: Review of Indonesian and Malaysian Affairs* 22 (1): 150–163.

Lambord-Salmon, Claudine, and Mira Sidharta. 2007. Traditional Chinese Medicine and Pharmacy in Indonesia: Some Sidelights. *Archipel* 74: 65–204.

Monnais, Laurence. 2019. *The Colonial Life of Pharmaceuticals: Medicines and Modernity in Vietnam*. Cambridge: Cambridge University Press.

Penerbit. 1932. Pendahoeloean. *Moeria* 1 (1): 1–3.

Pols, Hans. 2009. European Botanists and Physicians, Indigenous Herbal Medicine in the Dutch East Indies, and Colonial Networks of Mediation. *East Asian Science, Technology, and Society: An International Journal* 3 (2–3): 173–208.

Pols, Hans. 2016. *Jamu*: The Indigenous Medical Arts of the Indonesian Archipelago. In *The Bright Dark Ages: Comparative and Connective Perspectives*, ed. Arun Bala and Prasenjit Duara, 161–185. Leiden: Brill.

Pols, Hans, and E.S. Tyas Suci. 2021. The Contentious Social Positions of People of Chinese Descent in Indonesia. In *Mental Health in China and the Chinese Diaspora: Historical and Cultural Perspectives*, ed. Harry Minas, 181–196. Cham, Switzerland: Springer.

Porter, Roy. 1985. The Patient's View: Doing Medical History from Below. *Theory and Society* 14 (2): 175–198.

Reid, Anthony. 1993. *Southeast Asia in the Age of Commerce, 1450–1680, Vol. I: The Lands Below the Winds*. New Haven, CT: Yale University Press.

Reid, Anthony. 1998. *Southeast Asia in the Age of Commerce, 1450–1680, Vol. II: Expansion and Crisis*. New Haven, CT: Yale University Press.

Reporter. 1924. Dr. Saleh+. *Bataviaasch Nieuwsblad*, December 24, Part II, 1.

Reporter. 1927. Schadelijk en Gevaarlijk [Harmful and Dangerous]. *De Nieuwe Vorstenlanden*, January 10, 1.

Reporter. 1932. Bagaimana Seorang Boekan Thabib Menoeloeng Orang Sakit? [How can a non-physician help a sick person?] *Moeria* 1 (1): 3–6.

Reporter. 1934. Gevaarlijke Bedak en Loodwit-Ord: Is de Loodwit Ordonnantie Toepasselijk op Loodwithoudende Poeders? [Dangerous Powder and the Regulations on White Lead: Is the White Lead Regulation Applicable to Powders Containing White Lead?] *De Locomotief*, March 17. *Part V* I: 2.

Sai, Siew-Min, and Chang-Yau Hoon, eds. 2013. *Chinese Indonesians Reassessed: History, Religion and Belonging*. New York: Routledge.

Saleh, R. Moh. 1920. *Boekoe Nasehat* [Advice Book]. Bandoeng: Yo Bok Seng, Toko Marie.

Schijf, H., and B.A.M. The. 1992. Chinese Doctors in the Dutch East Indies: Social Mobility among an Ethnic Trading Minority in a Colonial Society. *Indonesia* 53: 33–50.

Schramm, Beda M. Chee. 1987. *Diagnosa dan Terapi Formula Penyembuhan dengan Metoda Alam Subur*. Semarang: Dahara Prize.

Sumardono, Asih, and Mark Hanusz. 2007. *Family Business: A Case Study of Nyonya Meneer, One of Indonesia's Most Successful Traditional Medicine Companies*. Singapore: Equinox.

Tagliacozzo, Eric, and Wen-Chin Chang, eds. 2011. *Chinese Circulations: Capital, Commodities, and Networks in Southeast Asia*. Durham, NC: Duke University Press.

Vorderman, A.G. 1889. De Chineesche Behandelingswijze van Keeldiphteritis [The Chinese Treatment of Throat Diphtheria]. *Geneeskundig Tijdschrift voor Nederlandsch-Indië* 29 (5): 559–649.

Yoeli-Tlalim, Ronit. 2021. *Reorienting Histories of Medicine: Encounters Along the Silk Route*. London: Bloomsbury.

Yoo, Genie. 2018. Wars and Wonders: The Inter-Island Information Networks of Georg Everhard Rumphius. *British Journal for the History of Science* 51 (4): 559–584.

13

Needham's Legacy in Clinical Research Revisited: Refashioning Acupuncture with Biomedicine

Wen-Hua Kuo

1 Needham's Legacy in Clinical Research Revisited

This chapter intends to be, in part, a continuation of my reflexive paper for the "A Second Look at Joseph Needham" forum in Isis (Volume 110, Number 1). Titled "An Ecumenical Medicine Yet To Come: Reflections on Needham on Medicine" (Kuo 2019), that paper offered an assessment of Needham's account of Chinese medicine, focusing on his thesis on the making of ecumenical science with a close look at acupuncture, on which topic he and Gwei-djen Lu together contributed a book (Lu and Needham 1980) during the early years of the global visibility of this East Asian healing art.

Medicine is certainly a field where there exist numerous therapeutics, yet an integration of them, whether theoretical or practical, is hard to

W.-H. Kuo (✉)
National Yang Ming Chiao Tung University, Taipei and Hsinchu, Taiwan
e-mail: whkuo@ym.edu.tw

A. Bala et al. (eds.), *Multicivilizational Exchanges in the Making of Modern Science*,
https://doi.org/10.1007/978-981-97-3541-9_13

achieve. In the United States, for example, Chinese medicine was recognized, along with Ayurveda, as a non-biomedical systematic medicine with its inclusion in 1999 by the National Center for Complementary and Alternative Medicine (NCCAM), part of the National Institutes of Health. Nonetheless, as the NCCAM was transformed into the National Center for Complementary and Integrative Health in 2014, its efforts shifted in part from how these alternative therapeutics work to how they can be integrated into biomedicine in a more scientific fashion. Perhaps due to this challenging nature, medicine was one of the subjects Needham most concerned himself with, though it was never his first priority in composing Science and Civilisation in China (SCC). He never even had the chance to see Part 6 of Biology and Biological Technology, the volume dedicated to medicine, which came out five years after his death.

Noting an increase in the scholarship on the history of medicine that has shifted away from Needham's positivist approach, in my Isis paper I called instead for a revival of his eclecticism in situating Chinese medicine as a living tradition, arguing that scholarship drawn from science and technology studies can facilitate the reorientation of Needham's project. This is particularly significant in the case of acupuncture, which, according to Paul Unschuld, is a therapy effective yet having no equivalent in biomedicine. As he once commented at the International Congress of Oriental Medicine in 2014, while acupuncture might remain independent after its integration into biomedicine, its underpinning science of systematic correspondence, based on the schema of yin-yang and five agents, would be replaced by chemistry, physics, and technology.

Having no intention of arguing against Unschuld, this chapter instead follows how the transition of acupuncture toward ecumenical medicine—its "due course", in Needham's terms—is taking place, and the possible role of Needham's viewpoint in attempts to make acupuncture scientific: that is to say, theoretically comprehensible and clinically applicable. As in my Isis paper, I argue as to how we should treat a legendary figure like Needham today, and why—whether his ideas have become outdated in interpreting the historical development of Chinese science or whether, in the case of acupuncture, they might become a practical guide for those who hope to preserve the therapeutic essence of this healing art while searching for empirical rationales on which it is based.

With the above concerns in mind, and echoing the quote from Huangdi Neijing at its beginning, this chapter takes a close look at Jaung-Geng Lin (林昭庚), lauded as the "father of defining the safe margins of needling depth in acupuncture" (針刺安全深度之父) and the first physician of Chinese medicine to be elected an academician at Academia Sinica (中央研究院院士). Departing from journalistic accounts that feature only Lin's personal merits, this chapter casts him and his achievements in a larger context that attempts to render this non-biomedical tradition scientific. It will show that, while inheriting some Needhamian concepts—whether wittingly or unwittingly—what has made Professor Lin and his team an undeniably successful research enterprise has been his professional training in 1970s Taiwan, his insights as a practicing acupuncturist, and his networking, which has turned acupuncture into an inviting field for research and a respectable clinical specialty.

2 Jaung-Geng Lin: A Practitioner, Researcher, and Professor[1]

Jaung-Geng Lin was one of the first medical students to receive an education in both biomedicine and Chinese medicine (between 1966 and 1973), and he has been active as an acupuncturist for more than four decades. As a senior practitioner of Chinese medicine, highlights of Lin's professional career include his participation in a diplomatic medical mission to Saudi Arabia (1979–1980) and his chairmanships of the National Union of Chinese Medical Doctors' Association (1998–2004) and of the Global Budget Payment System of the Chinese Medical Service Committee (1998–2001). Lin has also enjoyed a successful and productive academic career as a university professor. Receiving his degrees in Taiwan, Lin was the first PhD in acupuncture and has served in various positions at the China Medical College (中國醫藥學院,

[1] For a biography of Jaung-Geng Lin, please see Lin and Yeh (2012).

later China Medical University, hereafter CMC), his alma mater, and in governmental roles relating to Chinese medicine. He has authored more than 300 scholarly pieces and penned or edited more than 40 books on Chinese medicine. He is recognized as an expert who has contributed to global policies on acupuncture, through, for example, the Intergovernmental Committee for the Safeguarding of Intangible Cultural Heritage.

Given Lin's remarkable profile, his election as an Academician at Academia Sinica in 2022 (albeit after three attempts) was greatly encouraging. This was also the first time that non-academics became aware of this scholar and his achievements in their entirety. For the public, Lin's story reads like that of a living legend, who, though born into a modest background in a rural area of central Taiwan, completed a thorough training in medicine and chose to devote his life to Chinese medicine, a subject which was not even part of mainstream research. "To the utmost of one's ability" (盡己所能) is the motto which summarizes his research output—from acupuncture for pain relief, to the safety margins of acupuncture. He has been able to contribute to the emerging field of experimental acupuncture—according to his own account an integration of biomedicine and Chinese medicine—and plans to continue working on it, toward making the therapeutic tradition of acupuncture more empirically based and clinically evaluable.

Lin is unquestionably an extraordinary figure. Though this is not unusual among his biomedical colleagues, who can manage a career back and forth between universities and hospitals as both researchers and practitioners, what makes Lin stand out is the fact that, unlike Korean medicine, Chinese medicine in Taiwan had not until recently been institutionalized within hospitals or as a regular research discipline. In fact, it was only when National Yang Ming Chiao Tung University, one of the country's leading biomedical universities, was granted approval to establish its department of Chinese medicine in 2023 that it would provide training for next generation of Chinese medical doctors that featured the integration of clinical practice and interdisciplinary research. Against this background we can recognize that Lin, who has almost single-handedly created a research enterprise comparable to that of his colleagues in biomedicine, not to mention his administrative engagement in a professional association, requires a fair assessment of his achievements.

Even so, instead of repeating Lin's successes in terms of personal traits such as his incorrigible enthusiasm and diligence in research, this chapter hopes to trace his research interests and achievements, by which, like Needham in the 1960s, he aimed to make acupuncture universally acceptable through empirical investigations. I argue that, in the general trend of acupuncture research, Lin coincidently followed a path Needham might have expected to see—one that combined basic research and its clinical application, the medical classics and their modern interpretations, and therapeutic experiences and their accumulated effects. In the sections that follow, I will further set out that path.

3 Refashioning Acupuncture as an Experimental Clinical Discipline

As mentioned previously, Lin received training in both Chinese medicine and biomedicine, and he was not an exception. When in 1972 the CMC, Lin's alma mater, was granted approval to extend its training program in Chinese medicine to a full department, it offered only dual degrees; all its students had to learn both biomedicine and Chinese medicine to qualify to take their licensing exams. As a result, although these graduates held two licenses in medicine, many of them chose to use only that of biomedicine.

Having decided during his years at medical school to devote his career to acupuncture research, Lin nonetheless practiced biomedicine after graduation just like the rest of his cohort, working as a surgeon for three years. As he recalled in an interview, Lin waited for the opening of an acupuncture unit at the Veterans General Hospital (later Taipei Veteran General Hospital), one of the leading medical centers in Taiwan, in 1977. It was a decision of great importance. In 1971, a research committee on acupuncture (針灸委員會) had been officially established under National Science Council, followed by out-patient acupuncture clinics at a number of biomedical centers in 1974, and a Master's program on the study of Chinese medicine (中國醫學研究所) at the CMC. All of this created a perfect research environment that had every component Lin asked for to carry out empirical studies on acupuncture.

The key to starting such research, just as he had learned in biomedicine, was anatomy. An eager student in medical school and a trained surgeon, Lin was able to apply his anatomical knowledge to acupuncture, and the safe margin of needling depth (針刺安全深度) appeared as a boundary subject between acupuncture and biomedicine. Still a resident at Veterans General Hospital, in 1978 Lin started collecting cadavers of various ages, heights, weights, and of both sexes and worked on measuring the acupuncture points of the bladder meridian (BL11 to BL27; BL41 to BL52) on their backs.

The idea of this study, as shown in his Master's thesis (see Fig. 1), was not sophisticated; it could be described as an anatomical investigation on the depth of different places between the back and the abdominal cavity. However, from the viewpoint of acupuncture it was crucial, as the points chosen had therapeutic significance and, to some extent, clinical implications. More importantly, the way Lin measured the depth of these points was astonishing. With help from the hospital's pathology department, Lin was able to work to remove a cadaver's internal organs so that he could measure the points he chose from the inside out (to be exact, to insert a needle from the inner side of the abdominal cavity through to the anatomical table where the cadaver lay). The hybrid nature of his thesis could also be seen in its advisory committee, which included two biomedical professors—one in anatomy, and the other in charge of the hospital's acupuncture research committee.

Interestingly, Lin did not follow the biomedical research to move on to the study of the microanatomy of acupuncture points: this had been studied by the Chinese scholar Zhenguo Yan (严振国) and his team since 1975 (Lin et al. 2014, Chapter 6).[2] Aware that Yan had been systematically working on the anatomical location, including the depth from the skin surface, of each classical acupuncture point listed by the WHO using tissue embedding technology, Lin decided to skip histology and pathology and studied instead the needling depth of acupuncture

[2] A case for such approach can be seen in the textbook by H.C Dung, a renowned professor of anatomy and Yun-tao Ma, a professor in physical therapy (Dung and Ma 2000).

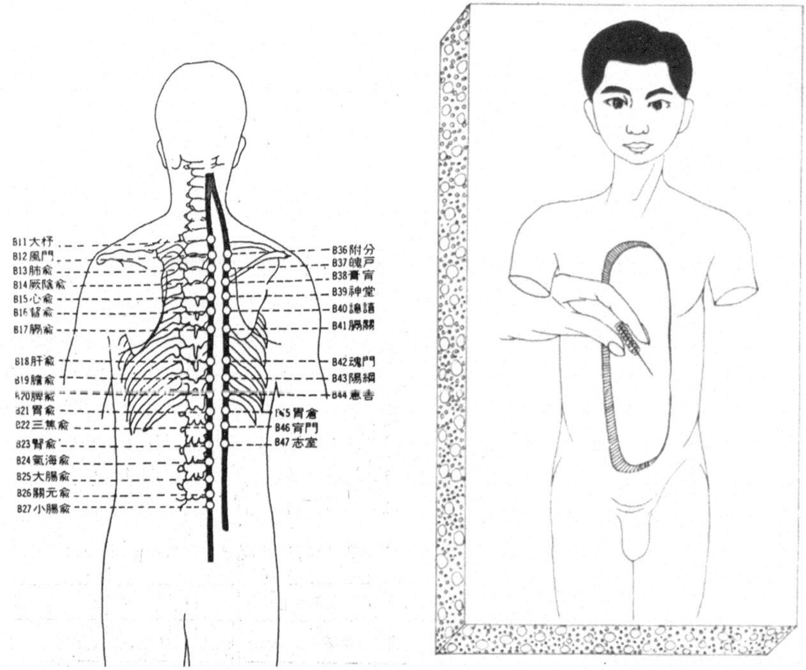

Fig. 1 Anatomical approach to approximating the needling depth of selected acupuncture points of the bladder meridian, from Lin's Master's thesis (Lin 1982)

points in living bodies. As an acupuncture practitioner, he knew very well that, although the way he measured the depth of acupoints on cadavers was new, the results, like those uncovered by Yan and his team, were far from the clinical reality. In retrospect, it proved to be a groundbreaking approach that combined biomedicine with acupuncture.

This time around, Lin's biomedical collaborators were radiologists, who were then testing the computed tomography (CT) machines newly installed at the Tri-Service General Hospital and Mackay Memorial Hospital. As an immature technology in the late 1980s, CT had not yet become an imaging tool able to make clear clinical contributions. Yet it was good enough for Lin to determine the small distances between the body surface and the inner side of the thoracic and abdominal cavities, while the cross-sectional aspects of the images fitted well with the acupoints on meridians that pass through the trunk, such as those on the stomach or spleen meridians. With help from his radiologist colleagues, Lin was able to collect CT images from 128 patients (97 females and 31 males, aged from 17 to 82 years old) for such analysis, and the results won him a PhD in 1991.

Even before finishing the collection of CT images, Lin, already an associate professor and attending acupuncturist at the CMC, could not wait to apply what he had found to an explanation of *deqi* (得氣), a therapeutic phenomenon of acupuncture that can be translated as "arrival of *qi*" (Lin et al. 2014, Chapter 6). Clinically speaking, *deqi* was usually characterized as a complex feeling of sour, numb, swelling, heavy, pain, hot, and cold when an acupuncture needle was placed at a point. What Lin attempted to establish can be understood as a physiological mechanism that correlated the needling depth of acupuncture points, their *deqi* effects, and clinical outcomes.[3]

At first glance, Lin's findings were not very surprising: there were no significant relationships between the needling depth of acupuncture points and *deqi* effects among patients with different sexes and body physique, while acupoints on the abdomen required larger depths in order to generate *deqi* effects. Nonetheless, considering these in the context of clinical practice, a larger plan for refashioning modern acupuncture emerged—Lin was about to set up a workable procedure of acupuncture that was empirically based and comparable to that of biomedicine. The following years witnessed this effort: Lin proposed a treatment course of acupuncture and a standard procedure for each session that featured *deqi* effects, and worked on therapeutic studies on individual acupoints.

Even years later, not everyone has appreciated this plan. On one occasion, in 2023, I saw Lin challenged on the idea of *deqi* effects as an indicator for measuring and monitoring acupuncture treatment, and he responded by citing his depth studies. For some professionals, the course Lin proposed seemed too simple and straightforward and risked missing the subtle therapeutic differences between individual patients.

---

[3] In fact, in one of his early studies on *deqi* effects, Lin tried to explain *deqi* effects with the electronic status of acupoint loci when being needled.

But what Lin was asking for, in my understanding, was a common, "bilingual" framework for acupuncture that could cradle concepts from both biomedicine and Chinese medicine.[4]

As historian of medicine Marta Hanson commented when Youyou Tu (屠呦呦) won a Nobel Prize for her work on anti-malarial herbs (Hanson 2015), medical bilingualism is the key to future medicine: "This medical bilingualism is a quality that current researchers mining the same fine line between the empirical knowledge of traditional medical traditions and the highest level of modern biomedical science would be lucky to share with Nobel Laureate Youyou Tu". And this was one of the main reasons, in addition to his connections to biomedical colleagues and institutions, that Lin could be considered for and elected as an Academician in the category of life science, where biomedicine (both basic and clinical) dominates.

4 Refashioning Acupuncture as a Research Institution

After finishing his residency at Veterans General Hospital, Lin returned to CMC in 1981 as an attending acupuncturist and lecturer and later became the director of its Acupuncture Research Center (針灸研究中心), a position he served from 1985 to 1999. Alongside over one hundred research articles penned and published, it was also during the period of his directorship that Lin gradually formed his working agenda to make acupuncture scientific.

As mentioned previously, Chinese medicine in Taiwan was considered an individual-based healing art, with those who wished to become acquainted with Chinese medicine going through an apprenticeship

[4] Interestingly, Lin's approach is so unique to be considered as "evidence-based" (Shizheng, 實證), which, according to the standards for biomedicine (such as those launched by Cochrane), is too basic and does not refer to any specific disease. For a literature review on evidence-based research on acupuncture in Taiwan, see Committee on Chinese Medicine and Pharmacy (2008). The concept of *Shizheng* and its relationships with experience and experiment is in my larger research project on the transformation of modern acupuncture and deserves more studies of their own.

under a senior physician. Even after the foundation of the CMC, an institution providing a professional education in Chinese medicine, many students spent their extracurricular time working outside of the university, learning directly from well-established acupuncturists. Although in 1973 the Chinese Traditional Medicine Research and Development Foundation (中國醫藥研究發展基金會, CTMRDF), an organization closely associated with the CMC,[5] was founded with a vision to integrate Chinese medicine with biomedicine, the establishment of a center for acupuncture research being one of its early projects, there was still much room for improvement. Compared to biomedicine, textbooks on Chinese medicine seemed to be out of date and its scholarship out of step with clinical practice.

How Lin changed that status quo with the transformation of the Acupuncture Research Center was, therefore, to write, edit, and revise books on modern acupuncture. In fact, as soon as he returned to the CMC Lin joined with other junior fellows to write *Acupuncture Science from the Perspectives of East and West* (中西針灸科學, Chang et al. 1982), contributing three chapters and introducing acupuncture anesthesia along with Cheng-Kuo Chang (張成國), who had been in the same cohort at medical college and a colleague at Veterans General Hospital. When Lin began his directorship, he started compiling papers on acupuncture. In the following years, two books were published in this light—*An Edited Collection on Acupuncture Research* (針灸研究論文專輯, 1985) and *Digest of Acupuncture Medicine* (針灸醫學文摘, 1988).

Setting aside the content and quality issues of the work included, we can perceive three common characteristics in these two books. The first is sponsors. As a part of the CMC, Lin knew to appreciate Lifu Chen and the school's leadership. But without specific financial support from the college for this project, Lin raised funds for publication from physicians of Chinese medicine, notably his classmates and leaders of local Chinese Medical Associations. Also an active practitioner, Lin gave these people credit for their contributions and mentioned their good relationship. A

[5] The main player of this foundation was Lifu Chen (陳立夫), a political figure who believed in the therapeutic merit of Chinese medicine and facilitated the transformation of the China Medical College.

second characteristic is the generational transition in acupuncture scholarship which they evidence. Although Lin was among the first graduates to receive professional acupuncture training in college, he carefully introduced his mentors, some of whom were not inclined toward scholarship, in the papers he collected.

Third, and most importantly, these books provided Lin with a stage to promote his own research on the safe margin of needling depth. His efforts here can also be clearly seen in the new edition of the *Compendium of Acupuncture and Moxibustion* (新針灸大成, 1988), a canonical text of the seventeenth century by Jizhou Yang (楊繼洲). The way Lin and his team treated this book was intriguing. Knowing that it had to be revised to meet the needs of modern acupuncture, but that any such revision would be time-consuming and idiosyncratic, Lin decided not to change a word of the original text. Instead, annotations—as was the age-old tradition for the Chinese classics—and a brief summary and commentary for each paragraph were provided separately. To make them useful for teaching purposes, Lin also added a résumé of each chapter. And what made this new edition new were two additional chapters. One was negative, aiming to correct parts of the text that were too opaque, too obsolete, or had lost their clinical meanings in practice. He also inserted his preliminary study findings on needling depth here. The other chapter was positive, with Lin displaying his vision of modern acupuncture—from locating acupuncture points and the names of meridians through to the settings for clinical practice, including treatment charts, a quick guide to adjunct acupuncture points (*peixue*, 配穴), and a working proposal to incorporate acupuncture into biomedicine.

Lin's effort to refashion acupuncture was rewarding. Still a fresh PhD, Lin already had accumulated years of research experience, which was presented in his *New Perspectives on Acupuncture and Moxibustion* (針灸學新論, 1991a). He was granted a certificate of professorship from the Ministry of Education (教育部部定教授) in August, 1992, the first in the history of acupuncture in Taiwan.

The professorship enabled Lin to be involved in the refashioning of acupuncture in graduate education. As a student and then a teaching fellow who had witnessed the development of graduate studies at the CMC, Lin knew it well through editing the *Digest of Chinese Medicine*

(中醫文摘, 1991b), a collection of 99 abstracts of master's theses from the Graduate Institute of Chinese Medicine there. Appointed as its associate director later in the same year, it became Lin's responsibility to further refashion acupuncture research by substantiating the scholarship it created, elevating its productivity as well as its quality to a level comparable to biomedicine.

Lin's strategy, in hindsight, was institutional, having to do with general trends in the 1990s to improve the professional status of Chinese medicine in Taiwan. The first and foremost was to establish hospitals of Chinese medicine and departments of Chinese medicine in biomedical hospitals. There had been a long history of such attempts. Back in the 1930s, Tsung-Ming Tu (杜聰明), the leading pharmacologist in colonial Taiwan, had proposed a Hospital of Chinese Medicine (漢方醫院). In the post-World War II years, Tu had planned departments of Chinese medicine in the hospitals of National Taiwan University and, later, of Kaohsiung Medical College. Even so, it was not until the foundation of the Chinese Traditional Medicine Research and Development Foundation that these ideas were able to be realized. In the early 1980s, public hospitals started founding departments of Chinese medicine, and the enactment in 1987 of the *Standards for Medical Institutions* (醫療機構設置標準) opened up the possibility of high-quality hospitals of Chinese medicine. It was in that same year that Lin was invited as a visiting staff member to the department of Chinese medicine at Taipei Municipal Ho-ping Hospital, which, with Lin's help when he served as chairman of the Taipei Chinese Medical Association, became Taipei City Hospital of Chinese Medicine in 1997.[6]

Another trend, corresponding with the first, was the reconstruction of graduate education in Chinese medicine. In 1991, National Yang Ming Medical College, which enjoyed close collaboration with Taipei Veterans General Hospital, founded its graduate institute of traditional medicine. Meanwhile, the China Medical College expanded its single PhD program on Chinese medicine to two—one on Chinese pharmaceutical sciences, and the other on acupuncture and clinical sciences. It also established

[6] When the accreditation system for hospitals of Chinese medicine was introduced in 2006, there were 25 such hospitals, plus 72 hospitals that had departments of Chinese medicine.

the Graduate Institute of Integrated Medicine (中西醫結合研究所) in 1999.

Along with the increasing number of Chinese medical physicians who were returning to university for advanced degrees, there was also a demand from the clinical side: from 2002 to 2006, Taiwan's Ministry of Health and Welfare launched its project on the promotion of general strengthening of clinical training on Chinese medicine (建構中醫整體臨床教學體系推動計畫). This was followed by a training program for Chinese medical physicians in medical institutions. Starting in 2009, hospitals that had departments of Chinese medicine had to set up two-year, post-graduate programs in internal medicine, gynecology, pediatrics, acupuncture, orthopedics, and traumatology (傷科), and a Chinese pharmacy with qualified teaching facilities. All of these required more physicians with graduate degrees.

It was alongside the trends outlined above that Lin built up his research enterprise. Among the numerous papers and books on which he worked with colleagues and students,[7] the one that best represents the social side of Lin's achievements was *An English-Chinese Dictionary of Medical and Pharmaceutical Terminology* (中西醫病名對照大辭典, Lin 2001), a five-volume, three-million-word megaproject that took nine years to accomplish. It was shortly after Lin was appointed as a member of the Executive Committee on the Promotion of Collaborative Research of Biomedicine and Chinese Medicine (中西醫合作研究推行委員會) that its chairman, Lifu Chen, assigned him this task. With his CMC colleague Wei Wang, he was supported by the National Research Institute of Chinese Medicine. The 16 physician editors, most of them CMC graduates with double degrees in biomedicine and Chinese medicine, and many of them still PhD students, were invited to join in. In order to be engaged more exactly with the International Classification of Diseases (ICD, the coding system used in Taiwan's national health insurance scheme), a revision project was launched immediately after the book was published (Lin 2004). While it took another four years to finish that second edition, its editorship also shifted to the next generation of

[7] From 1992 to 2019, Lin co-authored 417 research articles and penned or edited 45 books on various topics of Chinese medicine, ranging from history, the collection of scientific works, textbooks, handbooks, and clinical guidelines through to popular readings.

Chinese medical physicians. Among the 25 editors invited, many hailed from the Graduate Institute of Integrated Medicine at the CMC.

All of these efforts nicely echo sociologists Peter Keating and Alberto Cambrosio's analysis of biomedicine (2006), with "biomedical platforms", components of clinical practices, technology, and even administration and architecture interweaving in the making of medical knowledge. Seen in this light, what Lin realized was a platform for Chinese medicine. Implanted in institutions for biomedicine, he nonetheless created an enterprise in which he was able to mobilize components using the terms of Chinese or integrated medicine.

5 Toward an Ecumenical Medicine Traced Back to Needham

Knowing that Lin's career path explains his achievements in clinical research and practical service, some may wonder whether there has been any immediate influence from Needham's work. The answer may be surprising, but it is understandable—no particular connection can be traced back to Needham from Lin's research.

In fact, although the first five volumes of *Science and Civilisation in China* had been translated into traditional Chinese by the early 1970s (Chu 2020), Needham only visited Taiwan once during his tours of Asia, from September 16 to October 2, 1984. To outsource work on materials for Volume 6 of *Science and Civilisation in China*, Needham and Gwei-djen Lu arranged to visit the Veterans General Hospital and the CMC, two institutes where Jaung-Geng Lin trained and worked. Even so, there is no record of Needham and Lin having met, talked, or exchanged ideas on making acupuncture scientific. At the Veterans General Hospital, Needham and Lu were shown around an acupuncture clinic and presented with instruments for what would later be called "energy medicine"—apparatus relating to the then newly developed "Kirlian" photography, and a "*Ch'in*-value detector" (秦值儀). In return, Needham and Lu gave Dr. Chieh Chung (鐘傑), main inventor of the *Ch'in*-value detector, a copy of their book *Celestial Lancets* (Fig. 2).

Fig. 2 Dr. Chieh Chung explained his Ch'in-value detector to Joseph Needham and Gwei-djen Lu in their visit to Taipei Veteran General Hospital on September 18, 1984 (*Ming Sheng Daily* 1984)

In their brief stay in Taichung, Needham and Lu went to the CMC's museum of Chinese medicine, but the focus was mainly on *materia medica*. As he did elsewhere, Needham took notes on the people he met and the issues discussed, though it seems that little from those notes met with a chance to be used outside of the archive room at the Needham Research Institute, for the construction of which the Executive Committee of the Chinese Cultural Renaissance Movement (中華文化復興運動推行委員會), the host of Needham's Taiwan visit, donated US$5,050.

Lin, then still a junior researcher, nonetheless followed Needham's spirit and spent his career making acupuncture an empirically based, clinically reliable specialty in Chinese medicine. Unlike his contemporaries, who worked on the physiological mechanisms of acupuncture (such as the body's ching-lo system) or interpretive models of meridians, Lin, a veteran clinician, had as his main research focus the application of acupuncture, such as the correct therapeutic margins for acupoints and the measurement of acupoint locations according to body metrics, aspects which were not only clinically practical but also easily integrated into the clinical practice of biomedicine. In this sense, it does not matter whether Lin actually learned anything directly from Needham; what is important, in my opinion, is the fact that Lin's achievements present the possibility of an acupuncture that is universal and scientific, and this future medicine has something in common with Needham.

In my *Isis* paper, I questioned "whether there is any medicine that is universal, or what the vision of ecumenical medicine that Needham has opened up for us would look like" (120). For this inquiry I have presented in this chapter a case study from the perspective of science, technology, and society studies (STS), an interdisciplinary field that features holistic approaches to issues between and about science and society. Despite its originality, my analysis of Lin's work cannot cover all his achievements in global scenes, notably those in acupuncture analgesia,[8] nor his international involvements, such as journal editorships and the making of health policies and standards concerning acupuncture. It requires historical understanding of the early post-World War II years, when acupuncture was considered one of the "trading zones"—to borrow the term coined by the historian of science Peter Galison (1997)—to describe the production of scientific knowledge between engineers, physiologists, and biomedical physicians in Europe and East Asia toward making acupuncture a future part of medicine. It is hard here to claim all these efforts as "scientific"; nonetheless, their encounters with acupuncture and what they were pursuing across the dichotomies of Chinese medicine and biomedicine, of basic research and clinical application, as well as medicine and other scientific fields, in this light coincided with Needham's ambitious plan to discover scientific treasures in Chinese civilization, and they deserve serious studies of their own.

By way of conclusion, let me draw attention again to a larger issue on Needham's envisioning of an ecumenical science and how to approximate it. While there is a rich literature on this topic, I would just like to introduce a conversation on the development of science between Needham and Nathan Sivin, an expert on the history of Chinese medicine and later editor of *Science and Civilisation in China, Vol. 6: Biology and Biological Technology, Pt. 6: Medicine.* When asked about how to compare science (in this case alchemy) in different civilizations, Needham responded with the following rather lengthy statement:

[8] Such as his series studies with anesthetists at National Taiwan University Hospital and China Medical University Hospital on the acupuncture assisted anesthesia on patient undergone hysterectomy.

> My collaborators and I have long been accustomed to use the image of the ancient and medieval sciences of all the peoples and cultures as rivers flowing into the ocean of modern science. In the words of the old Chinese saying: "the Rivers pay court to the Sea." In the main this is indubitably right. But there is room for a great deal of difference of opinion on how the process has happened and how it will proceed. One might think of the Chinese and Western traditions travelling substantially the same path towards the science of today, that science against which, on the inductivist view, all ancient systems can be measured. But on the other hand, as Nathan Sivin maintains, they might have followed, and be following, rather separate paths, the true merging of which lies well in the future. …. In all such ways the thought-complex of traditional Chinese science may yet have a much greater part to play in the final state of all science than might be admitted if science today was all that science will ever be. Always we must remember that things are more complex than they seem, and that wisdom was not born with us. To write the history of science we have to take modern science as our yardstick—that is the only thing we can do—but modern science will change, and the end is not yet.[9]

While appreciating the cultural fabrics from which peculiar ways to interpret nature emerge, the ultimate goal for Needham, in the end, was to better science so that it might contribute to a new civilization together with society. The history of Chinese medicine, as well as of biomedicine, is the means to achieve this goal.

Acupuncture is exactly such a case. In *Celestial Lancets*, Lu and Needham indicated that "China is a living civilization, not a dead one, so that all her traditional arts and techniques are at this day in lively development; and acupuncture is by no means an exception" (163). The research effort Jaung-Geng Lin has made on acupuncture should be understood in this light. Wittingly or not, he has demonstrated a form of modern acupuncture in the style of Needham; he has done so by preserving the lively part of this healing art while mobilizing a generation of practitioners of Chinese medicine to move toward science.

[9] *Science and Civilisation in China*, Vol. 5 part 2, pp. xxvii–xxix.

This is also why, after over a quarter of a century since his death, we still need Needham's guidance for the advancement of science. In addition to remembering his authority in making visible the contribution of non-Western civilizations, we should also take serious consideration of him as a renowned scientist, who always gave top priority to the seeking of nature's truth. Let us closely watch Asian medicines—their theories, practices, and how they transform to cope with the modern world—and make Needham new again.

Acknowledgements I would like to thank the Joseph Needham Foundation for Science & Civilisation (李約瑟科技與文明基金會) and Needham Research Institute (李約瑟研究所), the main organizers of the forum "Needham's Dialogical Vision: Understanding Science as a Multi-Civilizational Outcome", which granted me an opportunity to reconsider Needham as a life scientist, a humanist, and civilization synthesizer. I am grateful to Nathan Sivin (1931-2022); his pioneering work on Chinese medicine guided me to consider acupuncture as a living tradition. My gratitude also goes to Professor Jaung-Geng Lin, who has kindly shared his research and offered his insights into acupuncture, and to my colleagues of Chinese medicine who have helped me in preparing this paper, in particular Dr. Fang-Pey Chen, the former Director of the Center for Traditional Medicine at Taipei Veterans General Hospital and its current Director Dr. Yen-Ying Kung. I would like to thank Dr. Po-Hsun Chen at the University of Manchester. Without his assistance on the archival material on Needham's 1984 Taiwan visit, this paper would have taken a very different direction. In addition to the Needham Forum, early versions of this chapter were presented to a workshop on the history of science on November 4, 2019, at Academia Sinica, Taipei, and at the annual meeting of the Society for the Social Studies of Science (August 18–21, 2020, in Prague and online), and the "Materiality of Medical Cultures in/between Europe and East Asia" Conference at Kiel University on June 10, 2023. I thank the participants of these occasions for their comments and suggestions.

References

Chang, Cheng-Kuo, 張成國, 林昭庚, 姜潤次, 王璧瞩, 孔傑, 潘隆森, 陈宏哲. 1982. 中西针灸科学.

Chu, Pingyi. 2020. Needham in Taiwan: Translating *Science and Civilisation in China* as Politics of Modernity and Identity. *East Asian Science, Technology and Society: An International Journal* 14 (2): 379–392.

Dung, H.C., 董厚吉 and Yun-tao Ma 马云涛. 2000. *Scientific Acupuncture for Health Professional* 科學性針刺療法 (in Chinese). 北京:中國醫藥科技出版社.

Galison, Peter. 1997. *Image & Logic: A Material Culture of Microphysics*. Chicago: The University of Chicago Press.

Hanson, Marta. 2015. Is the 2015 Nobel Prize a Turning Point for Traditional Chinese Medicine? A Conversation. Webpage retrieved at https://theconversation.com/is-the-2015-nobel-prize-a-turning-point-for-traditional-chinese-medicine-48643

Keating, Peter, and Alberto Cambrosio. 2006. *Biomedical Platforms Realigning the Normal and the Pathological in Late-Twentieth-Century Medicine*. Cambridge, MA: MIT Press.

Kuo, Wen-Hua. 2019. An Ecumenical Medicine Yet to Come: Reflections on Needham on Medicine. *Isis* 110 (1): 116–121.

Li and Yeh. 李良松, 叶海涛, eds. 2012. 针灸英杰—林昭庚博士. 北京:北京大学医学出版社。

Lin, Jaung-Geng 林昭庚. 1982. 探討人體背部膀胱經穴位直刺安全深度, 台中: 中國醫藥學院中國醫學研究所碩士論文。.

Lin, Jaung-Geng 林昭庚, ed. 1985. 針灸研究論文專輯, 台中: 中國醫藥學院針灸研究中心。.

Lin, Jaung-Geng 林昭庚. 1988. 新針灸大成. 台中: 中國醫藥學院針灸研究中心.

Lin, Jaung-Geng 林昭庚. 1991a. 電腦斷層掃描照相術探討人體胸背部各穴位安全深度之研究, 台中: 中國醫藥學院中國醫學研究所博士論文。.

Lin, Jaung-Geng 林昭庚. 1991b. 針灸學新論. 台中:中國醫藥學院針灸研究中心.

Lin, Jaung-Geng 林昭庚, ed. 2001. 中西醫病名對照大辭典 (An English-Chinese Dictionary of Medical and Pharmaceutical Terminology, in Five Volumes), first edition. Taipei: 衛生福利部國家中醫藥研究所.

Lin, Jaung-Geng 林昭庚, ed. 2004. 中西醫病名對照大辭典 (An English-Chinese Dictionary of Medical and Pharmaceutical terminology, in Five Volumes), second edition. Taipei: 衛生福利部國家中醫藥研究所.

Lin, Jaung-Geng, Yu-Qi Liu, De-Mao Li, Xiao-Neng Chen, and Chia-Ying Yu. 2014. *A Review of the History and Practice of the Needling Depth of Acupoints*. National Research Institute of Chinese Medicine, Ministry of Health and Welfare.

Lu, Gwei-Djen, and Joseph Needham. 1980. *Celestial Lancets: A History and Rationale of Acupuncture and Moxa*. Cambridge: Cambridge University Press.

Ming Shen Daily (民生報). 1984. Why are Old Technologies in China Unable to Recover from a Setback (中國古科技為什麼一蹶不振). September 19, p. 9.

14

Classical Chinese Medicine and the Needham Question

Keekok Lee

1 Introduction

This chapter recognises and acknowledges the distinction between treating the Needham Question as a comparative issue on the one hand and as a dialogical issue on the other.[1] However, it wishes to argue that the two are interconnected. A plausible case could be made that Needham's own understanding of the former may have limitations which undermine that of the latter. Hence a considerable amount of attention must be paid to the comparative issue, not for its own sake but from the necessity demanded by this interpretative standpoint.

I wish to thank Bala Arun and Raymond Lau for their insightful comments and suggestions for improvement. However, flaws still remain and for these I take full responsibility.

[1] For similar attempts, see Cohen (1997, 2010). See also Sivin (1982).

K. Lee (✉)
University of Manchester, Manchester, UK
e-mail: keekok.lee@manchester.ac.uk

A. Bala et al. (eds.), *Multicivilizational Exchanges in the Making of Modern Science*,
https://doi.org/10.1007/978-981-97-3541-9_14

Chinese medicine has been chosen for some detailed exploration as it is an achievement which Needham had not done sufficient attention to from the methodological vantage point, just observed. In modern Western medicine/biomedicine, Needham had also appeared to ignore the domain of epidemiology while focusing on clinical medicine. Through under-exploring these two areas, Needham had overlooked the opportunity for dialogical encounters between epidemiology in biomedicine on the one hand and between modern Western medicine/biomedicine and Chinese medicine on the other. In turn, such oversight (implicitly) led Needham to misrepresent Chinese philosophy-cosmology. Furthermore, it led him, ironically, to go down the road of Coloniality in spite of his overt "scientific oecumenism". According to the concept of Coloniality,[2] the yardstick of "scientific goodness" it implies (albeit subtly and invisibly in Needham's case) is that of the modern Western scientific mode and its methodology, so that the indigenous (in this case, the Chinese scientific mode) would only count as having a cutting edge over the modern Western scientific mode if that cutting edge could be explained within the latter's own terms. Indeed, what could be more ironical, as Coloniality may be said to embody in spirit the very opposite of dialogical encounters. However, this way of assessing Needham is not, of course, intended to deny that Needham was without positive insights. Furthermore, another ironical development has recently occurred in the twenty-first century which confirms his insight about the dynamic nature of scientific thinking, thereby seemingly providing yet another opportunity, through this thought experiment, for a dialogical encounter between biomedicine and Chinese medicine. It remains to be seen whether Needham could take advantage of it and under what terms.

In other words, this chapter, methodologically speaking, adopts an oblique way of addressing the main theme of this volume. It discusses the opportunities for dialogical encounters which Needham had himself missed because of certain flaws or limitations in his understanding of themes under the domain of comparative issues.

---

[2] For one detailed exploration, see Mignolo (2017).

2 Chronology of the Needham Question

Needham first raised it in 1942 in a very cryptic form: "Sci. in general in China – why not develop?" (Chinasage 2017). This was followed by more than a decade of research and investigation culminating in the *Science in Chinese Civilisation* series (1954–present).[3] In 1937, three Chinese biochemists came to Cambridge to work with him (during his first incarnation as a scientist). What he learnt from them about Chinese culture and civilisation fired his intellectual curiosity to such an extent that he learnt classical Chinese so that he could read its texts. Writing under a pseudonym, he said that he, Needham, had an epiphany; it was as if he "received from them some kind of liberation for which he has always been looking … it was clear that talking with his Chinese colleagues about their cultural background, the traditions of Chinese language and literature, he found something equal and opposite to all that in which he himself had been brought up, and something for that very reason of compelling fascination (Henry Holorenshaw and Needham 1973, p. 12)". He enlarged the Needham Question in *The Grand Titration* (1969, pp. 16 and 190).

Could one plausibly argue that the Needham Question, first articulated in 1942, had shaped Needham's reading of the Chinese texts and informed his interpretation of them, even if not always consciously? As such, it is not that the Needham Question emerged after decades of textual immersion; rather it had guided him from the beginning of his journey, which implies that he garnered certain contents from the texts he read which provided him with the evidence he was looking for to confirm his hypothesis.

[3] A relevant, though minor observation may be mentioned here. The series concentrates on Chinese technological achievements and has less to say about Chinese Science, in spite of its title.

3 Needham's Understanding of Chinese Philosophy-Cosmology. Is It Flawed?

Let us look at Needham's understanding of Chinese philosophy-cosmology as this issue is critical to one's understanding of Chinese medicine.

In Needham and Wang (1956, p. 504), he called the key Chinese concept "organic naturalism". In Needham (1969, pp. 20–21), he called it "organic materialism". In Needham and Wang (1956, p. 339), he wrote that modern (Western) science since its emergence in the seventeenth century, had "to assimilate field physics" as well as biology (since Darwin) in which "the philosophy of organism" played a key role. This philosophy of organism "stemmed from the bureaucratic society of ancient and medieval China". He concluded that "Chinese bureaucratism and the organicism which sprang from it may turn out to have been as necessary an element in the formation of the perfected world-view of natural science, as Greek mercantilism and the atomism to which it gave birth". It strikes me as quite an extraordinary account of the history of science from the vantage point of Chinese philosophy-cosmology whose key concept Needham appeared to identify in terms of the philosophy of organism enmeshed in Chinese bureaucratism.

I'm not a sinologist; I stand to being corrected. As far as my own limited understanding goes, Chinese bureaucratism was based on a form of Confucianism dating back to the early Han dynasty. Hanwudi 汉武帝 (156 BCE-87 BCE) finally defeated the Xiongnu (a Northern nomadic tribe which was constantly harassing the Chinese and raiding their territory); Han rule and society were consolidated. The emperor was worried that the dynasty might turn out to be as short-lived as the preceding Qin dynasty. He needed a new ideology to consolidate his newly established empire. One Chinese philosopher and official obliged. He was Dong Zhongshu/董仲舒 (ca. 176-ca.104 BCE).[4] He constructed for the emperor what became known as Han Confucianism. In my opinion, this philosophy differed from what had gone before (for roughly three

[4] For an excellent detailed account by a Chinese scholar, see Wang Liqun (2020) which, however, is available only in Chinese.

and a half centuries since the death of Confucius) in the following crucial way—Dong Zhongshu converted the traditional philosophical-cosmological framework, (implicitly) adhered to by Confucius and his followers which was ontologically dyadic in nature to one which was dualistic in outlook.[5] Dualism entails a hierarchical relationship while dyadism does not, as is shown by the familiar iconic image of the *Yinyangliangyitaiji tu* /阴阳两仪太极图 where the pair of polar contrasts, *yin* and *yang* are not hierarchically ordered. On the contrary, they are regarded as complementing each other to form a harmonious Whole (*he*/和). In Chinese medicine, health is understood in such terms and illness is defined as the lack/loss of equilibrium between *yin* and *yang* in the patient.

On this interpretation, Han Confucianism constructed under Dong's Dualist Thinking is an aberration from the Contextual-dyadic Thinking embedded in the history of Chinese culture which was not challenged by Confucius's teaching prior to Dong. Contextual-dyadic Thinking, as articulated here, comes from *Daojia* 道家/Daoist philosophy. The corpus of early *Daojia*/Daoist philosophical writing, according to Chinese scholarship, includes the *Yijing*/*I Ching*, the *Laozi*/*Daodejing*, the *Zhuangzi*, the *Huainanzi* to name some of the key texts. To grossly oversimplify matters, one could say that the key concept of these texts is about the relationship between *yang qi*/阳气 and *yin qi*/阴气. It follows that the core concept of Chinese philosophy-cosmology is *qi*/气.

What then is *qi*? Is it a kind of matter? If it is, then Needham would have got the materialism part of "organic materialism" right? Unfortunately, the concept is not as simple as Needham made it out to be. For one account of *qi*, see the *Zhuangzi* (attributed to Zhuangzi himself who lived c 350–300 BCE) which contains a chapter, Knowledge Roaming North. This passage either makes clear or implies several important theses, including the following:

[5] Modern Western philosophy teaches that Dualism originated with Descartes, yet Dong Zhongshu had long anticipated the famous Descartes in this regard. For detailed discussion, see Lee (2018, Chapters 6 and 10).

1. *Qi* is omnipresent in the universe and is the ontological category accounting for the myriad things/*wanwu* 万物 (all natural phenomena on Earth, biotic as well as abiotic).
2. As such, it cannot therefore be translated by "breath", "air", "gas", nor even straight-forwardly by "energy", which it often is; nor can it be translated by "matter".
3. The notion of *qi* points to a puzzling feature that it is perhaps both "energy" and "matter" simultaneously—may be it can be called "Emism", as suggested by this author.

From these grossly oversimplified and brief remarks about the concept of *qi*, one could conclude this historical and comparative section by saying that Needham's "organic materialism" is somewhat off the mark as the fundamental notion of the Chinese philosophical-cosmological framework is bounded by *qi*, a concept which embraces both matter and energy and cannot be cashed out simply as matter or materialism. It encapsulates Timespace, and is astronomy/physics-oriented rather than biology-oriented.

4 Needham's "Oecumenical Science": Is It Flawed?

This exploration uses Chinese medicine to test the claim that Needham's account of "Oecumenical Science" is flawed and to show where the flaw might lie. Let us first note that Lu and Needham had co-authored a book in the *Science in Chinese Civilisation* series on acupuncture (and moxa) published in 1980. Acupuncture, indeed, is an essential component of Chinese medicine as evidenced by the fact that its foundational text, the *Huangdi neijing*/《黄帝内经》, consists of two parts, *Suwen*/*Basic Questions* and *Lingshu*/*Spiritual Pivot*, with the latter dealing with acupuncture. It is interesting to observe that in their volume, Lu and Needham showed that they were greatly impressed by the discovery in the 1970s of acupuncture analgesic. Was their reaction prompted solely by the fact that this was a novel, modern application of an ancient technique? Or was there something more that they implied?

In 1967, Needham published "The Roles of Europe and China in the Evolution of Oecumenical Science". In it, he distinguished between what he called a "fusion point" and a "transcurrent point". He made it clear that a fusion point is harder to achieve in the case of Chinese medicine as shown by his Fig. 1. Needham implied that the more complex the organism, the longer it takes for fusion to occur, that is, if it ever does.

Let us scrutinise further his notion of "oecumenism" per se. Needham appeared truly "oecumenical" as he was agnostic as to which system of medicine had the cutting edge. In the case of acupuncture analgesic, it was Chinese medicine. However, on closer scrutiny, that "oecumenism" disappears and is replaced by Coloniality which, to recall, amounts to upholding that the modern Western scientific mode and its methodology constitute "proper"/"authoritative" knowledge, such that any indigenous knowledge claim (in this case, the Chinese) would only count as having a cutting edge over the modern Western scientific mode if that cutting edge could be explained within the latter's own terms. If it could not, then one would be sceptical of the indigenous claim to be a genuine cutting edge or even to dismiss it altogether as under-hand trickery when used by the crude and the unsubtle.

Back to Needham's example of acupuncture analgesic—it turned out that the modern Western scientific mode has an explanation for the phenomenon, which says that the needling leads to the production of opioids by triggering endogenous opioid mechanisms through stimulating high-threshold, small diameter nerves which act upon the spinal cord, brainstem and hypothalamic neurons. In other words, under Coloniality, the colonised system must be legitimated by the science of the colonists. By 1976, the modern Western scientific mode could provide an explanation for acupuncture analgesic in its own terms, not in terms of what acupuncture theory and practice have to say (that is in accordance with the Chinese scientific mode). Suppose, however, in 1976, the modern Western scientific mode did not know about the production of opioids and their mechanisms on the nervous system and the brain, then acupuncture analgesic could not be regarded as "cutting edge" but would have to be written off as fakery at worst, a suspect phenomenon or a puzzling state of affairs at best.

However, regarding the use of acupuncture as general therapeutic practice in Chinese medicine, in his 1967 paper, he was careful in pointing out that the outcome of such a therapy should not be written off as chicanery at worst or the placebo effect at best. He speculated that it could be because acupuncture works upon the immune system. However, as the modern Western scientific mode had not thrown much light on the mechanisms involved in the way it has with acupuncture analgesic, there was nothing more to be said about the subject. There is a question mark in his 1967 Fig. 1; it marks Needham's ambivalence as well as his hope that in the future, the modern Western scientific mode would deliver sufficient knowledge to render fusion possible.

Another way of interpreting that question mark in his 1967 Fig. 1 is possible: to read it as Needham's prescience about the changing character of science which I have already drawn attention to in Needham and Wang (1956, p. 339). Let us do a thought experiment concerning the findings of the modern Western scientific mode which have "become still more 'modern'" by today, that is, in the first two decades of this century. It has to be a thought experiment as Needham had already passed away in 1995. The Human Genome Project/HGP was completed in 2003, with implications for our understanding of the role played not only by genes but also by the environment and the relationship between these variables in shaping the individual's phenotype as well as in disease-generation. The triumphalism of Reductionist genetics is much more muted, today, as a result (see Noble 2006). The HGP together with big-data analysis development have succeeded in spawning new areas of investigations in medical research. For our limited purpose here, only metabolomics[6] would be briefly raised.[7]

Metabolites are small molecules derived from the metabolic process found in readily accessible tissue samples, such as cells or biofluids (blood/plasma/serum/urine/cerebrospinal fluid). They conveniently act as biomarkers as they reflect a response to normal or pathological processes which manifest themselves as functional, physiological,

[6] A whole suite of "omic" studies now exist which gives a collective characterisation and quantification of biological molecules, underlying the structure, function and dynamics of an organism.

[7] See Clish (2015) for a definition and its expected role in future medicine.

biochemical or molecular interactions. As they are part of the metabolic processes in an organism, these processes are dynamic, changing in response to changes in both the internal and external environments of the organism as well as reflecting how the entire organism is working. In this sense, they may be seen as giving a picture of the organism from the Wholist vantage point, rather than from a localised or Reductionist one which genomics might be recruited to support.[8] Metabolites-as-biomarkers, are expected to be found in those with a certain disease, such as Alzheimer dementia and those without. These can act as (a) a warning about the eventual onset of Alzheimer dementia in an individual, (b) as an indicator of how fast the disease may be progressing, and (c) whether a drug given to the patient is helping to retard the progression of the disease. That is why metabolites-as-biomarkers are seen as a powerful tool of individualised/precision medicine.

How would Needham (in our thought experiment) evaluate this "“modern”" development in twenty-first-century science and medicine? The exploration carried out would yield the following salient points:

1. Chinese medicine, since its inception, has upheld the notions of preventive medicine as well as personalised medicine (which biomedicine now calls "precision medicine").
2. As biomedicine had nothing to say about them until after the emergence of the HGP, two decades or so ago, the notions postulated by Chinese medicine were at best ignored or at worst ridiculed as "unscientific". Hence, Needham's ambivalence as shown in his 1967 Fig. 1.
3. Posthumously, one could argue that Needham's prescience that science does change had stood him in good stead.
4. The change arrived with the HGP and the emergence of metabolomics as an innovative domain of medical science and its relationship with biomedicine which can now set forth its notions

[8] Reductionism postulates that diseases can be caused by the existence of a single gene, when in reality single-gene disorders are very rare and, in any case, the expression of any gene is not dependent causally on the gene alone but on a whole host of other factors pertaining to the internal and external environments of the individual person, as in phenylketonuria.

of preventive medicine and personalised/precision medicine. Furthermore, its own theoretical account can cover Chinese medicine's analogous notions without the need to fall back anymore on Chinese medicine's own explanation for these phenomena.

If Needham had been still alive in the first two decades of this century, he could claim another victory for "oecumenism" in medicine. The cutting edge fittingly comes from biomedicine when it pioneered the measurement of metabolites as biomarkers in disease-diagnosis and disease-prognosis.

5. However, there is a price to be paid for this "oecumenism", as it requires biomedicine to validate the "goodness" of Chinese medicine's own claim staked as long ago as the emergence of that medicine in China itself more than two thousand years ago, that is, to by-pass/ignore Chinese medicine and its Chinese scientific mode or to regard it as inferior/sub-standard. This constitutes Coloniality. Is Coloniality an exercise in "scientific oecumenism"? Herein lies irony and paradox, surely.

5 One Missed Dialogical Encounter: Epidemiology

We are still going through with the COVID-19 disease/illness even more than three years after its emergence in 2020.[9] Indeed, in the last two thousand years of European and Chinese history, there have been numerous epidemics and even pandemics. In China, the earliest surviving comprehensive text known to Chinese scholarship on the practice and theory of epidemiology is the late Han text of Zhang Zhongjing 张仲景 (150–219 CE), the *Shanghanzabing lun*/《伤寒杂病论》.[10] In contrast, in the West, epidemiology as well as clinical medicine appeared

[9] WHO (5 May 2023) declared an end to COVID-19 as a public health emergency which, one must remember, does not equate with the claim that the disease no longer exists or that it no longer poses a global threat.

[10] See Lee (2018, Appendix 2).

in the nineteenth century; however, only clinical medicine was considered to be science while epidemiology played a "Cinderella" role to it.[11] Epidemiology, according to one view, began with the work of John Snow (1813–1858) who grappled with the cholera outbreak in London in 1854 when he dismantled the handle of the pump in Broad Street which he traced as the source of the contaminating factor in the cholera epidemic. The next game-changing stage of development may be said to occur in the twentieth century with the work of Austin Bradford Hill (1897–1991) and Richard Doll (1912–2005) which is said to have transformed epidemiological research, putting it on an impeccable scientific footing from the methodological standpoint—see Peto (2004–2005). Yet, the Nobel Prize in Medicine was not bestowed on it. Furthermore, in 2021 and 2022, in spite of being COVID-19 pandemic years, the Prize had, nevertheless, eluded epidemiology, confirming yet again its "Cinderella" status.

This deep-seated bias against epidemiology could perhaps be related to the fact that its model of explanation is not that which underpins clinical medicine. To grossly oversimplify, clinical medicine adheres to thing-ontology and a linear, mono-factorial model of causality—these are hallmarks of the modern Western scientific mode or the Newtonian paradigm of scientificity. It upholds the following theses: (a) the monogenic conception of disease—one disease-entity (the cause), one effect (the disease)[12]; (b) the ability to conduct Randomised Controlled Trials enabling researchers to seek for the one differentiating factor which accounts for why an individual falls ill and/or why an individual having fallen prey to a disease-entity could be cured, feats of scientificity which epidemiology cannot hope to aspire to and emulate.[13] All that epidemiology can offer are "common sense" strategies, such as lockdown,

[11] See Lee (2012, Chapter 12) for a full discussion why it was viewed as inferior to clinical medicine. See also Broadbent (2013)

[12] Every disease (cholera/HIV/candida/schistosomiasis/Creutzfeldt-Jakob) is caused by a single disease-entity, be it bacteria/virus/fungus/parasite/prion. Koch pioneered it via the Koch-Henle's Postulates.

[13] For a critique of this superiority, see Lee (2017b).

self-isolation, physical/social distancing, mask-wearing and so forth.[14] Not only are these strategies not "sexy", they also appear today in certain countries to attract the ire of those who subscribe to the ideology of extreme individualism from which vantage point, these measures are to be resisted as they smack of collectivism/authoritarianism. Epidemiology, alas, has got entangled with the intricacies and intrigues of geopolitics, and become politicised which makes some people overlook the possibility of evaluating these "mundane", non-pharmaceutical measures in terms of their efficacy and their scientific validity.

6 Summary of Arguments

1. Chinese medicine is selected for discussion in this chapter as it may be said to constitute *par excellence* an exemplar of a post-Newtonian paradigm of scientificity in the world history of medical theorising and practice. Perhaps, to call it the post-Newtonian paradigm of scientificity is not altogether apt; if so, it could be called the "Not-Newtonian paradigm of scientificity".

   In other words, this chapter invokes Chinese medicine as a methodological tool to pinpoint two different sets of limitations underlying Needham's reasoning owing to some limitations in his understanding of (a) the Chinese scientific mode and (b) modern Western medicine/biomedicine by his ignoring epidemiology.
2. Needham's neglect of epidemiology had arisen because he, like many others, appeared to have been over-impressed by the achievements of clinical medicine as pioneered by Pasteur and Koch in the nineteenth century. Their monogenic conception of disease permits clinical medicine to elevate the concept of the randomised controlled trials to the highest level of probity in the hierarchy of evidence (as shown in the Triangle of Causation).

[14] Vaccines are regarded as the "magic bullet" but it will be argued by this author (Lee 2023, forthcoming) that vaccines cannot be properly understood except as falling into both clinical medicine and epidemiology.

Grossly oversimplified, clinical medicine may be said to be underpinned by thing-ontology. A microbe, though small and invisible to the human eye is, nevertheless, a thing. It implies a model of causality which is mono-factorial (one cause, one effect) and its causal arrow is linear and unidirectional, like this:

$\rightarrow$

In contrast, epidemiology is not concerned with an individual thing (be it a microbe or a human being) but with the pattern of distribution in a population (of human beings) of a certain pathogen to which it has been exposed and its rate of transmission in such a population. Epidemiology, therefore, necessarily involves numerous factors in its model of causation, such as the age, the health (physical and mental) of the human population, their genetic endowment and so on. Minimally, the causal arrow is bi-directional or reciprocal, thus:

$\leftrightarrow$

In other words, the causal relationship between the various possible factors is more complex than that underpinning clinical medicine as it is multi-factorial, non-linear, bi-directional, synergistic, with feedback mechanisms both negative and positive depending on specific contexts. Ontologically, as epidemiology is concerned with the distribution of events, and as events involve processes, it is primarily, though not solely, focused on process-ontology rather than thing-ontology alone.

3. By ignoring epidemiology, Needham had, thereby, failed to notice that the theoretical underpinning of epidemiology appears to have more in common with Chinese medicine as the latter, too, focuses on process-ontology rather than simply on thing-ontology; its causal model, too, is multi-factorial, non-linear, bi-directional, synergistic with identifiable feedback mechanisms.[15] Given the flaws in his accounts of Chinese philosophy-cosmology and of "oecumenical science", Needham had also failed to notice that Chinese medicine (in its theory as well as practice) constituted a science of medicine, albeit one which does not conform to the Newtonian paradigm of

[15] For detailed exploration, see Lee (2012, Chapter 12; 2018, Chapters 2 and 3).

scientificity which could explain (at least in part) why Needham had overlooked Chinese medicine as a possibility for dialogical encounters, as its paradigm of scientificity is so different from that embedded in clinical medicine in biomedicine.

4. Needham appeared to have overlooked the fact that science (and therefore medicine in so far as it claims to be scientific) is not free-standing, existing in a social-cultural vacuum. Philosophy is acknowledged to be an important, if not the most important constituent of culture. He was right to point out that modern Western science is ultimately embedded in Greek atomism (which is a prime embodiment of thing-ontology). This tradition of thing-ontology was picked up again in the modern era, in the so-called billiard ball image of Newton's three laws of motion. In other words, thing-ontology provided the appropriate philosophical framework for the emergence of modern Western science and medicine/biomedicine, conforming to the Newtonian paradigm of scientificity.
5. The fundamental impulse behind Needham's posing of the Needham Question appears to be a misreading of Chinese philosophy-cosmology, which has nothing to do with biology or with Chinese bureaucratism as it is about the relationship between *yin qi* and *yang qi*, a subject matter which (if one were to use today's language of scientific categorisation) would fall into the academic disciplines of astronomy and physics.
6. One could plausibly argue that Needham had been led to pose the Needham Question because of such failings on his part. Let us use an analogy to explore this conclusion further. Consider philosophy to be the DNA of a cultural tradition. If there are two very different cultural traditions, if one were to dig deep enough, one should find their respective philosophies, their DNAs. The DNA of a cat is not and cannot be the same as that of a dog. It is absurd to complain why the DNA of the former does not give rise to a dog but a cat. Given their fundamentally different "philosophical DNAs", it follows that it would make no sense to pose the question as Needham did: why did the ancient Chinese not articulate modern (Western) science

and medicine, just as it would be equally absurd to raise its counterpart: why did modern Europeans not develop a medicine like Chinese medicine?

It follows that the Needham Question is misposed.

7. If Needham had not overlooked the achievements of Chinese medicine as a science, and if he had not been under-impressed by the achievements of epidemiology in biomedicine, he would have been in a far better position to seek out the opportunities for dialogical encounters between the two traditions, East and West.

   If these areas of exploration had not been overlooked by Needham, one could even argue that his account of "scientific oecumenism" would not, ironically, have been undermined by his (implied) Coloniality.
8. In other words, this chapter has argued that Needham's limitations in his understanding of the domain of comparative issues have undermined his ability to do justice to the domain of dialogical issues. The attention paid to the comparative issue domain in this chapter may appear to be excessive, but it is, nevertheless, essential to do so given the methodological approach it has adopted—it seeks to establish that Needham's lapses in the comparative domain had led to his overlooking opportunities for dialogical encounters between the two philosophical-cosmological traditions which underpin their respective conception of science/medicine.

It may be fitting to conclude by citing an example of a scientist who also lived in the twentieth century, whose life overlapped with that of Needham (1990–1995) and who immediately recognised the relevance of the Chinese paradigm of scientificity during the pioneering research he was conducting in what we now call quantum physics. That scientist is Niels Bohr (1885–1962). He was bestowed the Nobel Prize in physics in 1922 at the age of 37. Bohr invoked the notion of complementarity to account philosophically for the new physics which is really his attempt to capture *Yinyang* philosophy as exemplified in the pair of polar contrasts—*yin* and *yang*—forming a harmonious whole, namely, the *Yinyangliangyitaiji tu* as shown below (Fig. 1).

Fig. 1 *Yinyangliangyitaijitu*/the iconic *yinyang* symbol

This iconic image was chosen by Bohr as a central component of his coat of arms when he was given the Order of the Elephant in 1947 by the King of Denmark. He amplified its message by adding the phrase in Latin: *Contraria sunt complementa* (opposites are complementary). (See Bala 2016) When he died, he also caused his coat of arms to be engraved on his tombstone (Fig. 2).

Bohr, unlike Einstein, straightaway realised that the new physics is here to stay and that it is incompatible with

Fig. 2 Bohr's Coat of Arms

> the standard philosophical framework of atomism and mechanism which underpins so well Newtonian science. The world of sub-atomic phenomena appears not to conform to the boundaries laid down by such a framework, as such phenomena are not simply one moment particle (a thing) and the next wave, but that they are at once both wave and particle. If this, indeed, is quantum Reality, then that Reality appears to be incompatible with the whole tradition of Western philosophy constructed on thing-ontology, from Parmenides, Aristotle, Leucippus, Democritus, Descartes, Newton to the emergence of quantum physics. Not only is quantum Reality incompatible with thing-ontology, it appears to be incompatible also with the logic which thing-ontology entails, that is, it appears to violate the principle of excluded middle, one of three principles of thinking laid down by Aristotle and which have guided and still guide standard thinking. This principle, in brief, forbids one to claim that a state of affairs can be both p and not-p at one and the same time …. (Lee 2017a, Chapter Eight; see also Lee 2017b, Bala 2016)

Bohr said one must look East, to the Daoist philosophical texts, such as the *Laozi/Daodejing* (which he had read in his youth, having been given a copy by his tutor) for a different philosophical-cosmological framework to do justice to the duality, simultaneously wave and particle, he had discovered in the new physics. Unlike Needham, he did not hold the *yinyang* philosophical-cosmological worldview as "suspect" or "problematic" in any way until he "validated" it through his own research. He obviously knew of its existence and implied he regarded it with respect in spite of it being obviously different from the Western Newtonian philosophical-cosmological worldview. In other words, unlike Needham's "scientific oecumenism" which was undermined by Coloniality, in the case of Bohr, in so far as one would wish to argue that he leant on that notion, his "scientific oecumenism" is certainly free from Coloniality.

Note that although Needham and Wang (1956, 339) wrote *en passant* that modern (Western) science since its emergence in the seventeenth century, had "to assimilate field physics" in the twentieth century, he pursued the matter no further and hence missed a significant opportunity of a dialogical encounter with Chinese (Daoist) philosophy, an opportunity which Bohr recognised straightaway and immediately seized upon.

References

Bala, Arun. 2016. *Complementarity Beyond Physics: Niels Bohr's Parallels*. New York: Palgrave Macmillan.

Broadbent, Alex. 2013. *Philosophy of Epidemiology*. Basingstoke: Palgrave Macmillan.

Chinasage. 2017. The Needham Question. Available at: The Needham Question (chinasage.info). Accessed 14 June 2022.

Clish, Clary B. 2015. Metabolomics: An Emerging But Powerful Tool for Precision Medicine. Cold Spring Harbour Mol|Case Stud. October 1 (1): a000588.

Cohen, H. Floris. 1997. *Joseph Needham's Grand Question, and How to Make It Productive for Our Understanding of the Scientific Revolution*. Paper Delivered at the 20th International Congress of the History of Science, in Belgium, 20–26 July 1997. Published in *Science and Technology in East Asia*, Vol. 9: *The Legacy of Joseph Needham*, ed. A. Arrault and C. Jami, 2001.

Cohen, H. Floris. 2010. *How Modern Science Came Into The World: Four Civilizations, One 17th-century Breakthrough*. Amsterdam: Amsterdam University Press.

Lee, Keekok. 2012. *The Philosophical Foundations of Modern Medicine*. Basingstoke: Palgrave Macmillan.

Lee, Keekok. 2017a. *The Philosophical Foundations of Classical Chinese Medicine: Philosophy, Methodology, Science*. Lanham: Lexington Books.

Lee, Keekok. 2017b. Bohr, Quantum Physics and the *Laozi*. *Australasian Philosophical Review* 1 (3): 298–304.

Lee, Keekok. 2018. *Classical Chinese Medicine: Theory, Methodology and Therapy in Its Philosophical Framework*. Newcastle: Cambridge Scholars Publishing.

Lee, Keekok. 2023. *Exploring The Interconnected Complexities of COVID-19*. Newcastle: Cambridge Scholars Publishing.

Lu, Gwei-Djen., and Joseph Needham. 1980. *Celestial Lancets: A History and Rationale of Acupuncture and Moxa*. New York: Cambridge University Press.

Mignolo, Walter. 2017. Part 1 (Activism and Trajectory) and Part 2 (Key Concepts). Interviews conducted by Alvina Hoffman, and editor at E-IR.

Available at: https://www.e-ir.info/2017/01/17/interview-walter-mignolopart-1-activism-and-trajectory/;www.e-ir.info/2017/01/21/interview-walter-mignolopart-2-key-concepts/. Accessed 19 February 2022.

Needham, Joseph. 1967. The Roles of Europe and China in the Evolution of Oecumenical Science. *Journal of Asian History* 1 (1): 3–32.

Needham, Joseph. 1969. *The Grand Titration: Science and Society in East and West*. Toronto: University of Toronto Press.

Needham, Joseph. 1973. Medicine, *Science and Civilisation in China* series, *Volume 6. Biology and Biological Technology, Part VI*, ed. Nathan Sivin. Cambridge: Cambridge University Press.

Needham, Joseph, and Ling Wang. 1956. *History of Scientific Thought* in *Science and Civilisation in China* series, Vol. 2. Cambridge: Cambridge University Press.

Noble, Denis. 2006. *The Music of Life: Biology Beyond Genes*. Oxford: Oxford University Press.

Peto, Richard. 2004–2005. Mortality in Relation to Smoking: 50 Years' Observations on Male British Doctors. *British Medical Journal* 328: 1519.

Sivin, Nathan. 1982. Why the Scientific Revolution Did Not Take Place in China—Or Didn't It?" *Chinese Science* 5: 45–66. (Revised in 2005). Available at: https://www.jstor.org/stable/pdf/43290352.pdf?refreqid=excelsior%3A9b3326382f73318e6987253ddae47230&ab_segments=&origin=&acceptTC=1. Accessed 4 December 2022.

Wang, Liqun 王立群. 2020 (31/03). Lecture 10 in the series Reading the *Shiji* in The Lecture Room/百家讲坛, entitled 董生对策/*Dong sheng duice*. Available at: https://tv.cctv.com/2010/03/31/VIDE1355519749981648.shtml?spm=C95797228340.P3XSH3djsU2B.0.0. Accessed 1 December 2022.

WHO. 2023. WHO Chief Declares an End to COVID-19 as a Global Health Emergency, May 5. https://news.un.org/en/story/2023/05/1136367. Accessed 1 May 2023.

Part V

Modes of Inquiry in Dialogue

15

Mathematics in India: Pluralism and the Possibility of Dialogue

Sundar Sarukkai

In this chapter, I want to analyse the challenges mathematics—as a discipline (or set of sub-disciplines) and as a set of practices—poses for a meaningful dialogue across scientific cultures. Unlike the example of technology, where transmission and circulation of knowledge are linked through some physical processes, the universe of mathematics offers no easy medium for a dialogue when mathematical cultures are so diverse.

The case of mathematics presents a classic test case of the possibility of dialogue of intellectual traditions that are based on different foundations, metaphysics and objectives. Perhaps the most important challenge to the possibility of understanding the 'species' of mathematics from the Indian, Chinese and the African traditions comes from the presumed universalism of mathematics. Contemporary discussions in decoloniality also point to the consistent problem of accepting mathematics (and associated with it, the sciences) with very different foundations and assumptions.[1]

S. Sarukkai (✉)
Bangalore, India
e-mail: ssarukkai@gmail.com

[1] See Boaventura de Sousa Santos (2018).

A. Bala et al. (eds.), *Multicivilizational Exchanges in the Making of Modern Science*,
https://doi.org/10.1007/978-981-97-3541-9_15

This has led to an interest in what is called ethno-mathematics but within the practice of mathematics these approaches are never accepted. Whether they should be or not is not the concern of this chapter. Here, I am using the example of Indian mathematics (or as is sometimes preferred—mathematics as practised in ancient and mediaeval India) to discuss some of these challenges for a dialogue across different models of mathematics and suggest drawing upon the growing field of pluralism in mathematics as a way of negotiating these obstacles to a dialogue.

1 Universality of Mathematics

Arguably, mathematics is the most powerful and best-known example of a universal knowledge system. But what is universal about mathematics? The ontology of mathematics consists of the 'objects' of mathematics, like numbers and sets. Universality of ontology is nothing more than the acceptance of this set of mathematical objects 'universally'. The most important impact of universalism in mathematics is with respect to its epistemology. Mathematical knowledge, such as $2 + 2 = 4$, is seen as universally true in that this statement is true across cultures (and even species!). Mathematical truths are also seen to be atemporal (so do not change with time) and eternal. They are the best exemplars of a universal knowledge system. Mathematical theorems and results from its many sub-disciplines are also similarly seen to be universally true. But what do these claims really mean? Are they 'proved' to be universal and if so, proved by systems that are 'inside' mathematics or 'outside'? Is this 'universality' more a claim of self-consistency of a discourse? If mathematical knowledge is indeed universal, then any invocation of 'cultures of mathematics' runs into problems. Other than the set of mathematical objects, mathematics as a discipline is characterised by operators. Does the universality of mathematics also presume the universality of these operators and operations? To add to this complexity, mathematics is unique in that its writing strategies, including the production of symbols, are different from other discourses. Can we imagine, in principle, a system of language and signs, with different ideas of operations and different sets of 'objects' that could correspond to the dominant mathematics of today?

The universality of mathematics entails a reality to this domain—a reality that has become the metaphysical bedrock of modern science. Claims to the universalism of the natural sciences are to a great extent derived from the presumed universality of mathematics. One of the most powerful expressions of this view is the claim that mathematics is the 'universal language' of science. Moreover, following Galileo's description of nature as an 'open book' written in the 'language of mathematics', the history of modern science, deeply influenced by physics, has become essentially entwined with mathematics. Even in disciplines like biology there is an indebtedness to mathematics even though it is not as straightforward as in physics. But there are open questions on this relation between science and mathematics as far as their respective claims to universality are concerned. In a 'perverse' argument, the argument known as the Quine-Putnam argument reverses this history by arguing that the success of science is the ground for believing in the reality of mathematical entities! Moreover, there is an essential difference between science and mathematics since science is about the empirical world and mathematics is a formal linguistic system. So, in what sense can the empirical be universal? Other philosophical terms that capture the distinction such as the necessary and the contingent, essential and the accidental, only point to the difficulty in associating a naive sense of universality to the sciences based on the universality of mathematics. Mathematics as necessary, essential and science as contingent is one expression of this problem. Finally, if we look at mathematics as a linguistic system for after all it is a system of signs, then the question of translation of mathematics raises different sets of questions about its universality.[2]

[2] For more on this, see Sarukkai (2001).

2 What is Mathematics Really?

Thus, we need to go back to the question of what mathematics is. There is no clear answer to this question even within the domain of mathematics. There are different views about mathematics such as the following: Realist, Naturalist, Structuralist, Formalist, Constructivist, Type of language (where language can be seen as not representational but symbolic, and mathematics understood as essentially having to do with certain types of symbolic manipulations). Each one of these views gives a different meaning for mathematics. A realist, Platonist definition of mathematics describes mathematical objects as being 'real', although a reality that is not defined by spatiotemporal characteristics. Other views move away from ascribing a foundational ontology to mathematical objects but look at it in terms of explicit constructions.

As Hellman and Bell (2006, p. 1) point out, "As soon as one asks questions concerning fundamentals, such as "What is mathematics about?", "What makes mathematical truths true?", "What axioms can we accept as unproblematic?", and notoriously, even "What are the acceptable logical rules by which mathematical proofs can proceed?", we find we have entered a mine-field of contentiousness". For them, plurality in the way we understand mathematics is part of the practice of mathematics and is an empirical fact. They argue that a "plurality or multiplicity of approaches to central questions of truth and proof is simply an observable fact".

Thus, it is reasonable to look at multiple interpretations of mathematics as is being done now by a growing body of work on pluralism of mathematics. One immediate connection to cultures is that cultures are often seen to be fundamentally pluralistic in character. In the case of mathematics, there are different types of pluralism. These different types can be classified as foundational, methodological, epistemological and alethic (Friend 2014). Friend (2014, p. 24) argues that there is "insufficient evidence to think that there is a unique foundation for mathematics…several, together inconsistent, foundations for mathematics" leading to pluralism of foundations. The pluralist "takes seriously the possibility that there are several, together inconsistent, foundations for mathematics".

Pluralism in Perspectives is the view that there is no best, absolute perspective of mathematics. Friend argues that "Each mathematical theory, which is powerful enough to give a perspective on quite a lot of mathematics, will have its philosophical sins and virtues. There is insufficient evidence to think that there is an absolute perspective that is best either philosophically or mathematically" (ibid., p. 25). Pluralism in Methodology is the argument that there are many methodologies across the different branches of mathematics. The open nature of searching for methods for mathematics means that methods from other disciplines can also become useful for mathematics. Given that the notion of proof has often been used as an important rhetorical strategy for the universalism of mathematics, it is useful to note that there is also Pluralism of Proof such as the analytical, computational, diagrammatic, and so on, leading to a view that is "tolerant towards proof techniques". Similarly, one also finds a strain of pluralism when mathematics is evaluated through the category of success. Pluralism is an acknowledgement of the fact that there are different ways by which one can measure success in mathematics. As Friend points out, "The pluralist not only distinguishes himself from other positions in the philosophy of mathematics, he is inspired by other positions. In particular, the pluralist retains lessons from the realist, the naturalist, the structuralist, the formalist and the constructivist" (ibid., p. 2).

Some of these observations have emanated within the field of mathematics earlier too. Hardy's description of proof in mathematics as 'rhetorical devices', Hilbert's approach to the foundation of mathematics, and the intuitionist and constructivist approach to mathematics all have the seeds of pluralism. Formalism is arguably how most mathematicians view mathematics and Friend suggests that the approach of formalism is close to pluralism. She accepts five characteristics of formalism from Detlefsen—invert priority of geometry over arithmetic; rejects classical mathematical notions of proof—so 'proof as a purely formal exercise' and based on non-interpretative role of language; mathematics primarily as syntactic; and finally the position that freedom of interpretation is what gives "genuine knowledge of mathematics" (ibid., p. 82). The real question is what the scope of the 'freedom of interpretation' could be and how this can be deployed in the analysis of other types of mathematics.

One can thus view pluralism in mathematics as the acceptance of different theories on mathematics, on a multiplicity of views about its various practices as well as its manifestations. Davies (2005) makes this point strongly when he observes that "…the view that classical mathematics, constructive mathematics, computer assisted mathematics and various forms of finitistic mathematics can co-exist. I revive Carnap's dictum that one must decide the framework of discourse before questions about existence and truth make sense; see Carnap [1950]. In different frameworks the answer to a question may be different, but this in no way implies that one or the other is 'right'. This position is strongly anti-Platonistic".

One of the pluralistic ways of understanding mathematics is useful for finding common ground with Indian and Chinese mathematics. Naturalism is the doctrine that finds an essential relation between mathematics and the sciences. Drawing on Maddy, who "treats mathematics as a science in its own right", Friend argues that naturalism leads to pluralism. She also notes that Quine's well-known argument about inferring an ontological commitment to mathematical objects based on the success of science does not cover all of mathematics but only that part of mathematics that is indispensable to science. This distinction between the domain of mathematics as such and its limited use in the sciences (and this limited use in a large measure validates the universality of the whole domain of mathematics!) remains a genuine problem in any of the theories about mathematics. For example, Sarukkai (2003) adds to this argument by showing how the 'same' mathematics can model physical realities that are completely contrary to our own and thus mathematics seems to be quite "indifferent" to the demands of our physical world. This is part of the puzzle that becomes famously known as the 'unreasonable effectiveness of mathematics'. This point is particularly important in a discussion of Indian mathematics because all mathematics in the Indian (and Chinese) case is 'applied'. For example, Mumford points out that negative numbers were very easily accepted as part of mathematics in India and China since they were essential for accounting practices, whereas in Europe it took much longer to accept them given their metaphysical association with negative ontology (Mumford 2010). That is why the approach by Maddy and extension of it by Friend is relevant for understanding what is 'mathematics' in Indian mathematics.

In this context, it is useful to consider the role of 'applied mathematics' and what it says about the nature of mathematics itself. First of all, what does it mean to apply mathematics? Is it applied to the world directly or to descriptions of the world?[3] It is reasonable to believe that applying mathematics is not of one 'kind'. There are different cultures and modes of applying it. It is relevant to consider what Frege himself thought about the applicability of mathematics. As he notes, "[I]t is applicability alone which elevates arithmetic from a game to the rank of science". Maddy's approach of viewing mathematics as a science or the Asian cultures of mathematics which grounds it in human cognition and interactions are plural manifestations of what it means to apply mathematics. I would argue that the applicability of mathematics illustrates its different foundational structures. Moreover, it may seem as if all this concern about pluralism comes only from foundations of mathematics and not from its daily practice. However, pluralism in daily practice is best seen in applied mathematics as well as in mathematical theories which are not seen as 'successful'.

We can evaluate pluralism in each one of these features of mathematics. For example, even the larger argument that mathematics is the 'language of nature' can be shown to have multiple manifestations. As I have argued elsewhere (Sarukkai 2003), mathematics cannot be 'one' language but only a family of languages. The languages of topology, calculus, arithmetic, set theory, etc., are not the same but share some commonality, although it needs work to see what the types of similarities could be. Once we look at these languages in terms of their difference, varieties of pluralism can be identified. Moreover, language is related to human practice, both cognitively and socially. Thus, explicit association of mathematics with language brings in the role of the human body, thereby questioning the assumed universality of this discourse. Recent work on embodied mathematics draws further attention on the locality of mathematical discourse. Even analytical approaches to understanding mathematics as rule-following, most famously articulated by

[3] See Sarukkai (2005); see also Steiner (1998).

Wittgenstein, also suffer from a reduction of the Platonist assumption to embodied cognitive acts. All these are good reasons to more critically examine the assumed universality of mathematics.

The entrenched power of universality of mathematics constitutes a problem for a dialogical vision. Given this, my argument is that we can use approaches such as pluralism in mathematics to argue for a more inclusive understanding of mathematics across cultures.

3 Indian Mathematics and Pluralism

I will begin with the term 'Indian mathematics', a term that has provoked some criticisms since it seems to point to a different type of mathematics. For those who subscribe to a universal understanding of mathematics, such a term is a problem and they attempt to overcome it by using the phrase 'mathematics in India'. Both these usages are common, and in what follows, I will use 'Indian mathematics' to invoke the possible plural dimensions of mathematics.

Firstly, is Indian mathematics a species of mathematics? One could argue that ancient Indian mathematics dealt with the subject matter of modern mathematics such as numbers, geometry, operations related to addition, convergence and algebra. Numbers were obviously ubiquitous in ancient systems but the way numbers were understood differed as we will see below. However, many unique differences characterise Indian mathematics. It has been argued that the structure of this mathematics was influenced by an algorithmic approach and not an axiomatic one such as that in Euclid's geometry. There are various implications of an algorithmic approach towards understanding the metaphysics of numbers and the operations on them. Since the metaphysics of Indian mathematics was not influenced by theology as in the Greek tradition, it is difficult to find a theological metaphysics of mathematical entities in this tradition. In contrast, Platonism functions as a kind of theological doctrine, postulating mathematical entities as existing in a non-spatiotemporal manner, reminiscent of a transcendent being. The algorithmic approach makes mathematics an essential tool for acting and doing-in-the-world and not as an extra-wordly entity that magically makes sense of the world. For example, the metaphysical problems surrounding the ideas of irrational numbers and negative numbers that plagued western mathematics do not arise in the Indian (or Chinese) context at all, and these numbers were used commonly in mathematical practice in these cultures. As Mumford points out, "In every culture, one of the main reasons for the development of arithmetic... is the need of merchants to keep accounts" and the practice of accounting in these cultures led them to accept negative numbers without worrying about their metaphysical foundations (Mumford 2010). Negative numbers were accepted in Europe only around 1685 by John Wallis and then Newton. Until the seventeenth century, Europe resisted negative numbers. Mumford suggests that perhaps it was due to Euclid's influence as well as the presence of imaginary numbers.

In the first millennium, astronomy dictated the course of Indian mathematics. Mathematics had a different relationship with language and the techniques of writing. We could say that mathematics in these cultures arose through an engagement with the world through the process of counting, measuring, describing, and so on. Thus, abstraction of a particular kind that characterises modern mathematics is not the only point of origin of this discourse. An important implication of this approach is that the so-called applicability problem is not really a problem for this view of mathematics. The earliest extant text, the *Sulvasutras,* describes many results in arithmetic and geometry but with a different approach to deriving these results (Dani 2010). How does one interpret these different expressions and modes? So also about the rich work on combinatorics, again expressed in completely different ways from a study of "Sanskrit prosody" and "musical patterns", which led to the discovery of Fibonacci sequences (much before Fibonacci) as well as the general study of recursion relations (Sridharan et al. 2010). One can work out the 'mathematics' in these texts by converting them to modern symbolic notation as is done in algorithms to solve algebraic equations such as the Pell's equations (Dutta 2010). So also the pioneering work in calculus, including series summation, infinite series, calculation of instantaneous velocity, the series expansion for π and many other results often associated with European mathematics (Ramasubramanian and M. D. Srinivas

2010; Divakaran 2010). The way in which these results were derived and the prose nature of the text might make this seem like a discourse different from mathematics. But the depth of all this work must make us ask whether these approaches offer another pluralistic definition of mathematics instead of arguing whether these cultures possessed mathematics or not.

There are other important differences that need to be understood more deeply before one can claim universal status for the mathematics that we know of now. The relation between language and mathematics is a profound one. Historically, mathematics has always been distanced from natural language. Much of the rhetoric about mathematics is about how mathematics is 'precise' unlike natural language. The dissociation from natural language also helps to dissociate mathematics from human action since languages are understood to be a social process produced by a community of speakers. Even the writing of mathematics is a way to distance itself from verbal languages. Mathematics as a symbolic system also moves away from convention-based ideas of meaning (meaning as associated by convention with words) towards more necessity-based ideas of meaning.[4] The rejection of meaning itself as a category within mathematics is another way to project its special praxis. But in spite of all these moves, is it possible to dissociate mathematics from natural languages? Symbols are not necessarily different from words and the way meaning functions in mathematics suggests the possibility that the relation between mathematics and natural language is much closer than is thought. Symbols in mathematics are not just rules for manipulating them but also rules for creating them (Sarukkai 2002). Through these writing strategies mathematicians discover new patterns, new properties and new operations. In the Indian system, there are different types of linguistic representation of mathematical entities. In the system of *Bhutasamkhya*, numerals are represented as words and in the Katapayadi system they are represented by consonants of Sanskrit. For example, the number 2010 in Katapayadi is written as *natanara* (na = 0, ra = 2, ta = 1) and when written in this way it also has a particular meaning

[4] In this, it shares an important feature with Indian logic. For more on this, see Sarukkai (2005b).

which in this case is 'actor-man' (Sarma 2012). The implication of this form of writing is that one can do mathematical calculations as verbal constructions which may be more helpful to remember, reproduce etc. As another idiosyncratic example, consider one of the most important Indian mathematicians, Bhaskara of the twelfth century. In his monumental work, *Bijaganita*, there is a discussion of various aspects of algebra written in verse form. In the section on 'six operations with unknowns', the "unknown quantities are represented by the names of colours, which are manipulated arithmetically in accordance with the following rules" (Plofker 2009, p. 193). The verse form and its associated vocabulary are very different from modern mathematics, and while one can make correspondences between them, it is also useful to look at the implication of the original formulations to discover other foundations for mathematics. That is possible only if we begin with the pluralistic foundations of mathematics.

The use of other terms for symbols is a particular critique of the idea of symbols and what they stand for. To make sense of this argument, we need to look at what distinguishes Indian logic in its use of symbols. We should remember that symbols play a dual role in modern mathematics—they seem like meaningless, abstract terms but constantly play on meanings inherent through the forms of writing, more like pictorial forms (Sarukkai 2005, 2005a). Moreover, today's notations that are used in mathematics are largely from the seventeenth century onwards. More systematic use only began in the nineteenth century such as the picturisation of unit vectors, vectors with the symbol of arrow on top, symbols for the factorial, and so on. (It is instructive to remember that even the ubiquitous = sign was only from 1557, the symbol of infinity in 1655, and even the ever-present π was accepted in its present form only in 1706). Symbols play a very important role in mathematics for they are not just passive representations of mathematical entities and operators. Techniques of symbolisation produce new mathematics. We know well the importance of tensor notation for the development of tensor calculus and for the general theory of relativity. Seen from this perspective, Indian mathematics with its multiple formulations of representing mathematics can be seen to produce not just alternate formulations but also mathematical knowledge that is dependent on these representations.

In this context, it is worth noting that the deep influence mathematics had on western philosophy is completely absent in the rich Indian philosophical schools. This is an ironic fact: although Indian mathematics was well advanced in its time as were the many philosophical schools, the latter did not think it necessary to privilege mathematics or the knowledge that it produced unlike the Greek/European tradition. The only way we can make sense of these differences in Asian traditions of mathematics (as well as African and other traditions) is to build upon the formulations that led to the growth of these mathematical traditions in the first place. While there is a tendency to rewrite these ideas of mathematics in ways similar to modern mathematics, there is also something that is lost in this rewriting. Why privilege the rewriting to show similarities while ignoring the implications of the foundational ideas that led to those mathematical developments in the first place? In this project of showing equivalences, the notion that mathematics is universal has played an important role. If we now begin with the pluralistic character of mathematics in its foundations, practice, epistemology, etc., is it possible to reinterpret many of the other mathematical traditions as independent and different forms of doing mathematics? In any case, the possibility of a dialogue across these diverse traditions can either be accomplished by reducing one system to the language of the other or to retain plural perspectives and see them without such reduction. I believe that the latter approach can be very fruitful for this dialogue.

References

Carnap, R. 1950. Empiricism, Semantics, and Ontology. *Reviews of International Philosophy* 4: 20–40.

Dani, S.G. 2010. Geometry in the Sulvasutras. In *Studies in the History of Indian Mathematics*, ed. C.S. Seshadri. Delhi: Hindustan Book Agency.

Davies, E.B. 2005. Pluralism in Mathematics. *Philosophical Transactions: Mathematical, Physical and Engineering Sciences* 363 (1835): 2449–2460.

Divakaran, P.P. 2010. Notes on Yuktibhasa: Recursive Methods in Indian Mathematics. In *Studies in the History of Indian Mathematics*, ed. C.S. Seshadri. Delhi: Hindustan Book Agency.

Dutta, Amartya K. 2010. Kuttaka, Bhavana and Cakravala. In *Studies in the History of Indian Mathematics*, ed. C.S. Seshadri. Delhi: Hindustan Book Agency.

Friend, Michele. 2014. *Pluralism in Mathematics: A New Position in Philosophy of Mathematics*. Dordrecht: Springer.

Hellman, Geoffrey, and John L. Bell. 2006. *Pluralism and the Foundations of Mathematics*. Minneapolis: University of Minnesota Press.

Mumford, David. 2010. What's so Baffling about Negative Numbers? A Cross-Cultural Comparison. In *Studies in the History of Indian Mathematics*, ed. C.S. Seshadri. Delhi: Hindustan Book Agency.

Plofker, Kim. 2009. *Mathematics in India*. Princeton: Princeton University Press.

Ramasubramanian. K., and M. D. Srinivas. 2010. The Development of Calculus in India. In *Studies in the History of Indian Mathematics*, ed. C.S. Seshadri. Delhi: Hindustan Book Agency.

Santos, Boaventura, and de Sousa. 2018. *The End of the Cognitive Empire*. Durham: Duke University Press.

Sarma, S.R. 2012. The Katapayadi System of Numerical Notation. *Revue D'histoire Des Mathématiques* 18: 37–66.

Sarukkai, Sundar. 2001. *Translating the World: Science and Language*. Lanham: University Press of America.

Sarukkai, Sundar. 2003. Applying Mathematics: The Paradoxical Relation between Mathematics, Language and Reality. *Economic and Political Weekly* 38 (35): 3662–3670.

Sarukkai, Sundar. 2005. Revisiting the 'Unreasonable Effectiveness' of Mathematics. *Current Science* 88 (3): 415–423.

Sarukkai, Sundar. 2005a. The Use of Symbols in Mathematics and Logic. In *Essays on the Foundation of Mathematics and Logic*, vol. 2, ed. Giandomenico Sica. Italy: Polimetrica Publishers.

Sarukkai, Sundar. 2005b. *Indian Philosophy and Philosophy of Science*. Delhi: Motilal Banarsidass.

Sridharan, Raja et al. 2010. Combinatorial Methods in Indian Music. In *Studies in the History of Indian Mathematics*, ed. C.S. Seshadri. Delhi: Hindustan Book Agency.

Steiner, Mark. 1998. *The Applicability of Mathematics as a Philosophical Problem*. Cambridge: Harvard University Press.

16

Explaining the Rise of Modern Science: A Dialogical Perspective

Arun Bala

1 Three Modes of Nature-knowledge

In his recent study *How Modern Science Came into the World: Four Civilizations, One 17th-Century Breakthrough*[1] the historian Floris Cohen undertakes to account for the rise of modern science as the emergence and fusion of three distinct approaches to natural phenomena he labels

I wish to acknowledge my deepest gratitude to Bernard Lightman and Raymond W.K. Lau for their invaluable comments and suggestions which helped to enrich and tighten the arguments made here.

[1] An abbreviated version of this study was later published as *The Rise of Modern Science Explained: A Comparative History*, Cambridge University Press, 2015. This title makes more explicit Floris Cohen's comparative approach to the history of the origins of modern science.

A. Bala (✉)
Institute for the History and Philosophy of Science and Technology,
University of Toronto, Toronto, Canada
e-mail: arun.bala@yahoo.com

A. Bala et al. (eds.), *Multicivilizational Exchanges in the Making of Modern Science*,
https://doi.org/10.1007/978-981-97-3541-9_16

as modes of nature-knowledge[2]—mathematical realism, kinetic corpuscularism,[3] and fact-finding experimentalism. Cohen claims that two of these three modes of nature-knowledge—kinetic corpuscularism and mathematical realism—are rooted in different traditions of ancient Greek science. Kinetic corpuscularism emerged through a radical transformation of ancient atomism of Epicurus popularized in Athens, that he refers to as Athens-plus. By contrast mathematical realism is seen by him as a transformation of the abstract mathematical orientation to disciplines such as optics, astronomy, statics, and hydrostatics mainly associated with Greek science in Alexandria. He therefore refers to it as Alexandria-plus. He takes the third tradition of fact-finding experimentalism to have grown out of another transformation of a distinctive orientation to nature emphasizing both accurate description and practical orientation that emerged *sui generis* in late Renaissance Europe.

Cohen links these new modes of nature-knowledge to pioneering figures of the Scientific Revolution—kinetic corpuscularism to René Descartes and Isaac Beeckman, mathematical realism to Galileo Galilei and Johannes Kepler, and fact-finding experimentalism to Francis Bacon, William Gilbert, and William Harvey. In this chapter I will examine only Descartes, Galileo, and Bacon as separately exemplifying these different modes of nature-knowledge. Nevertheless, my arguments could, with slight modifications, be extended to the rest.

Let us begin with Descartes and his new concept of atomism that went beyond the ancient Greek tradition. The ancient atomists held the world to be composed of empty space and atoms that only moved in straight lines unless they collided with other atoms, although Epicurus had allowed them to make unpredictable swerves in order to allow for

---

[2] Cohen uses 'modes of nature-knowledge' to refer to 'consistent ranges of distinct approaches to natural phenomena, which may differ in several dimensions. Their scope may have been comprehensive, with a view to deriving the whole wide world from first principles, or deliberately partial. The way in which knowledge was attained may have been predominantly empiricist or chiefly intellectualist. If any practices went with a given mode of nature-knowledge, these may have been observational, experimental, instrumental, etc. Knowledge may have been sought for its own sake or with a view to achieving certain practical improvements' (Cohen 2010, p. xxviii).

[3] Cohen (2010) deploys the term 'corpuscularianism' but in this chapter we have elected to use 'corpuscularism', since this is the more familiar term referring to the corpuscular philosophy adopted by many thinkers in the seventeenth century.

free will in humans. (Cohen 2010, p. 7 & p. 223). Descartes radically transformed this classical conception by introducing the notion that God had originally imparted to all material corpuscles a fixed quantity of motion that they could pass to one another as they collided through contact interactions, but without change in the total amount of motion. This allowed Descartes to formulate conservation laws that could be used to predict the later state of a system of interacting corpuscles from its earlier state, and to treat the motion of large composite bodies in terms of the motions of their constituent particles. It led to a radical change in the way natural phenomena came to be explained. Classical atomists had focused on the geometric shape, type and hardness of atoms to explain their behavior. By contrast Cartesian kinetic corpuscularism explained the behavior of particles in terms of their *dynamic* motions and interactions. Cohen sees this shift from the structure of the particles to their motions to explain the behavior of material bodies as revolutionary. It made it possible for Descartes to include a mathematical approach to atomism that ancient Greek atomism had not succeeded in accomplishing (Cohen 2010, p. 233). Nevertheless, he also considers it to be a latent developmental potential in ancient atomism which remained unrecognized until modern times. Hence, he labels the change from Athenian structural atomism to Cartesian kinetic corpuscularism as a shift from 'Athens to Athens-plus'.

Cohen also takes the second mode of nature-knowledge—realist mathematics—to develop from another latent potential inherent in ancient Greek thought. Claudius Ptolemy (100–170 CE) from Alexandria had developed powerful mathematical models to deal with both optical phenomena and predict planetary motions. But in order to achieve mathematical precision his optics treated light rays as emanating in a cone out of the eyes, and his astronomy invoked equants which were abstract points in space not centered on physical bodies around which planets moved in circles. Ptolemy recognized there were concerns about whether his assumptions represented actual processes in the world, but dismissed such realist concerns. He saw his task as 'saving the phenomena' where mathematics was merely an abstract tool for computational purposes that could predict observed events. He did not require that the rays from the eyes and the equants in the heavens possess any concrete physical representation in the world.

The revolutionary change that Galileo and Kepler made was to develop a theory of planetary motions and optics where mathematical relations applied to real entities in the world—a shift from ancient abstract mathematics to modern mathematical realism. Cohen maintains that although this change existed as a latent possibility in the ancient Greek tradition, it only came to be realized in modern Europe. He labels this transformation from the abstract mathematical Alexandrian mode of nature-knowledge to realist mathematics as 'Alexandria-plus' (Cohen 2010, p. 160).

Cohen sees the third mode of nature-knowledge as a transformation of something that first emerged in Renaissance Europe with artist-artisans such as Leonardo da Vinci (1452–1519). It was associated with a penchant for knowledge that involved both accurate description and served practical action. This can be illustrated by Leonardo's painstakingly precise drawings of the wings of birds in flight and his efforts to use that knowledge to build flying machines. A century later this mode of nature-knowledge came to be developed further by thinkers such as Bacon, Gilbert, and Harvey into a mode of fact-finding experimentalism designed to discover new phenomena never before found in nature—what we now see as the experimental method of modern science.

In this chapter I argue that Cohen's three modes of nature-knowledge are better explained not as *latent* developments of two modes of nature-knowledge from ancient Greece and one born in Renaissance Europe, but as outcomes of encounters with influences which entered Europe from the outside—the powerful computational techniques carried by the Indian number system combining with the atomic theory that made possible Cartesian kinetic corpuscularism, the realist mathematical optics of Ibn al-Haytham that inspired mathematical realism in optics and astronomy in early modern Europe, and the numerous technologies that flowed from China through the corridors of communication established by the Mongol Empire which inspired fact-finding experimentalism.[4] Although the impact of Indian numbers, Arabic optics and Chinese technologies on Renaissance Europe is widely recognized their profound implications become far clearer after we adopt Cohen's conception of three modes of nature-knowledge that combined to produce modern science—albeit by going beyond the comparative context within which he confines them and adopting a dialogical perspective that traces their roots beyond Europe.

2 Indian Number System and Kinetic Corpuscularism

In order to appreciate the impact of Indian mathematics on Cartesian kinetic corpuscularism I will begin with the observation by the historian Alexandre Koyré that modern atomism cannot be considered simply a revival of the ancient Greek theory:

> The atomism of the ancients, at least in the aspect presented to us by Epicurus and Lucretius – it may be that it was different with Democritus, but we know very little about Democritus – was not a scientific theory, and though some of its precept, as for instance, that which enjoins us to explain the celestial phenomena on the pattern of terrestrial ones, seem to lead to the unification of the world achieved by modern science, it has never been able to yield a foundation for development of a physics; not even in modern times: indeed, its revival by Gassendi remained perfectly sterile. The explanation of this sterility lies, in my opinion, in the extreme sensualism of the Epicurean tradition; it is only when this sensualism was rejected by the founders of modern science and replaced by a mathematical approach to nature that atomism – in the works of Galileo, R. Boyle, Newton, etc. - became a scientifically valid conception, and Lucretius and

[4] Here we need to distinguish fact-finding experimentalism from fact-finding empiricism. The latter is a turning toward actual observation and away from ancient Greek rationalism of Plato and the qualified rationalism of Aristotle in approaching natural phenomena. The discoveries made of new geographies, new peoples, new plants, animals, and food crops through the voyages of Iberian discovery had encouraged a new turn toward greater trust in the empiricism of the senses. For more on this see Barrera-Osorio (2010) and Canizares-Esguera (2006).

> Epicurus appeared as forerunners of modern science. (Koyré 1957, p. 278 note 7).

The distinctive achievement of Descartes, rightfully recognized by Cohen, is that by moving away from the sensualism—as Koyré puts it—of ancient atomism with its appeal to the structural properties of atoms to explain phenomena, and shifting to the kinetic properties of corpuscles, we can offer a mathematical account of atomic behavior and the phenomena associated with them. However, what Cohen has failed to recognize is that the shift also involved a change in the nature of the mathematics deployed—the mathematics of Descartes was not simply that of geometry (closely linked to the sensual properties of atoms, such as shape, smoothness, and angularity in the ancient Greek tradition Koyré refers to), but algebra which had only recently entered Europe. In his *Discourse on the Method of Rightly Conducting Reason and Seeking Truth in the Sciences* Descartes contrasts the method of geometry and that of algebra and why he hopes to bring them together:

> As to the geometrical analysis of the ancients and the algebra of the moderns, they cover only highly abstract matters and seem not to be useful for anything; besides which ancient geometrical analysis is always so closely tied to the study of figures [*i.e. geometric shapes*] that it can't exercise the intellect without greatly tiring the imagination; and modern algebra is so confined to certain rules and symbols that what emerges from this is not a science that develops the mind but rather a confused and obscure art [*i.e. set of techniques*] that entangles it. (Descartes 1637)

He then proceeds to show how the situation can be remedied by borrowing the best in geometrical analysis and algebra and combining them so that the defects of one become corrected by the other. His great insight was that both deal with proportions between objects and that it was best to study such proportions in general by expressing geometrical relations in algebraic symbols, and algebra relations through geometric figures. This was, of course, the Cartesian discovery of co-ordinate geometry in which geometrical straight lines, circles, ellipses, and other figures could be represented by algebraic equations such as $y = ax + b$ for a straight line, and $x^2 + y^2 = r^2$ for circles and so on.

The importance of the Cartesian algebraic approach for kinetic corpuscularism can be appreciated when we contrast it with the orientation of Galileo whose mathematics only adopted the ancient Greek geometric approach. Galileo writes:

> Philosophy is written in this grand book, the universe, which stands continually open to our gaze. But the book cannot be understood unless one first learns to comprehend the language and read the letters in which it is composed. It is written in the language of mathematics, and its characters are triangles, circles, and other geometric figures without which it is humanly impossible to understand a single word of it; without these, one wanders in a dark labyrinth. (Galileo 1623, pp. 237–238)

Although Galileo speaks of the book of nature as written in the language of mathematics it is really only the language of geometry. But, as noted by Ladislav Kvasz in his essay *Mathematical Language as a Bridge Between Conceptualization of Motion and Experimental Practice*, the language of geometry is too concrete to be the language of modern science. Modern science is not based on triangles, circles, and other geometric figures but upon functions and differential equations since the concrete language of synthetic geometry cannot make place for either universal laws or interactions between bodies. This new mathematical approach is traced by Kvasz to Descartes—it was his interest in interactions between bodies that pioneered the search for universal laws that regulated such interactions. It leads Kvasz to argue that in contrast to the approach of Galileo, and in resonance with most of modern science, Cartesian physics is algebraic. He concludes:

> Cartesian physics takes from algebra the universality of its laws. Even though these laws are formulated in ordinary language, they have the same kind of universality as algebraic formulas. The language of algebra enabled Descartes to create the first description of interaction. He did it by introducing the notion of the quantity of motion, and formulating the law of conservation of the quantity of motion. It is important to realize that the quantity of motion is an algebraic quantity and the law of its conservation is an algebraic equation. (Kvasz 2015, p. 236)

However, this 'algebra of the moderns', as Descartes refers to it in contrast to ancient Greek mathematics of geometry, did not emerge de novo in modern Europe. It came through the Arabs, especially through the work of Muḥammad ibn Mūsā al-Khwārizmī (780–850), a mathematician of the Abbasid Caliphate, and scholar in the House of Wisdom in Baghdad. His writings on the Indian numerals entitled *On the Calculation with Hindu Numerals* introduced the decimal positional number system to the Western world, and his *The Compendious Book on Calculation by Completion and Balancing* presented systematic solutions to linear and quadratic equations. He is often considered one of the fathers of algebra, but his work built on the earlier achievements of Indian mathematicians. The Indian scientist and historian of science Roddam Narasimha argues that the Indian number system had been the basis for the development of a sequence of number-centric mathematical disciplines quite distinct from the geometric tradition of the Greeks:

> [W]ell before 1600, Indian mathematicians had a concept of mathematics of which algebra, equations, computation and calculus were integral parts. Each was connected with the others, and the methods used for deriving them often contained arguments that are familiar today. Nevertheless, they differed fundamentally from European mathematics in their approach to the subject; this approach was number-centric rather than axiom-proof or geometry-centric. (Narasimha 2016, p. 127)

It is also notable that Narasimha includes calculus as one of the disciplines that emerged from the number-centric orientation of Indian mathematicians. Indeed, Indian mathematicians of the Kerala school, which flourished from about 1350–1550 associated with a lineage of mathematicians including Madhava, Nilakantha, and Jyesthadeva, had known many of the infinite series expansions for trigonometric functions as well as some of the notions of modern calculus before their re-discovery in Europe in the seventeenth century by Newton and Leibniz.[5]

[5] See Joseph (2018), especially Chapter 9 on 'Exploring Transmissions: A Case Study of Kerala Mathematics'.

Fig. 1 Page from *Margarita Philosophica* (1503) by Gregor Reisch showing power of Indian number calculation against the abacus

Let us now address the question: Why are the algebraic methods centered on the number-centric approach so congenial to Cartesian kinetic corpuscularism. One answer is already provided by Kvasz—it allows us to use equations to deal with the conservation of motion laws to understand the dynamic development of systems. But there is also another deeper reason—the number-centric orientation facilitated an atomic mode of analysis that went beyond the physical atomism of ancient Greek thought. To see this, we have to first recognize that there is a close link between algebra and arithmetic in the Indian decimal place number system with zero. Every number can be represented in the form of a power series—e.g., $1729 = (1 \times 1000) + (7 \times 100) + (2 \times 10) + (9 \times 1)$. This could also be written in reverse as $9 + 2.10 + 7.10^2 +$

1.10^3 as a power series of ten. This is easily extended to series like $a_0 + a_1 x + a_2 x^2 + a_3 x^3 + \ldots$. which is a power series of an unknown number symbolized as 'x'. This was a generalization Indian mathematician did go on to make to arrive at the concept of a general power series. Just as a number can be represented as a power series based on the number ten, so is a power series a representation of a number on a base that could be any number x. This paved the way for representing trigonometric functions as power series—precisely the route taken by Indian mathematicians of the Kerala School. The basic principle adopted in every such representation is that any number or function can be represented as the sum of parts. For rational numbers the sum involves a finite series; for irrational numbers an infinite series. Nevertheless, the idea that the whole can be calculated as a sum of parts is built into the arithmetical number system, and the algebraic system was discovered as its generalization, by Indian mathematicians. Significantly the same idea plays an important role in integral calculus, where the area within a figure is treated as the sum of infinitesimal parts—an idea the Indian mathematicians of the Kerala School developed although they did not reach the modern concept of a limit function.[6]

The notion that a whole can be treated as the sum of its parts is crucial for the development of seventeenth-century physics and is the distinctive contribution of the Cartesian analytic method. Many of the fundamental concepts and laws of physics discovered at that time embody this additive principle—the mass of a body is the sum of the masses of each of its parts; the momentum of a body is the sum of the momenta of each of its parts; the volume, force, and distance moved of a body is the summation of the volumes of its parts, the different forces on it, and the separate small intervals it traverses. Thus, the mathematical system inherited from the Indians provided a powerful instrument for the atomic mode of analysis central to the development of seventeenth-century science. This atomism did not apply to physical bodies alone—it also applied to the properties of these bodies like volume, mass, length, momentum, force, and so on. Even time could be seen as the sum of separate temporal intervals. If we had to find a term to characterize this flexible atomic notion

[6] For a comprehensive survey of Kerala mathematics see Joseph (2018).

then the expression 'mathematical atomism' could serve to distinguish it from the narrower physical or material atomism associated with Hellenic thought. The number theory of the Indians embodied mathematical atomism and served as the mathematical foundation to study a vast range of properties of bodies and their behavior. Even the integral calculus could be seen as the outcome and natural extension of the mathematical atomism embodied in the Indian number system since it computed areas, volumes, and so on by dividing them into tiny infinitesimal parts which are then summed together. This is precisely why calculus came to be a natural extension of numeric methods, first in India and later in Europe.

Hence when Koyré argues that the sensualism of ancient atomism could only be fruitful after it became possible to take a mathematical approach he is quite right. But what we now have to see is that the simple shift from Athens to Athens-plus—from geometric to kinetic corpuscularism—would not, as Cohen supposes, have done the job. It needed the combination of kinetic corpuscularism with the articulated algebraic developments of the Indian number system.

3 Arabic Optics and Realist Mathematics

A similarly fructifying transformation occurred as a result of the transmission of Ibn al-Haytham's optical theory to Europe. Ibn al-Haytham had designed his optics to replace the abstract mathematical theory of Ptolemy with a realist mathematical model. In order to appreciate his accomplishment, consider the state of mathematical optics as it ended with the Greeks prior to the time Arabic scholars were to take up the subject. Ptolemy, following the Platonic view, assumed in his *Optics* that radiation emanates from the eyes in the form of a cone—the visual cone—and that perception occurs when the rays of the cone get interrupted by an external opaque object. The observer identifies the object—its shape, size, and location—by the pattern and location of the intercepted rays. One can imagine this as similar to the way a blind person reaches out with a stick into the world to identify objects by the information relayed backed by the stick as it explores the surfaces it comes into contact.

Al-Haytham repudiated Ptolemy's theory on the grounds that it was physically implausible. He argued that we cannot suppose that there were real rays emanating from the eyes because this would mean that perception can occur in the dark since the source of illumination is the eye of the observer. Moreover, he argued, an emanation from the eye cannot, by its very nature, damage the eye, but since bright objects can injure the eye, the injury must be caused by something without. Furthermore, if perception is the result of an emanation from the eye, the material from the eye must fill the whole of space that can be reached by vision—including the sphere of the fixed stars. Such a claim, he maintained, is incredible.

In place of Ptolemy's emanation theory al-Haytham offered an alternative theory able to appropriate the visual cone of the emanation theory, and therefore the mathematical power associated with it, without also implying its physical implausibility. He suggested that light travels as rays into the eyes from opaque illuminated objects. Al-Haytham's approach is familiar to students today in the ray diagrams they are taught when they learn about the optics of vision. By taking the visual field as the base and the center of the crystalline eye-lens as the vertex al-Haytham managed to get the visual cone which made it possible for Ptolemy's theory to exploit the mathematical features of visual perception (Lindberg 1992, p. 311). However, in contrast to Ptolemy who saw the cone as involving emissions reaching out from the eyes into the world, al-Haytham conceived of the cone as formed by rays from the object entering the eye and focused onto the lens so that each point of the image in the lens receives a single ray from any point on the object. This is exactly what happens in a pinhole camera—an instrument al-Haytham studied experimentally in the process of articulating his theory.

His mathematical realist alternative to Ptolemy's optical theory also led al-Haytham to turn his attention to Ptolemy's astronomy. In his *Al-Shukūk ʿ alā Batlamyūs*, translated as *Doubts Concerning Ptolemy* published at some time between 1025 and 1028, Al-Haytham criticized Ptolemy's Planetary Hypotheses:

> Ptolemy assumed an arrangement (*hay'a*) that cannot exist, and the fact that this arrangement produces in his imagination the motions that belong to the planets does not free him from the error he committed in his assumed arrangement, for the existing motions of the planets cannot be the result of an arrangement that is impossible to exist... [F]or a man to imagine a circle in the heavens, and to imagine the planet moving in it does not bring about the planet's motion. (Sabra 1978, p. 121 note 13)

Although he failed to develop a convincing mathematical realist astronomical model, he did inspire subsequent astronomers to make the attempt. He set in motion a chain of research activity within what has come to be known as the Arabic Maragha school of astronomy that ultimately led to a geo-centric astronomical model that did not deploy equants. The mathematical techniques deployed to accomplish this task—the Tusi couple and Urdi lemma—were also used by Copernicus when he developed his heliocentric theory similarly inspired to formulate a mathematical realist model of the cosmos without equants.

This raises the question of the extent to which Galileo was influenced by al-Haytham when he adopted a mathematical realist orientation. Indeed, the importance of al-Haytham's influence in the time of Galileo is evident in Johannes Hevelius' (1611–1687) *Selenographia lunae sive descriptio* or *The Study of the Surface and the Relief of the Moon* published in 1647. It was one of the first lunar atlases with great influence on European astronomers at the time and a major reference text for over a century. What is significant for us is that the frontispiece of the *Selenographia* carries the image of both Ibn al-Haytham and Galileo.[7] Ibn al-Haytham appears on the left with a geometric pattern in his hand and an inscription below with his name that indicates he represents rational thought. Galileo appears on the right holding a long

[7] Figures 1, 2 and 3 are in the public domain here:

https://en.m.wikipedia.org/wiki/File:Gregor_Reisch,_Margarita_Philosophica,_1508_(1230x1615).png (Fig. 1).
Houghton Typ 620.47.452—Selenographia, title - Ibn al-Haytham—Wikipedia (Fig. 2).
Houghton EC.B1328.620ib—Novum organum scientiarum—Novum Organum—Wikipedia (Fig. 3).

telescope with his name and inscription that he represents observation (i.e., empiricism).

The reason why Hevelius includes these two figures in his work is obvious. He treats the moon no longer as a smooth polished surface but as having a contour of craters and hills. His atlas of the moon is constructed by combining Galileo's approach of observation through the telescope with al-Haytham's theory of how light forms images and shadows. It is the same theory that Galileo used when he did his famous drawings of the moon to prove it was not simply a smooth sphere.

It is also important to recognize that although Galileo did not develop a theory of the telescope, we cannot ignore the significance of al-Haytham's theory to making possible the telescope. Around 1230–1235 Robert Grosseteste had used the theory to show how to magnify images so that letters could be read at great distances. Roger Bacon also mentions the magnifying properties of transparent lenses in his treatise *Opus Majus*. In 1271 Witelo writes *Perspectiva* which largely incorporates much of al-Haytham's theory of optics. In fact, Geoffrey Chaucer connects Ibn al-Haytham, Witelo, and Aristotle in one of the great works of English literature *The Canterbury Tales* written around 1387–1400 in the lines:

> They spoke of Alhazen and Vitello.
> And Aristotle, who wrote, in their lives,
> On mirrors strange and on perspectives,
> As all they know who've read their published word.[8]

These developments are also inseparable from the invention of spectacles around 1285–1300. It leads economic historian Joel Mokyr to write:

> The greatest advances before Johannes Kepler's celebrated essay, *Expounding the Optical Part of Astronomy* (1604), were made by Alhazen (al-Haytham, early 11th century), who studied curved mirrors and lenses and first established that light travels from the source to the eye and not

[8] See *The Canterbury Tales: The Squire's Tale*, lines 232–235 on the mirror.

Fig. 2 Frontispiece of *Selenographia*—Hevelius' Moon Atlas

vice versa. And yet from a technological point of view, the first successful application was the emergence of eyeglasses in the 1280s. Without some underlying epistemic base, the probability of this technique emerging was low indeed. Given, however, that this knowledge emerged, the eventual occurrence of even better spectacles (correcting for myopia in addition to presbyopia), telescopes and microscopes was more likely. (Mokyr 2003, pp. 55–56)

Clearly when Galileo builds his telescope in 1609 it has to be seen in the light of a developmental sequence that began with and was set in motion by al-Haytham's theory of optics. This sequence also influenced Kepler and we know that when he built his refracting telescope two years after Galileo's he was quite familiar with al-Haytham's discoveries. In his study *Theories of Vision from al-Kindi to Kepler* historian of science David Lindberg writes:

> Alhazen's theory of vision was enormously influential. It was powerfully defended and had the additional advantage of being essentially without competitors, for no other theory could pretend to the same universality. … Directly, or indirectly, his *De aspectibus* inspired much of the optics that occurred between the thirteenth and seventeenth centuries, and among his followers we must number Roger Bacon, Witelo, John Peckham, … But Alhazen's theory was more than influential; it was the source of fundamental conceptions on which Kepler based his theory of retinal image. (Lindberg 1976, p. 86)

This strongly suggests that what Cohen sees as the mathematical realism of Galileo is more naturally traced to the influence of al-Haytham's optical theory, and we must add possibly his mathematical realist critique of Ptolemy. This is a far simpler explanation than Cohen's view that Galileo's realist turn is probably due to a number of fortuitous factors such as the large number of practitioners of Alexandrian science in Renaissance Europe, inspiration from the ambiguous realism of Copernicus' book, and Galileo's Paduan exposure to the practical crafts. After all Cohen himself admits that these factors do not compel Galileo's realist turn (Cohen 2010, p. 213).

Hence what Cohen terms the transformation from Alexandria to Alexandria-plus must be directly traced to the impact of al-Haytham's optics and his critique of Ptolemaic abstract mathematical astronomy. What Galileo achieved, as al-Haytham did not, was to produce a mathematical realist account of projectile motion which later came to be united with a mathematical realist planetary theory in the higher synthesis of Newtonian cosmology.

4 Chinese Technologies and Fact-Finding Empiricism

The third major mode of nature-knowledge is traced by Cohen to originate in the desire for accurate description and the interest in the practical crafts evident in artist-artisans such as Leonardo da Vinci. It became transformed by Francis Bacon into fact-finding experimentalism. However, to see this trend as emerging spontaneously in Europe would be to overlook the impact of the combination of Arabic optics and Chinese technologies on Renaissance Europe—the first in nurturing the desire for accurate observation and description, and the second an interest in practical crafts. Consider first the desire for accurate description. It seems reasonable to assume that it was inspired by the new aesthetic perspectivism of artists which Leonardo expresses thus:

> Those who are in love with practice without knowledge are like the sailor who gets into a ship without rudder or compass and who never can be certain wither he is going. Practice must always be founded on sound theory, and to this Perspective is the guide and the gateway; and without this nothing can be done well in the matter of drawing. (Leonardo 1970, p. 24)

Indeed, this aesthetic orientation had an impact on many areas of knowledge including anatomy as noted by Butterfield:

> Finally, the Renaissance brought a greater insistence on observation and a refinement of observational skill. It is perhaps not an accident that the first branch of science transformed by improved observation was that of anatomy, the science of the painter, the one restored by Vesalius, in whom the mind of the artist and the mind of the scientist seem almost to have been fused into one. (Butterfield 1957, pp. 52–53)

However, this perspectivism uniting both aesthetics and science is grounded in Ibn al-Haytham's optical theory. In his seminal study *Florence and Baghdad: Renaissance Art and Arab Science* Hans Belting documents how this happened. He delineates how the pictorial theory of mathematical perspective in Florentine Renaissance art directly adapted the visual theory in Ibn al-Haytham's *Book of Optics* to the making of images. It leads him to conclude that perspective has a double history—initially as visual theory in Arabic science and subsequently as pictorial theory in Renaissance art when Arabic mathematical optics was co-opted to serve Western art. Belting concludes that the aesthetic culture of the Renaissance was created by the historical encounter with Arabic science.

Thus, the impact of al-Haytham's optical theory cannot be confined to the domain of theoretical science alone. It also led to a profound shift in the perceptual sensibilities of Europeans. It did this by transforming the notion of what it is to make unbiased and accurate observations. Whereas Greek art was concerned with the representation of ideal types or forms, and Christian art in Europe was directed toward creating symbolic representations of spiritual ideas and events, the new optical science gave birth to a theory of perspectivism in which the artist could represent objects as they appeared to the eye of the observer. To represent the world as it is—or as it appeared to the eye of the observer—without the interposition of idealizations or spiritual projections—became a real possibility and excited Renaissance artists. The world now could be considered to have become transparent to the human visual gaze.

The notion of exact and accurate observation of nature which inspired the artists—along with the confidence given by the al-Haytham theory that such observations did not deceive because they were underpinned by a correct understanding of how the human eye perceived the world—also had an impact on science. It generated the confidence that accurate and unbiased observation of nature was both desirable and possible in science as in art. This is at the heart of the penchant for accurate description in map-making, anatomical drawings, and moon-relief drawings as we saw, that constitutes one dimension of Cohen's third mode of nature-knowledge which emerged in Renaissance Europe.

However, the third mode also has another dimension—an interest in the practical crafts that Cohen also associates with Leonardo da Vinci. Indeed, Cohen contrasts Leonardo's approach with the Chinese approach to nature embodied in its organic materialist natural philosophy:

> The difference at stake here is between Leonardo's studies aimed at emulating bird flight and the Chinese infatuation with making huge kites

> fly with the winds. Joseph Needham has sought to capture the overall contrast in terms of the phrase *wu-wei,* which occurs so often in texts in the Lao Tzu and Chuang Tzu lineage. On this view, China's world-view of organic materialism was marked by an attitude of acceptance of the world. The ultimate aspiration of China's rule, the achievement of harmony in microcosm and macrocosm alike, could be attained best by abstaining from improper action. No mere passivity lay in the aim to live in harmony with nature; yet the Chinese posture was constrained by *wu-wei* – by a reluctance to 'go against the grain of nature'. Behind the Chinese striving for harmony lay an acceptance of the world as it is given to us, in marked contrast to how European civilization aimed in action *and* in thought to intervene in the world and manipulate it for human ends. As diagnosed by Needham, Europe's singular dynamism is reflected in the peculiarly interventionist nature of its empiricist undertakings. [Cohen 2010, p. 138]

It is difficult to agree with Cohen that Leonardo's approach is very different from the Chinese. Both seek to learn from nature by following its way—the Chinese by following the way of the winds and Leonardo the wings of birds. Actually, the discoveries that impressed Bacon did not draw their inspiration from Leonardo but were, as we now know but Bacon did not, Chinese discoveries. In a famous oft-quoted passage Bacon writes:

> It is well to observe the force and virtue and consequences of discoveries. They are to be seen nowhere more conspicuously than in those three which were unknown to the ancients, and of which the origin, though recent, is obscure and inglorious: namely, printing, gunpowder, and the magnet. For these have changed the whole face and state of things throughout the world, the first in literature, the second in warfare, the third in navigation; whence have followed innumerable changes; insomuch that no empire, no sect, no star, seems to have exerted greater power and influence in human affairs than these three mechanical discoveries. (Bacon 1620, Book 1, Aphorism 129)

Indeed, even though Bacon saw the origins of the three great discoveries as obscure, he did connect them with the age of discoveries when

Fig. 3 Cover page of Francis Bacon's *New Organon*

European thought had broken out of the narrow confines of the Mediterranean Sea. His influential methodological treatise *Novum Organum, sive Indicia Vera de Interpretatione Naturae* (New Organon, or True Directions Concerning the Interpretation of Nature) was first published in 1620. On its title page depicting a galleon passing beyond the Pillars of Hercules, seen in Greek mythology as bestriding the two sides of the Strait of Gibraltar, into the wider ocean beyond. By that time the Atlantic, Indian, and Pacific Oceans had been connected by Iberian navigators. It led Bacon to place his House of Solomon where knowledge was actively pursued in all human disciplines in the Pacific Ocean off the coast of Peru.

Given Bacon's wider contextualization of the new knowledge he was calling for, and the impression Chinese technologies of gunpowder, printing, and compass upon his thinking, it is curious that Cohen should argue that Chinese organic materialist mode of nature-knowledge did not contain any developmental potentiality for modern science. I would like to argue that it was precisely Chinese natural philosophy which led to the discoveries that inspired Bacon. To see why we have to recognize that Chinese organic materialism is well placed to nurture creativity and invention when we deal with natural systems. Since it suggests understanding a thing in terms of its context it promotes a tinkering orientation in dealing with natural systems to see how the behavior of one part would change when other parts are altered. This is precisely the trial-and-error process of the experimental method—one that involves changing the context of an object in order to understand how its behavior is thereby altered. What is distinctive about Baconian science is that it carries this tinkering orientation to nature much further. It recommends the institution of new contexts never previously present in nature to, as Bacon puts it, 'torture nature' or 'put nature in constraints not seen before'. This is the distinctive coercive orientation of modern science which would not have been encouraged within the framework of the Chinese organic materialist worldview whose major task is to understand the behavior of a thing in terms of its correlations with other things, and not in isolation from all possible contexts. Bacon's experimental method goes beyond mere description of nature in its own contexts—it neither seeks to imitate the path of winds nor that of wings—but pursues the discovery of new facts by violating natural contexts. In short, it replaces the contextual experimentalism of the Chinese with the coercive experimentalism of modern science. If we adopt Cohen's notion of latent development potential, we could even label Bacon's experimental method as 'China-plus' since it was the inspiration provided by Chinese technology and the tinkering orientation which produced it, that led Bacon to fact-finding experimentalism.

5 Explaining the Rise of Modern Science: Toward a Dialogical History

Cohen claims that it is only by adopting a comparative approach that we can understand the rise of modern science in Europe—one that also shows why it failed to emerge in civilizations such as Islam and China which achieved quite advanced developments in science before the modern era. He attributes the failure of modern science to emerge in China to its organic materialist nature philosophy which lacked the development potential of Athenian and Alexandrian modes of nature-knowledge. Similarly, he considers Arabic civilization, despite inheriting the ancient Greek heritage, to suffer mainly because it did not develop the ethos for accurate description and practical orientation embodied in the third mode of nature-knowledge which emerged in late Renaissance Europe.

However, Cohen's comparative approach suffers in two respects when contrasted with the dialogical approach here proposed. First, it requires him to offer a multiplicity of different sociocultural explanations for why the Athenian and Alexandrian heritages failed to trigger either the mathematical realist turn or the kinetic corpuscular turn when they were transplanted to the Islamic world, Byzantine civilization, early medieval Europe, or late Renaissance Europe. A dialogical history does not have the same problem for it explains the emergence of the three modes of nature-knowledge in Europe, and their failure to emerge elsewhere, far more simply. Kinetic corpuscularism became possible only after Descartes combined the mathematical atomism inherent in the Indian number system with a dynamic interpretation of ancient Greek atomism, and mathematical realism became a possibility only after al-Haytham showed how it could be achieved in optics. Even the third mode of nature-knowledge seen as spontaneously arising in Renaissance

Europe as Bacon's fact-finding experimentalism is ultimately traceable to the dual impact of Arabic optics and Chinese technology.

The dialogical approach also explains why modern science did not emerge in any of the other civilizations—China and India did not have the Alexandrian or Athenian heritage as well as Arabic optics; the Islamic world did not have the Athenian heritage although it did turn the Alexandrian heritage in a mathematical realist direction in optics. But this did not inspire a representational theory of aesthetics and science as in European Renaissance art, since the making of images was forbidden in Islam. Consequently, the dialogical approach offers a simpler and more integrated explanation for why the transplantation of Greek science into a number of civilizations did not result in revolutionary transformations, and why such revolutionary changes did not occur in China, India, or the Islamic world.

Moreover, Cohen himself acknowledges the significance of the dialogical approach when he writes of the historically creative force of cultural transplantation:

> Civilizations may clash; they may also cross-fertilize. Processes of transformation and of what I shall from here on label 'cultural transplantation' offer the single most potent boost to novelty and creativity history knows of. An influx of foreign people, foreign ideas, foreign practices may under propitious circumstances greatly enhance chances for novel things to happen to ideas or habits worn out or petrified or grown stale in their original setting. [Cohen 2010, pp. 45–46]

However, in the case he is discussing he is concerned with how the transplantation of Greek ideas into Europe triggered new thinking that ultimately led to modern science. What he fails to appreciate is that Europe was also a recipient of ideas from other Eurasian civilizations—China, India, and the Islamic-Arabic world. These transmissions were at least as crucial, if not more so, in shaping the approaches to understanding nature we now associate with modern science.

Nevertheless, the notion of modes of nature-knowledge that Cohen has introduced to explain the rise of modern science offers a powerful analytical tool for writing a global history of the origins of modern science. This is not possible if we adopt Cohen's comparative perspective that traces these modes of knowledge into European history alone. A dialogical perspective leads us toward quite a different evaluation. We come to recognize that the Indian number system, Arabic optics, and Chinese technologies carried modes of nature-knowledge which came to constitute integral features of modern science. The mathematical atomism of the Indian number-centric tradition and the mathematical realism of Arabic optics is now a part of scientific approaches to nature. Even if the coercive empiricism of early modern science went beyond the contextual empiricism of Chinese science, contextual empiricism is now becoming acknowledged to be relevant for understanding natural self-regulating processes in the sciences associated with ecology and complexity theory. We must conclude that Floris Cohen's concept of modes of nature-knowledge furnishes a powerful analytic tool for articulating a global account of the growth of modern science, provided we relocate it within a dialogical rather than a comparative history of science.

References

Bacon, Francis. 1620. *Novum Organum. Translated as The New Organon: Aphorisms Concerning the Interpretation of Nature and the Kingdom of Man, collected in James Spedding*, ed. Robert Ellis and Douglas Heath. The Works of Francis Bacon (1857).

Bala, Arun. 2006. *The Dialogue of Civilizations in the Birth of Modern Science.* New York: Palgrave Macmillan.

Bala, Arun & Prasenjit Duara (eds.). 2016. *The Bright Dark Ages: Comparative and Connective Perspectives.* Leiden: Brill.

Barrera-Osorio, Antonio. 2010. *Experiencing Nature: The Spanish American Empire and the Early Scientific Revolution.* University of Texas Press.

Belting, Hans. 2011. *Florence and Baghdad: Renaissance Art and Arab Science.* Belknap Press of Harvard University Press.

Butterfield, Herbert. 1957. *The Origins of Modern Science, 1300–1800.* New York: The Free Press.

Canizares-Esguerra, Jorge. 2006. *Nature, Empire, and Nation: Explorations of the History of Science in the Iberian World*. Stanford, California: Stanford University Press.

Cohen, H. Floris. 2010. *How Modern Science Came into the World: Four Civilizations, One 17th-Century Breakthrough*, Amsterdam University Press.

Descartes, René. 1637. *Discourse on the Method of rightly conducting one's reason and seeking truth in the sciences.* Trans. Jonathan Bennett (2017), see https://www.earlymoderntexts.com/assets/pdfs/descartes1637.pdf.

El-Bizri, Nader. 2022. The Confluences of Science, Philosophy, and Art: Ibn al-Haytham's Optics and the European Perspectiva Traditions. In *From Wisdom to Data: Philosophical Atlas on Visual Representations of Knowledge*, ed. José Higuera Rubio et al., 65–68, 112–117. Porto: University of Porto Press.

Galileo Galilei. 1623. *The Assayer*. In *Discoveries and Opinions of Galileo*, trans. Stillman Drake (1957), 229–280.

Hevelius, Johannes. 1647. *Selenographia, lunae sive descriptio* or *The Study of the Surface and the Relief of the Moon.*

Ibn al-Haytham. 2023. *The Optics*, trans. A. I. Sabra, Books I–III (London: Warburg Institute, 1989) and Books IV–V. London: Warburg Institute.

Joseph, George Gheverghese. 2018. *A Passage to Infinity: Medieval Indian Mathematics from Kerala and Its Impact*. Ltd: Sage Publications Pvt.

Koyré, Alexandre. 1957. *From the Closed World to the Infinite Universe.* Baltimore & London: Johns Hopkins University Press.

Kvasz, Ladislav. 2015. Mathematical Language as a Bridge Between Conceptualization of Motion and Experimental Practice. In *A Bridge Between Conceptual Frameworks Sciences, Society and Technology Studies*, ed. Raffaele Pisano, 229–248. Dordrecht: Springer.

Leonardo da Vinci. 1970. *The Notebooks of Leonardo Da Vinci* (vol. 1), ed. Jean Paul Richter. Dover Publications.

Lindberg, David C. 1976. *Theories of Vision from al-Kindi to Kepler*. Chicago: University of Chicago Press.

Lindberg, David. 1992. *The Beginnings of Western Science: The European Scientific Tradition in Philosophical, Religious, and Institutional Context, 600 BC to AD 1450.* Chicago: University of Chicago Press.

Mokyr, Joel. 2003. Why Was the Industrial Revolution a European Phenomenon? in Supreme Court Economic Review, *The Rule of Law, Freedom, and Prosperity*, 10: 27–63. Chicago: The University of Chicago Press.

Narasimha, Roddam. 2016. Barbarous Algebra, Inferred Axioms: Indic Rhythms and Echoes in the Rise of Western Exact Science. In *The Bright Dark Ages*, ed. Bala and Duara, 120–142. The Netherlands: Brill.
Needham, Joseph. 1954. *Science and Civilization in China.*, vol. 1. Cambridge: Cambridge University Press.
Sabra, A.I. 1978. An Eleventh-Century Refutation of Ptolemy's Planetary Theory. In *Science and History: Studies in Honor of Edward Rosen*, ed. Hilfstein Erna, Czartoryski Paweł and Grande, Frank D, 117–131. Studia Copernicana, XVI, Ossolineum, Wrocław.
Saliba, George. 2007. *Islamic Science and the Making of the European Renaissance*. Cambridge, MA: MIT Press.

17

Webs, Trees and Knowledge: Bunzo Hayata's Eastern Perspectival Model of Nature

Gordon McOuat

1 Introducing the Dynamic System of Bunzo Hayata

This chapter will explore a story of cross-bordering, of hybridization, and networks of exchange—but in a distinctly bio-political sense. It will examine the under-explored place of Japanese science in a period of its phenomenal growth—that is, the first few decades of the twentieth century—where European models of imperialism and science were encountered, exchanged, emulated, reformulated, nativised, resisted and ultimately refashioned, creating a whole new hybridized form. That form decidedly rejected certain forms of so-called European science and its epistemologies, while at the same time retaining and refashioning key concepts, entities and elements of that very science. It occurred in the midst of the Japanese own experience with the modernist project of 'scientific imperialism', and may have set the landscape for much of

G. McOuat (✉)
University of King's College, Halifax, NS, Canada
e-mail: gordon.mcouat@ukings.ca

A. Bala et al. (eds.), *Multicivilizational Exchanges in the Making of Modern Science*,
https://doi.org/10.1007/978-981-97-3541-9_17

late Japanese science, even as it 'universalized' under post-World War II occupation.

In 1921, in the very centre of the first Japanese experience with 'scientific imperialism', there appeared, attached as a rather long appendix to the final volume of an exhaustive ten volume study of the island flora of Formosa, a work on the 'Natural Classification of Plants according to the Dynamic System'. Its author, the famed Japanese botanist Bunzo Hayata (1874–1934), attempted a grand and general reform of the whole taxonomic enterprise in the light of the rise of the new experimental genetics and the associated decline of the naturalist project. Although couched in terms of laboratories and genes, Hayata's reform reached far deeper than genetics. Negotiating a heady recombination of elements from orthodox genetics, classical German morphological idealism ala Goethe and overarched by an appeal to Eastern metaphysics and ways of life, Hayata's 'new Dynamic system' aimed to free the life sciences from not only their European Darwinian entanglements, but from the very ground of hierarchical classification retained in the logical structure of Western classification and its own 'view from nowhere'.[1] Hayata's 'new system' was a full-frontal attack on what he termed 'Western thought' in general.[2] The 'dynamic system' promised to replace, all at once, the whole cloth of nested sets of natural objects inherited from Western notions of classification, logic and ontology (all the while retaining those sets as 'special cases' or particular perspectival moments) with a more general, multivocal, way of ordering the world—'dynamic', because 'there is no fixed criterion or criteria for the natural system', the order, rank, place and very nature of the entities would change from the perspective of the taxonomist.[3] Here Western thought, Western biology, in Hayata's view, was but a limited moment in a generalized Eastern, Buddhist, encounter with the fecundity and multifarious nature of the world and our place within it.

Hayata's philosophical foundation rehearsed a systematic idealism unheard of since the grand systems of the post-Enlightenment science.

---

[1] See Nagel (1986).

[2] For the logical premises of Linnaean classification, see the classic work: Cain (1958). For a controversial *long duré* discussion of taxonomic philosophy, see Atran (1990).

[3] Hayata (1928c, p. 1885).

At the core of its universalist ambitions, Hayata's system took irreducibly local perspectives seriously. The new 'Dynamic System' made the startling claim that there was a peculiar 'Eastern' way of seeing the world and that this Eastern view resisted the 'historicism', progressivism and 'paternalism' of the West. Here in the 'dynamic system', social thought and an approach to nature were combined, actively resisting the 'universalism' and perspective-free 'objectivity' required by the received view of Western science.[4]

Interestingly, in spite of its enormity and its self-professed parochialism, the new 'dynamic system' of Bunzo Hayata quickly, albeit only briefly, caught the imagination of the Western life sciences. For a time—a very short time—Hayata's system promised European and Anglo biologists an optimistic and exciting new way of examining their foundations—that is, living entities and their relations—and a way of transcending received finite systems, offering a bridge between naturalism and the new science, and a transcendence of limited perspectives and endless unsolvable disputes. Here Hayata's system crossed many borders—albeit rather uneasily. To its Western enthusiasts, his genetics looked familiar, although presented in a peculiar way. Its noble attempt at sublimating all received taxonomies into one dynamic system seemed on the right track, especially in a period marked by a Babel of natural systems vying for biologists' attention.

2 Inventing Formosa: 'Scientific Imperialism' and Making New Kinds of Beings

Bunzo Hayata, graduate of the Science School of the University of Tokyo (established on April 12, 1877), and eventual Director of its Botanical Gardens, first visited the island of Formosa (Taiwan) in 1900 with a grant from the government of Japan to collect and describe the entire flora population of the island.[5] Like most natural scientists and bureaucrats

---

[4] See, for example, Nagel (1986).

[5] For further biographical material, see Hiroyoshi Ohashi (2009).

of the newly 'modernizing' Meiji-period Japan, Hayata had received his formative scientific training in the European, mainly Germanic, tradition, and turned to his naturalist work enthused by the successes of German education and science.[6]

By the turn of the century, the Japanese—relative latecomers to the worldwide imperial project—were embarking on a rapid catch up, modelled on a powerful and passionate imitation of their European rivals and teachers, especially the Germans.[7] Closely emulating the European model, proponents of this new Japanese-style imperialism claimed to be wholly grounded on 'scientific principles', albeit with an Eastern twist. For the Europeans and Japanese alike, scientific imperialism demanded at its foundation an inventory and a classification. Like the West, Japanese imperialists would begin their project with the formation of exhaustive scientific inventories of their recent and growing spheres of influence.[8]

Japan's first colony, the newly acquired island of Formosa, would form the first experiment in 'modernized' Japanese colonialism—an experiment in scientific imperialism and ordering the world built upon the total inventory of natural history. In 1895, as a result of the treaty of Shimonoseki after the Chino-Japanese War, China ceded Taiwan (Formosa) and the Pescadores Islands to Japan. Japan occupied the legal capital, Taipei, in June of that year. The local Chinese population revolted and for several months Taiwan declared itself an independent republic. The revolt was soon overcome by Japanese army and

[6] See Bartholomew (1989) and Minoru (1965). Germany also provided advisors to the new imperial government of Japan, especially in the field of medicine, such as Leopold Mueller (1824–1894), Julius Scriba (1848–1905) and Erwin Baetz (1849–1913). The Japanese constitution was greatly influenced by the German legal scholars, Rudolf von Gneist and Lorenz von Stein. See Spang and Wippich (2006).

[7] By 1887, Japanese foreign minister could claim, in a memorandum, that 'in my opinion, what we must do is to transform our empire and our people, make the empire like the countries of Europe and our people like the peoples of Europe. To put it differently, we have to establish a new, European-style empire on the edge of Asia'. Cited in Myers and Peattie (1987, p. 64).

[8] For the idea of a colonial 'inventory' science, see Zeller (1987). Endersby (2008) describes the intricacies of setting up a colonial collection.

naval forces.[9] Immediately, Japan embarked on their program of intensive 'scientific colonialism'—the term made popular in Wilhelmina Germany by the colonial reformer, Bernhard Dernburg (1865–1937) and borrowed by the newly appointed civil administrator of Formosa, Dr. Goto Shimpei (1856–1921).[10] Like Dernburg, Goto Shimpei had studied biology and medicine in Germany, familiarizing himself with the intricate ties between the political and the biological in that newly ascendant European imperial power.[11] Goto brought his experience as the former head of the Department of Health in the Home Ministry of Japan and declared that that Formosa must now be governed on 'biological principles'. This meant, primarily, the imperial utilization of the soil, its treasures, the flora, the fauna and above all the human resources, in favour of the economy of the colonizing nation—an economic natural history combined with a robust 'biopolitics', so to speak. In the case of Formosa, the new imperial masters lost no time, embarking on a rapid and often stark modernization through the assimilation of the local population into Japanese culture and technology, coupled with the exhaustive inventorying of the entire island, living and non-living.[12] Here naturalists were the front line of the conquering imperial forces, just as they were in the European model. The British had begun the work on their own 'botanical survey of India' at roughly the same time, following on Hooker's monumental 'Flora of British India' (1872–1897) and during the early years of the Formosian survey, Japanese botanists repeatedly exchanged with their European colleagues, visiting the centre

[9] This is probably the reason why Hayata gives the date of Japan's acquisition of Formosa as being 1897. See Hayata (1911–1921), Vol. 10, p. 1. In 1898, there was a proposal to sell the island to France because it was so expensive to pacify. See Tsurumi (1967).

[10] For German 'scientific colonialism', see (Duignam and Gann 1972, Naranch and Eley 2014).

[11] Goto Shimpei had studied bacteriology and medicine in Germany before returning to become Japan's chief medical officer in 1892. For a recent assessment of the colonial regime, see (I-Te Chen, 1995). For biological metaphors in Japanese imperialism, see Myers (1972, p. 435). For Bernhard Dernberg, see Duignan and Gann (1972, pp. 179–189). For Shimpei, see Lo (2002).

[12] The rather chilling accounts of the systematic and scientific 'pacification' of the aboriginal population are to be found in the official *Report on the Control of the Aborigines in Formosa*, Bureau of Aboriginal Affairs (1911). The Japanese colonial government were proud of their achievements and were quick to publish the scientific results in all the major European languages.

of the British colonial collection at Kew to compare notes, specimens and styles of inventory.[13]

3 Troubles in Paradise

In 1900, the 26-year-old Hayata entered the College of Science of Tokyo Imperial University to train under Matsumura and others of the Meiji/European tradition, becoming an assistant in charge of the Formosan collections.[14] In 1900 surveyors of Formosa, including Hayata, were sent by the imperial regime to conduct a complete inventory of the new colony, dispatching the collections back to the University of Tokyo for proper classification. Hayata took up the charge with unmatched zeal, beginning with collections from the more passive north and coastal regions, and then heading inland to the more dangerous and unexplored regions.[15] By 1904, the colonial government had established the Bureau of Forestry with Hayata charged to oversee the survey of the entire vegetation. The conditions were difficult. The British had already encountered the formidable island and its rebellious inhabitants, restricting their own excursions exclusively to the coastal regions. Dr. Augustine Henry's late nineteenth-century 'List of Plants from Formosa', while listing 1446 species, complained of problems with climate, sanitary conditions and ever-present 'head hunters'.[16] Under the new Japanese rulers, pacification of the aboriginals and the creation of a total inventory of the island proceeded apace with T. Makino and T. Owatari collecting in the north in 1896, followed by more extensive surveys over the whole

---

[13] For a discussion of the first of multiple visits by Hayata's collaborator, Jinso Matsumura, see Royal Botanic Gardens (1907, p. 28). Hayata soon followed, starting in 1911, and visiting the great herbaria at Kew, Paris and Dahlem, Berlin. See the introduction to the Hayata (1911–21), Vol. 1, p. 3ff.

[14] Biographical details of Hayata's life are to be gleaned from the brief notes of Genkei Masamune (1934), and dispersed throughout Hayata's own writings.

[15] Elmer Merrill and Egbert Walker list some eighty-eight taxonomic papers by Bunzo Hayata. See Merrill and Walker (1938).

[16] Henry (1896). See 'An Abstract of the reasons submitted for the award of the Prince Katsura Commemoration Prize to Bunzo Hayata for his studies on the Flora of Formosa', published by the Authority of the Imperial Academy, in the Official Gazette of 10th May, 1920.

island by Y. Yashiro and others. Hayata's ambitious goal, with support from the new administration, was to push inland, and publish a complete descriptive flora of the island over the next 15 years.[17] The result was his monumental work of great ambition but unfulfilled dreams: the *Icones Plantarum Formosanarum.*

Hayata's plan for an exhaustive flora immediately met unforeseen obstacles. In spite of the enthusiasm regarding a total inventory of the island and the success at moving inland, the project was continually plagued by tensions regarding the aims of the colonial government and restrictions on the size of Hayata's grant. Taking economic reform as priority, the colonial government pumped for inventories emphasizing 'useful' plants that would yield industrial and agricultural results.[18] Hayata's own meticulous descriptions of each and every species demanded much more time and space. As early as the second volume of the Icones published in 1912, Hayata's government grant was cut forcing him to curtail his original intention to describe all the species of plants on the island in fifteen volumes, one per year. Later volumes in the series reveal a hurriedness and anxiety (and some redundancy). By volume 10, Hayata's Icones Plantarum Formosanarum had all the appearance of a 'concluding number of the series', pre-emptively including a general index to the series plus the studies he published while preparing his work on the Icones. Volume 10 was indeed to be the final number in the series. It is in that final volume that Hayata published his remarkable paper on the natural system of classification, synthesizing and critiquing what he had learned from the German tradition in the face of the incredible new abundance of being encountered in the inventory of Formosa (Fig. 1).

---

[17] Hayata (1911–21), vol. 10, p. ii.

[18] Under 'scientific colonialism', the increase in rice production alone was phenomenal. In 1901–1910, Formosa produced 571,000 tons of rice, while in 1921–30 the yield increased to 879,000 tons. The increase in sugar production was even greater: 82,000 tons in 1901–1920, 498,000 tons in 1921–30. The number of hectares involved in sugar production alone increased from 30,000 to 115,000. See Beasley (1897, p. 130). For the use of the police force in this increase in production, see Myers and Chung (1964). For the new colonial Governor-General, Sakuma Samata (1905–15), use of military force to open the mountainous forest regions for economic exploitation, see Roy (2003), pp. 50ff.

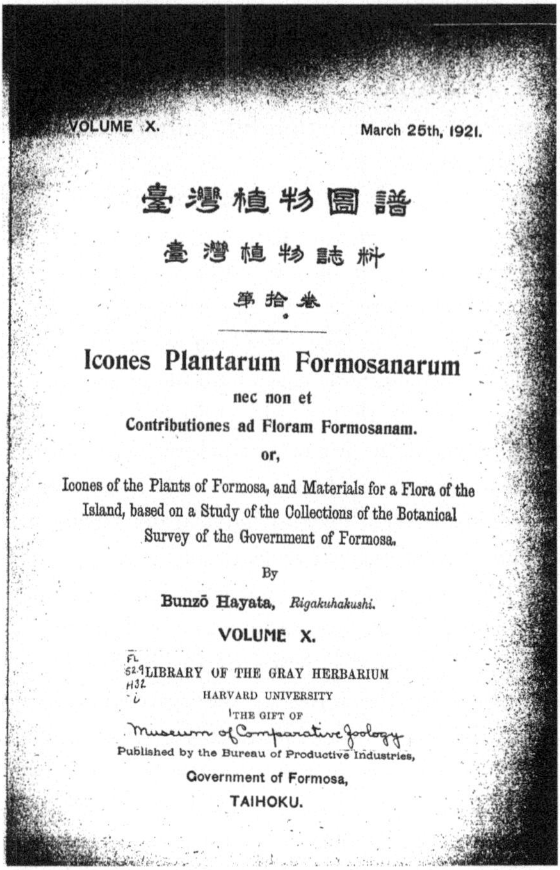

VOLUME X. March 25th, 1921.

臺灣植物圖譜

臺灣植物誌料

第拾卷

Icones Plantarum Formosanarum

nec non et

Contributiones ad Floram Formosanam.

or,

Icones of the Plants of Formosa, and Materials for a Flora of the Island, based on a Study of the Collections of the Botanical Survey of the Government of Formosa.

By

Bunzō Hayata, *Rigakuhakushi.*

VOLUME X.

Published by the Bureau of Productive Industries,

Government of Formosa,

TAIHOKU.

Fig. 1 The Gray Herbarium copy of Hayata's *Icones Plantarum Formosanarum* (Museum of Comparative Zoology, Harvard University)

4 Natural Systems and the Incredible Overabundance of Being

Economic and political restrictions were not the only impediment in the way of publishing an exhaustive flora of Formosa. The very nature of the object presented insurmountable problems. Hayata had already encountered the irreducible multifariousness of the relationship between plants. While his first reports from his working of the Formosa flora were deeply informed by phylogenetic methodology—that is, the search for primitives and non-overlapping nested sets of classification—he soon recognized the limitation of such approaches. His early study of the *Maratticeae* family was grounded in the traditional method of identifying the 'primitive type' of the various genera by the relationship of leaf and stellar structure. But he had already noticed the possibility of multiple relations and entailments of numerous different non-overlapping sets, considering the particular perspective and attention of the botanist. For example, the newly dubbed *Protomarattia* closely resembled the *Marattia* in the reproductive organs while, on the other hand, it could be closely related to *Archangiopteris* in its vegetative organs.[19] Still, at this stage, Hayata imagined that this dual relation meant that it somehow represented a 'primitive' of those two genera. But, the more he looked, the more trouble he saw in the phylogenetic/historicist project. As the mountainous regions were 'pacified' and brought under his purveyance and collections from those regions amassed, Hayata discerned great affinities between plants at similar 'levels' in China, the Malaysian archipelago and peninsula, and in Formosa. Since they were related morphologically, but not directly descendent, Hayata mulled over their historical and classificatory relationship. Here the purely phylogenetic classification wobbled.

The first volume of Hayata's *Icones* begins easily enough with an enumeration of plants, references to species plus a key to the families, genera and species with their respective localities and geographical distributions. The promise was to continue with this plan throughout the series. Here Hayata's taxonomic arrangement followed that of George Bentham and Joseph Hooker in their *Genera Plantarum*, the chief imperial flora of the British Empire.[20] Bentham and Hooker's arrangement had been standard in Japan since the first introduction of binomial nomenclature in 1884, and here Hayata was following convention.[21] Yet, for Hayata, Hooker and Bentham's work soon seemed wholly inadequate in the face of the innumerable kinds of plants in the bountiful space of

---

19 Hayata (1919, pp. 87–88).

20 Bentham and Hooker (1862–1888). See Endersby (2008).

21 Matsumura (1884). See Kimura (1977).

Formosa.[22] Any linear system of classification seemed to fall far short of the enormous reality of tropical life. The number and relations between plants seemed far too great to be held together by one particular scheme.

By the third volume of his *Icones*, Hayata lamented that new materials with which I had been loaded down since 1910, and especially the two collections made by myself in my two excursions to the island (in 1912) had been so numerous that it has required my whole time to work up even the first part of them.[23]

And so Hayata was forced to alter his original plan. The next seven volumes in the series have a sense of urgency, and even anarchy, about them. Species are described, sometimes repeated. The fourth volume presents floral sketches of some of the species already described in earlier volumes and the original descriptions are repeated with the apology: 'as I desire to give a clear conception of my species to anyone who examines them to see whether they are to be recognized as genuine new species or to be reduced to known species'.[24] By the fifth volume, 'the original description in each case is not repeated, as the space is here more limited than it was in the preceding volume'.[25] Increasingly, and in earnest, Hayata turned to general questions of classification, of getting the whole system in order. Recognizing that his grant was likely to run out soon, Hayata devoted his next volume to pulling the whole thing together and to an exposition of the general problems with taxonomy that had been gnawing at him as he produced his exhaustive flora.

The relations of plants on Formosa had finally proven to Hayata that any attempt to set up the relations and descriptions of plants on one preconceived fixed system was doomed to failure. In Volume 10 he presented his theoretical views on natural order and his attempt to set up a wholly new 'natural system of classification' grounded not in European phylogenetic trees and hierarchical sets, but in an ever moving web of 'dynamic' interconnections between biological types informed by the new genetics and a particular 'Japanese' way of seeing. The system,

---

22 Darwin himself had encountered the same fecundity in the tropical jungles of South America. It appeared as if there were no limits to the variation, the kinds and the shear bulk of things.

23 Hayata (1911–21), vol. 3, p. i.

24 Hayata (1911–21), vol. 4, p. i.

25 Hayata (1911–21), vol. 5, p. v.

its background methodology and associated metaphysics, is presented in the last volume of *Icones* in two associated papers: 'An Interpretation of Goethe's Blatt in his "Metamorphose der Pflanzen", as an Explanation of the Principle of Natural Classification'[26] and 'The Natural Classification of Plants according to the Dynamic System'.[27]

5 Resurrecting Archetypes Without History

In his first paper on Goethe's *Blatt*, Hayata opens his discussion of a new natural system by resurrecting the notion of 'typical form' so important to Goethe's Romantic notion of archetype and morphology. It is not surprising that Hayata begins there. Goethean morphology was experiencing a certain renaissance in the years surrounding the turn into the twentieth century, especially among Hayata's own German mentors. It was the great phylogenist Ernst Haeckel himself who gave Goethe a boost (he was matched by his late nineteenth-century rivals, Rudolph Virchow and Hermann Helmholtz).[28] Haeckel's morphology found much resonance with Goethe's emphasis on form and archetype and on what Haeckel saw as Goethe's 'monism'. Yet, Haeckel had given this morphology his own very non-Goethean phylogenetic twist. Against the original Romantic version notion of ideal type, Haeckel's *phylogenetic* archetype was now to be considered an original form *in time*. In Haeckel's scheme, the evolutionary history of an organism and its associated classification was an unfolding in time of that Goethean *Ur-type* through progressive developmental manifestations. Here phylogeny and ontogeny are closely united, both tracking historical/developmental aspects of the organism and searching for the *Ur-type* as ancestral form.[29] It is this historicism that Hayata now resisted.

26 Hayata (1911–21), vol. 10, pp. 75–95.

27 Hayata (1911–21), vol. 10, pp. 97–335.

28 Haeckel (1866). See Virchow (1861); Helmholtz (1892).

29 See Gould (1977). The importance of a historically sublated Goethean morphology cannot be overestimated. See, for example, the parallel studies by E.S. Russell (1916), and D. Thompson (1917).

Hayata's first move in constructing an alternative 'Eastern' biology and dynamic natural system was to explode this phylogenetic reading of Goethe and return to what Hayata thought was a more authentic understanding of Goethe's idealist system. For Hayata, Haeckel's nineteenth-century interpretation was the result of a particular European 'historicist' thinking infected by European fixation on lines of descent, inheritance and pure individualism. All readings of Goethe, Hayata noted, attend to the grounds of the organism and the system: the archetype—in the case of botany, the *Urblatt*. For Goethe the fundamental form (in plants it is the '*Blatt*', the leaf) is not the fully formed leaf, but the *Urblatt*, the archetypical leaf that forms the fundamental basis or, rather, 'plan' of all leaves and all the other parts of the plants. 'Although there are many kinds of vegetable organs, they are after all modifications of one and the same organ—*Blatt* which becomes, according to different circumstances, a foliage leaf, or a sepal, or something else…'[30] Following Haeckel, late nineteenth- and twentieth-century European commentators on Goethe interpreted the *Urformen* in the light of evolutionary historicism: the archetype or *Urpflant* is something all the plants share at bottom, and they are revealed, or 'evolved' through a development of the type through various stages and articulations.

Hayata questioned this assumed developmentalism by arguing that *Ur-types* are not developmental and individualist, but rather combinatorial and holistic. The original types exist as the multitude of fundamental morphological possibilities, existing 'since the beginning of time', combining and recombining in infinite ways. The organism is not a single being, but a collective of these 'generative qualities', or, as he somewhat peculiarly calls them, 'genes'—here Hayata translated the *Wesen* in Geothe's well-known phrase, '*lebendige selbständige Wesen*' as 'genes'. For Hayata an organism is not to be conceived as a particularly isolated individual. Rather, it has close contact throughout the rest of the biological world by 'sharing' in these fundamental forms. These forms 'participate' in the formation of the individual in the manner of genes making up living individuals.

30 Hayata (1911–21), vol. 10, p. 77.

Hayata's use of the term gene may seem idiosyncratic, but it was not unusual in a period when genes were deeply grounded in their original conception as timeless, factorial, almost Platonic forms, brought together in combinatorial relationship with other genes to form the foundation of the sets of characteristics that make up organisms and their relationships.[31] To bolster his combinatorial ahistorical notion of genes and organisms, Hayata drew upon Wilhelm Johannsen's original radical concept of the gene as completely independent of the life-history of the organism. Johannsen, who coined the term 'gene', would write:

> The genotype conception is thus an 'ahistoric' view of the reactions of living beings – on as far as true heredity is concerned. This view is analogue to the chemical view, as already pointed out; chemical compounds have no compromising ante-act, H_20 is always H_20, and reacts always in the same manner, whatsoever may be the 'history' of its formation or earlier state of its elements. I suggest that it is useful to emphasize this 'radical' ahistoric genotype conception of heredity in its strict antagonism to the transmission—or phenotype—view.[32]

Johannsen's ahistorical view of the gene pointed Hayata directly to the work of the great American geneticist, T.H. Morgan, and his theory of the gene as developed in 1917, wherein Morgan reiterates genes as fundamental unchanging, abstract, static units or factors, combining and recombining to produce all the myriad possibilities of possible existences.[33] For Hayata, morphological manifestations are a result of combining and separating, rather than development (or evolution) over time. Hayata's fundamental units are themselves timeless and unchanging. Different combinations and relations between these fundamental units manifest different types. The anti-historicist implications are striking:

---

[31] For an overview of the 'gene' as platonic form and fundamental unit, see Keller (2002).

[32] Johannsen (1911, p. 139).

[33] See T.H. Morgan, (1917). For a background on the history of the gene in this period, see Raphael Falk (2001), Carlson (1966) (especially Chapter 4) and Churchill (1974).

> an innumerable number of species of organic beings have existed from the eternal past and will exist to the eternal future; they unite with or separate from one another, and produce many different organisms by different combinations of the genes; or they change by themselves, as the genes change. Thus, they come from the eternal past, changing their forms incessantly, and will continue to change forever.[34]

Reworking of the 'beads on a string' metaphor coined by the British botanist, Joseph Hooker,[35] Hayata imagines this world of relationships as a spectacular network or web (Fig. 2):

The organism thus lies at the nodes of an infinitely large and ever-changing net. Each change in the net is reflected in a change in the individual, each change in the individual's relations reverberates through the net. Yet, for Hayata, the net remains substantially identical, conserved in its substance, infinity and its individuals. Each bead is along a string, a node of interconnecting threads. The beads manifest phenotypical characteristics by the relationship and combinations. Moreover, for Hayata (taking a certain gloss on his reading of Morgan), genes are *reflexive*: each phenomenal representation of each bead is a reflection from one perspective, one thread taken from the many. The most important idea here is not historical development but continuity and substance.

Although borrowing heavily from the factorial notion of a-historical genes, for Hayata this combinatorial point of view is particularly manifest in the *Eastern* way of seeing the organism. For Hayata, Goethe's theory, grounded upon Spinoza's theory of an eternal 'substance' revealing itself in various 'modes', is markedly similar to that of the

---

34 Hayata (1911–21), Vol. 10, p. 94.

35 'The Cohorts may thus be fancifully likened to the parti-coloured beads of a necklace, joined by a clasp, the beads touching at similarly coloured points of their surfaces. The position of each bead on the necklace is determined by the predominance of colours common to itself and those nearest to it; whilst the number and proportion of the other colours which each bead presents, indicates its claims to be placed elsewhere in the necklace; in other words, such colours represent the cross affinities which the Cohorts display with others remote from the position they occupy'. Hooker (1873, p. 994). Hayata makes no reference to Hooker's use of the bead metaphor.

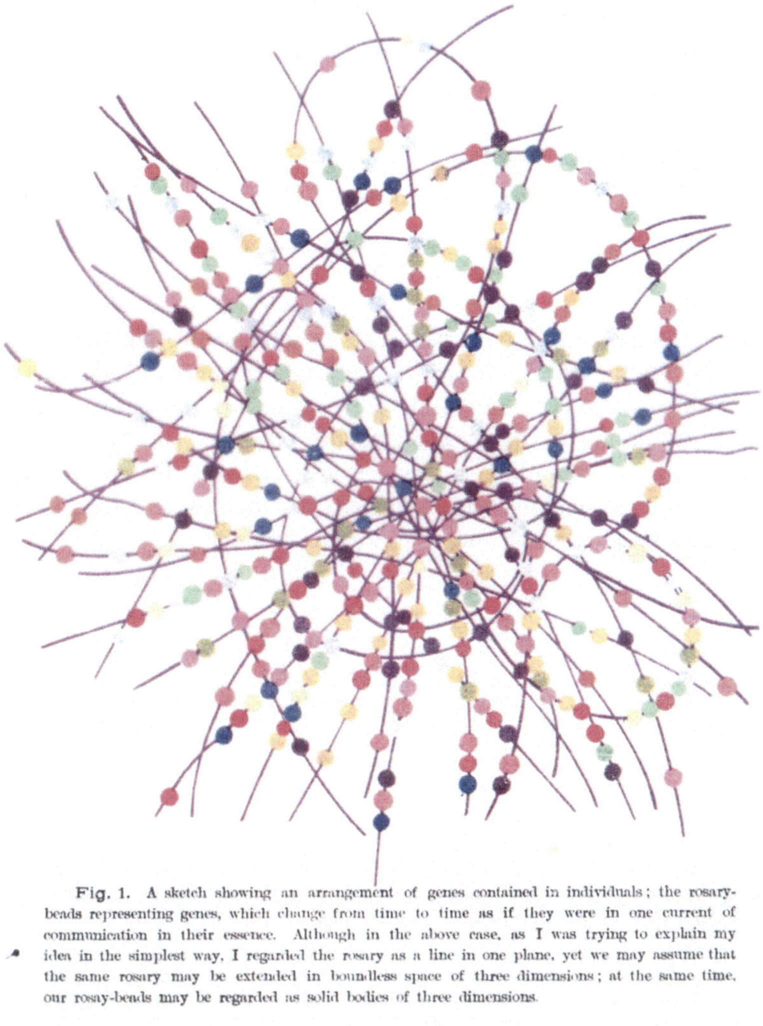

Fig. 2 Hayata's beads in a network illustration from 'An Interpretation of Goethe's *Blatt*' (From Hayata, 1921). The caption begins: 'A sketch showing an arrangement of genes contained in individuals; the rosary beads represent genes, which change from time to time as if they are in one current of communication with their essence'

Tendai Buddhism of Japan.[36] But, for Hayata, it is only in Tendai Buddhism that the fully formed natural system can be revealed. In Tendai's doctrine of the 'Middle Path', sentient beings are not viewed as characters of one quality, but rather as a collection of different qualities (or factors) which are sometimes latent and at other times apparent, according to the circumstances conditioning the inevitable causal nexus.[37] The result is a pluralism grounded upon a notion of unchanging substance—an 'Indra net' of connections.

> The universe is like a boundless net with innumerable millions of crystalline beads, each on a mesh of a different colour, each reflecting the images of the other beads, and each consequently presenting different hues, according to the position of the observer. The beads present different hues, according as they are observed from this point or that. It is, however, only in their phenomena that they are different; in their real entities, they are all and ever the same beads.[38]

---

[36] Tendai Buddhism was introduced to Japan from China in the middle of the eighth century, uniting with the indigenous Shinto religion. In this hybrid form, Tendai had remarkable appeal among the upper classes in Japanese society. See Ziporyn (2004).

[37] In an obvious appeal to his Western audience, Hayata interprets Tendai Buddhism through the works of his friend and colleague at the university in Tokyo, Anezaki Masaharu (1878–1949), professor of Japanese studies in Harvard before World War I. See, for example, Masaharu's lectures of 1913–1915 in Masaharu (1930).

[38] Hayata (1911–21), vol. 10, p. 84. Here Hayata is drawing directly on the notion of Indra-nets in the ancient Buddhist texts. See the Avatamsaka-sutra (c. 300 CE), as translated by Francis H. Cook: 'Far away in the heavenly abode of the great god Indra, there is a wonderful net which has been hung by some cunning artificer in such a manner that it stretches out indefinitely in all directions.... A single glittering jewel [is suspended] at the net's every node, and since the net itself is infinite in dimension, the jewels are infinite in number.... If we now arbitrarily select one of these jewels for inspection and look closely at it, we will discover that in its polished surface there are reflected all the other jewels in the net, infinite in number. Not only that, but each of the jewels reflected in this one jewel is also reflecting all the other jewels, so that the process of reflection is infinite'. In Cook (1977, p. 2). See also Odin (1982).

6 Nonhierarchical Systems

If the beads in Hayata's web are relative in their appearance, sometimes showing, sometimes hidden, but always present, then the true method of examining living beings entails tracing along these lines, revealing and concealing the innumerable contacts between all living beings. This is the metaphysics lying at the basis of his new system. In his next long paper on 'The Dynamic System', Hayata examined the influence such a view would have upon the nature of natural systems and taxonomy. It is here that he presented his new system of classification.

If such pre-existing factors come together in an indefinite number of possibilities, Hayata noted, then there are innumerable ways in which organisms can be combined. Western taxonomists, according to Hayata, have been concentrating only on one or two in order to build their nested sets of taxonomic categories. They have ignored other, equally justifiable, ways to classify. And so, Hayata presented his boldest assertion. There are, he claimed, indefinite, even infinite, ways to classify each individual organism. That is, there are indefinite possibilities of nested sets. It all depends on the 'perspective' of the taxonomist and the particular tracing along the threads. Western taxonomists have concentrated on the particular hierarchical, historicist picture because this happens to be their limited perspective, informed by their particular station in life and in society. Evolutionary theory itself is but one limited Western perspective.[39]

Like genetics, the natural system cannot be historical and phylogenetic. For Hayata, this is a fundamental error of Darwinian biology and Western thought in general. For him the Darwinian system is based on a misconceived social model: phylogenetic trees based on selection theory begin their error by borrowing analogies from British and European

39 In his discussion of the various non-reducible 'natural systems' being imported from the continent, the nineteenth-century philosopher, William Whewell introduced his notion of 'consilience of inductions', where one irreducible system is used to give independent evidence of the divisions of the other. See Whewell (1858, pp. 83–96). This is one solution to the multiplicity of possible classifications and hierarchy and it is one that Hayata rejects.

patriarchal trees of inheritance. Within biology, phylogenetic classification is simply analogous to such a family tree, with origins, branching points and unilinear development (Fig. 3).

Although certainly more articulated and multi-dimensional than the simple family tree, this representation remains fundamentally linear.

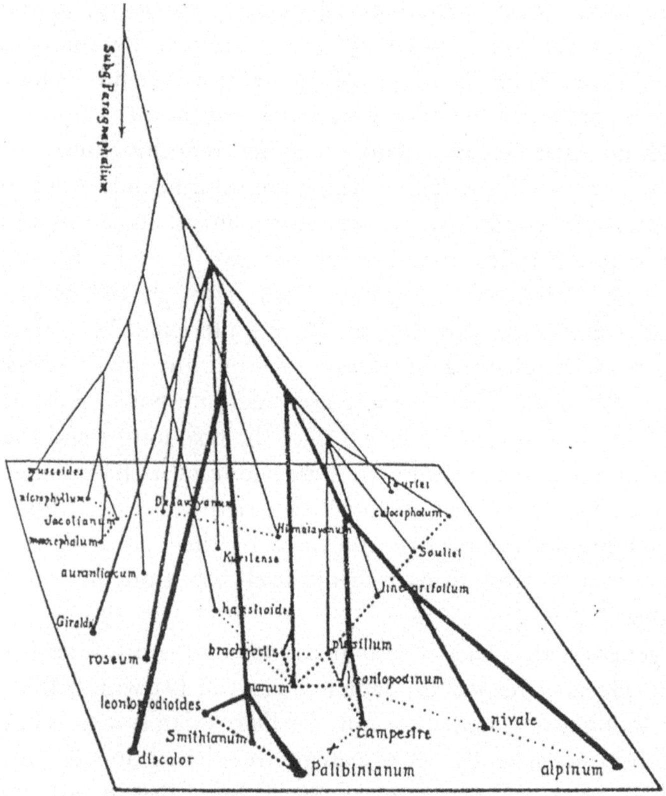

Stammbaum und gegenseitige Stellung der Arten der Sektion *Alpina*.

Fig. 5. Genealogical tree and mutual ("gegenseitige") position of species of *Leontopodium* section *Alpina* (from Handel-Mazzetti, 1927). Thickness of vertical lines indicates dimensionality; taxa joined by horizontal lines with thick dots are linked by intermediates; taxa joined by lines with small dots are clearly related but lack living intermediates; crosses denote hybrids between taxa.

Fig. 3 Phylogenetic tree of *Alpina*, from Handel-Mazzetti, (1927), reprinted in Stevens (1984, p. 192)

Here all organisms are traceable through ancestors to an original branching through time. The membership, then, of exclusive sub-categories are a necessary expression of their historical relationship. But, asked Hayata, is the original analogy valid? No: in a deep sense it is one-sided, 'patriarchal', tracing back to one exclusive site of origin, one track of inheritance and entailment. In a theme he would take up in his later work, Hayata pointed out that a 'matrilineal' society would trace the tree along different lines, to a different origin, producing different relations between orders and sub-orders. Here the tree would be radically different. Although the individuals remain the same, their arrangement does not. In a matriarchal genealogy, Charles Darwin is just as much a Wedgewood as he is a Darwin. The choice, then, for Darwin's lineage is between whether we choose a matriarchal or patriarchal line. Neither, *pace patriarchs*, is any more 'natural'.

However, in one way these systems are similar: that is, the 'genealogy is usually considered to be something like a tree with spreading branches, the ancestor being here compared to the trunk, and the descendants to the branches'.[40] The branches represent exclusive, nested sets of organisms. But, for Hayata, the tree is a result of an abstraction. We are looking at one side, or another, of the true, real nature of descent—that of both parents: 'It is clear that this graphical method is altogether useless when descent through both parents is taken into consideration'.[41] If we trace through a few generations, the relationships begin to look more like a web than a tree—Hayata's web.

In short, for Hayata all other systems of classification are merely one-sided attempts to use one, or some, character-threads as key. In reality there are indefinitely many possible links as there are combinations of factors.

> All systematizers regard the natural system as a static one with a definite form and believe that all species, genera or families have their fixed natural positions, so as to be arranged between this and that, according to their natural relations. My idea is quite different from this current opinion. I

[40] Hayata (1928a, p. 1868).

[41] Hayata (1928a, p. 1868).

> regard the natural system as a dynamic one, changing with the view of the systematizer and subject to alteration, according to the way in which it is considered, and I believe that none of the species, genera or families has a fixed natural position but has changeable positions, subject to alteration according to the criteria for comparison.[42]

On this foundation, Hayata then moves to build his new grand system.

Hayata perspectivalism permits him to start his new system from any received place—they are all legitimate in their individual perspective. He suggests that we take any classification system. Since we can be sceptical about any one in particular, it does not really matter which one we use, as long as that system identifies the individuals we recognize in the organic world. For his own exposition of the new system, Hayata chose the popular phylogenetic system of Adolf Engler (1844–1930) 'as it is one of the most widely used by systematisers'.[43] He could have chosen the alternative systems of Charles Bessey or many of the others contending for the status of 'natural system'.[44] For demonstration purposes, Engler's system was popular and ready to hand. And so Hayata begins to hybridize natural systems. In Hayata's representation of his new system, the series and families of Engler's classification system are arranged as if on a central spine of the new Dynamic System. On both sides of the central spine—or framework—are arranged all the series or families that are thought to have 'some sort of relation' to the series or family on the central spine (leaving ample space for the addition of families or series whose relation we do not yet know). These other sorts of relations are often revealed in rival classification systems. The effect is like highlighting but one strand of the above-mentioned ontological web, abstracting it from the rest, representing it on a vertical line and then representing all the other relations as horizontal cross members. It should be pointed out that this spinal arrangement is not to be reified as

[42] Hayata (1911–21), vol. 10, p. 98.

[43] Hayata (1911–21), vol. 10, p. 157. See Engler (1897).

[44] For a discussion of the alternate schemes of Engler and Prantl, Bessey and Arber, see Naik (1984, pp. 95–126). Indeed, Charles Bessey's rival system was a reworking of the Bentham and Hooker system, realigned by evolutionary criteria.

a fundamental 'ground' of the new dynamic system. Any system could suffice. Rather, the spine is only a reference point, and could be replaced by any of the orthogonal relations. As such it is 'dynamic' (Fig. 4).

9. THE DYNAMIC SYSTEM OF NATURAL CLASSIFICATION APPLIED TO THE ANGIOSPERMS, WITH ENGLER'S SYSTEM AS A FRAMEWORK.*

—— *Gymnospermæ* Subdivision **ANGIOSPERMÆ** ——

—— *Dicotyledoneæ* Class MONOCOTYLEDONEÆ ——

—— *Spathifloræ* Series I. ***Pandanales*** *Synanthæ* *Principes* ——

—— *Pandanaceæ* **1 Typhaceæ** Sparganiaceæ ——

—— *Araceæ* *Palmæ* **2 Pandanaceæ** Sparganiaceæ Typhaceæ ——
—— *Cyclanthaceæ* ——

—— *Araceæ* *Gramineæ* **3 Sparganiaceæ** Pandanaceæ Typhaceæ ——

—— *Triuridales* *Liliifloræ* Series II. ***Helobiæ*** *Spathifloræ* *Microspermæ* ——
—— *Farinosæ* *Ranales* ——

—— Hydrocharitaceæ **4 Potamogetonaceæ** Scheuchzeriaceæ ——
—— Alismataceæ *Araceæ* Najadaceæ Aponogetonaceæ ——

—— *Ceratophyllaceæ* **5 Najadaceæ** Potamogetonaceæ ——
—— Hydrocharitaceæ Scheuchzeriaceæ (=Juncaginaceæ) ——

—— Scheuchzeriaceæ Alismataceæ **6 Aponogetonaceæ** Potamogetonaceæ ——

—— Potamogetonaceæ **7 Scheuchzeriaceæ** (=*Juncaginaceæ*) Alismataceæ ——

—— Scheuchzeriaceæ Aponogetonaceæ **8 Alismataceæ** Butomaceæ Potamogetonaceæ ——
—— *Ranunculaceæ* Hydrocharitaceæ *Triuridaceæ* ——

—— Hydrocharitaceæ **9 Butomaceæ** Alismataceæ *Triuridaceæ* ——

—— Butomaceæ Potamogetonaceæ **10 Hydrocharitaceæ** Alismataceæ *Bromeliaceæ* ——
—— *Iridaceæ* *Amaryllidaceæ* *Burmanniaceæ* *Mayacaceæ* Najadaceæ ——

—— *Dicotyledoneæ* Series III. ***Triuridales*** Helobiæ *Microspermæ* ——

—— *Alismataceæ* *Orchidaceæ* **11 Triuridaceæ** *Burmanniaceæ* *Dicotyledoneæ* ——
—— *Butomaceæ* ——

—— *Farinosæ* Series IV. ***Glumifloræ*** *Liliifloræ* ——

—— *Juncaceæ* **12 Gramineæ** Cyperaceæ *Sparganiaceæ* ——

Fig. 4 The dynamic system as presented in Hayata (1911–21, Vol. 10, p. 159)

Here Hayata imagines his 'natural system' along the lines of Roget's thesaurus (especially the revised edition of 1918). The centre spine around which other systems of relations are constructed becomes just another relation among all the 'synonyms'. The central spine is purely arbitrary. Rather, we trace through the systems 'like running the hands along the web of beads, following first one thread and then another'. As soon as one system was complete, we may look in another direction. There is no logical necessity of exclusive natural sets. Perspective is all that matters.

Hayata's graphical system of the comparison of just two families is accompanied by 39 pages of explication of the revealed relationships between members of these two families. He imagines the dynamic system as being like a stack of index cards, moving, exchanging places, constantly checking against the other relations in the system.

7 From Mathematics to the Nationalization of the Natural System

In his later works, Bunzo Hayata increasingly turned to a refined notion of genetics and combinatory analysis as the model upon which a dynamic system would be built. If the first instance of the system included the relationships of the members of just two families, Hayata imagined that the relationship between three or more families to be infinitely more complicated and beyond the simple method expounded in the *Icones*. In a remarkable piece, published in *Berichte der Deutschen botanische Gesellschaft* in 1931, Hayata enrolled combinatorial matrices and mathematics in an attempt to work out the mathematical relations of the possible number of combinations given in a hypothetical dynamic system.[45] The combinatory nature of his genetics results in a natural system of utmost complexity, foreshadowing the kind of multivariant factoral analysis only attainable by large computer systems of the numerical taxonomists of the latter half of the twentieth century.

[45] Hayata (1931).

Here we reproduce one of Hayata's matrices concerning all the possible combinations of just four elements (Fig. 5).

The mathematics has now become rather abstract and complex and represents a turn away from the more vivid representational and philosophical models of his earlier dynamic system. Yet, while turning to abstract mathematics, Hayata now pressed home the *socio-political* import of his system. In his inaugural work on the 'Dynamic System', Hayata presented a sociological account of the 'error' of Darwin: Western scientists had grounded their phylogenetic retelling of Linnaean categories on a received metaphor of patriarchal inheritance within families. For Hayata, the Englishman Darwin was the worst offender—his historical explanations and 'trees of life' were built on an *English* imperial model. Against the European model, Eastern systems follow models of ancestral participation and matrilineal descent. By the late 1920s and 1930s, Hayata pushed this political, anti-Western, stance to the limit, culminating in his papers on the dynamic system presented at the Third Pan-Pacific Science Congress held in Tokyo in 1926.[46]

Following their participation in the international Pan-Pacific Science Movement of the early twentieth century, and especially after its successful Second Congress held in Australia, Japanese modernizers pulled out the stops in the effort to showcase the ascendancy of Japanese science in the world theatre.[47] (Even the great historian of science, George Sarton recognized this ascendancy as a caution against being 'hypnotized' by Eurocentric science.)[48] The Third Congress was to be a vehicle for the display by its Tokyo hosts. Hayata was commissioned to write both the history of Japanese botany, but also to elaborate his understanding of the new methods of biological systematics.

---

[46] For recent studies on the Pan-Pacific Congress, see the work of Japanese historian, Chihiro Yamanaka, e.g., Yamanaka (2021).

[47] In preparation for the 1926 Congress, the National Research Council of Japan published a heavy volume on 'Japanese science, past and present', which included work on Japanese botany by Hayata. Shinjo (1926). For background to the Pan-Pacific Science movement and its Second Congress, see MacLeod and Rehbock (1988).

[48] Sarton (1929). Sarton went further to note that 'the destinies of the new world, intellectual as well as material, are more likely to be shaped in the Pacific rather than the Atlantic region', p. 88.

(1) $a_0 b_0 c_0 d_0$, (2) $a_0 b_1 c_0 d_0$, (3) $a_0 b_2 c_0 d_0$, (4) $a_1 b_0 c_0 d_0$, (5) $a_1 b_1 c_0 d_0$, (6) $a_1 b_2 c_0 d_0$, (7) $a_2 b_0 c_0 d_0$, (8) $a_2 b_1 c_0 d_0$, (9) $a_2 b_2 c_0 d_0$, (10) $a_0 b_0 c_1 d_0$, (11) $a_0 b_1 c_1 d_0$, (12) $a_0 b_2 c_1 d_0$, (13) $a_1 b_0 c_1 d_0$, (14) $a_1 b_1 c_1 d_0$, (15) $a_1 b_2 c_1 d_0$, (16) $a_2 b_0 c_1 d_0$, (17) $a_2 b_1 c_1 d_0$, (18) $a_2 b_2 c_1 d_0$, (19) $a_0 b_0 c_2 d_0$, (20) $a_0 b_1 c_2 d_0$, (21) $a_0 b_2 c_2 d_0$, (22) $a_1 b_0 c_2 d_0$, (23) $a_1 b_1 c_2 d_0$, (24) $a_1 b_2 c_2 d_0$, (25) $a_2 b_0 c_2 d_0$, (26) $a_2 b_1 c_2 d_0$, (27) $a_2 b_2 c_2 d_0$, (28) $a_0 b_0 c_0 d_1$, (29) $a_0 b_0 c_0 d_2$, (30) $a_0 b_1 c_0 d_1$, (31) $a_0 b_1 c_0 d_2$, (32) $a_0 b_2 c_0 d_2$, (33) $a_0 b_2 c_0 d_1$, (34) $a_1 b_0 c_0 d_1$, (35) $a_1 b_0 c_0 d_2$,

(36) $a_1 b_1 c_0 d_1$, (37) $a_1 b_1 c_0 d_2$, (38) $a_1 b_2 c_0 d_1$, (39) $a_1 b_2 c_0 d_2$, (40) $a_2 b_0 c_0 d_1$, (41) $a_2 b_0 c_0 d_2$, (42) $a_2 b_1 c_0 d_1$, (43) $a_2 b_1 c_0 d_2$, (44) $a_2 b_2 c_0 d_1$, (45) $a_2 b_2 c_0 d_2$, (46) $a_0 b_0 c_1 d_1$, (47) $a_0 b_0 c_1 d_2$, (48) $a_0 b_1 c_1 d_1$, (49) $a_0 b_1 c_1 d_2$, (50) $a_0 b_2 c_1 d_1$, (51) $a_0 b_2 c_1 d_2$, (52) $a_1 b_0 c_1 d_1$, (53) $a_1 b_0 c_1 d_2$, (54) $a_1 b_1 c_1 d_1$, (55) $a_1 b_1 c_1 d_2$, (56) $a_1 b_2 c_1 d_1$, (57) $a_1 b_2 c_1 d_2$, (58) $a_2 b_0 c_1 d_1$, (59) $a_2 b_0 c_1 d_2$, (60) $a_2 b_1 c_1 d_1$, (61) $a_2 b_1 c_1 d_2$, (62) $a_2 b_2 c_1 d_1$, (63) $a_2 b_2 c_1 d_2$, (64) $a_0 b_0 c_2 d_1$, (65) $a_0 b_0 c_2 d_2$, (66) $a_0 b_1 c_2 d_1$, (67) $a_0 b_1 c_2 d_2$, (68) $a_0 b_2 c_2 d_1$, (69) $a_0 b_2 c_2 d_2$, (70) $a_1 b_0 c_2 d_1$, (71) $a_1 b_0 c_2 d_2$, (72) $a_1 b_1 c_2 d_1$, (73) $a_1 b_1 c_2 d_2$, (74) $a_1 b_2 c_2 d_1$, (75) $a_1 b_2 c_2 d_2$, (76) $a_2 b_0 c_2 d_1$, (77) $a_2 b_0 c_2 d_2$, (78) $a_2 b_1 c_2 d_1$, (79) $a_2 b_1 c_2 d_2$, (80) $a_2 b_2 c_2 d_1$,

(81) $a_2 b_2 c_2 d_2$.

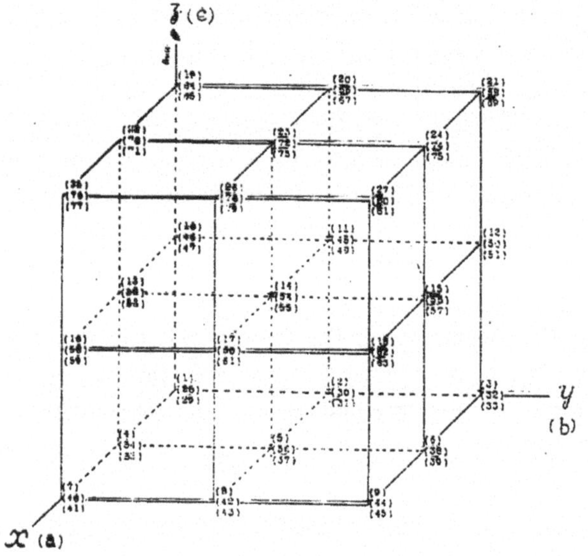

Fig. 5 From Hayata (1931, pp. 336–7). (The numbers represent possible sorts of 4 genes, a, b, c, d, and are then represented on a 3-dimensional axis of possible relations)

In terms of the stated aims of the Tokyo conference—'to promote feelings of brotherhood and strengthen the bonds of peace among Pacific peoples'—Hayata launched his full-frontal assault on the Darwinian theory and its logical roots. Simply, for Hayata, anything associated with Darwinism—natural selection, adaptation and phylogeny—thwarts those very aims. Hayata reiterated his notion of a genealogical 'net' replacing the branching phylogenies of his European colleagues. But now the notion of 'participation' was seen as a political statement: hybridization, mutual intermingling, cooperation and the substantial perspectivalism of Eastern thought are solid alternatives to the racism and social-Darwinism endemic to the West. Simply, Darwinism is the culmination of a faulty Western metaphysics: 'mistaken genealogy, the illusion of progress, and the malicious feeling of pleasure at the griefs of others'.[49] Selection theory and evolution were thus perceived as downright dangerous, and linked inexorably with Western imperialism: 'the natural selection theory when accepted incites hostility among different peoples, and when practiced results in the extermination of tribes'.[50] Hayata's presented his 'participation' theory and dynamic system as much more attune with the needs of indigenous peoples, and the more 'holistic' sensitivities of Eastern colonialism. The dynamic system was now utterly politically allied in an increasingly polarized world.

8 Border Crossing Back to the West (or, How to Leave Out the Metaphysics)

Yet, surprisingly, in spite of this political baggage, Hayata's 'dynamic system' promised much to Western naturalists and experimental geneticists going through a crisis of confidence and the shock of the new science. For a brief time, after its introduction to the West by geneticists and taxonomists, Hayata's 'Dynamic System' seemed to offer a way out of

[49] Hayata (1928b, p. 1869). Here Hayata would echo the interpretation given by his spiritual teacher, Anezaki Masaharu, who would claim in 1925 that the West represents 'aggression', the east 'passivity'. See Anezaki Masaharu (1925).

[50] Hayata (1928b, p. 1871). Hayata conveniently ignored the role of Japanese colonialism in the methodological 'pacification' of the indigenous tribes of Formosa.

the interminable problems with finite natural systems and the struggles over the 'death of Darwinism'.[51] For geneticists it promised a way into systematic taxonomy by directly linking the study of genes with systematic classification in a non-historicist way. For taxonomists, it seemed to offer an ecumenical approach to their field, at once preserving and revolutionizing the practice of composing natural systems.

In Germany especially, Hayata's system met with great interest and excitement. In 1923 survey in the German flagship journal, *Die Wissenschaften*, Kurt Krause (1883–1963), who began his collaboration with the phylogeneticist Engler at the Berlin-Dahlem botanical gardens, introduced a detailed elaboration of Hayata's new form of classification to Western audiences, dubbing it a first step ('ein Schritt') in a revolution of systematic taxonomy and hoping it would be applied across the board and into the animal kingdom.[52] Krause happily flagged Hayata's reliance on Goethe, but as a taxonomist of the Englerian stripe, he still worried about what appeared to be Hayata's 'equalization of all characters' ('Gleichstellung aller Merkmale'), a problem he saw associated with any and all web classifications. Krause's phylogenetic teachers had taught him that not all characters are created equally, and to distinguish between ancestral and recent characters, weighing them accordingly. For Krause, Hayata seemed to pump for a levelling equality among all the characteristics of the organism, uninformed by history or physiological, adaptive and morphological considerations. Krause couldn't imagine such an equality, and so Hayata's first introduction to the West came with a philosophical caution.

In turn, American and English naturalists found promise in this new way of ordering nature in the midst of their vigorous change to the new sciences.[53] William B. Crow, one of the strongest advocates of Morgan's new genetics, looked promisingly at Hayata new system.[54] In his long and detailed survey of the problems of phylogeny, taxonomy and genetics

[51] For the "death of Darwinism" in the decades around 1900, see Bowler (1983).

[52] Krause (1923, p. 63). Hayata approved of most of Krause's interpretation. See Hayata (1928b, p. 1880).

[53] Turrill (1937, pp. 563–6); Crow (1926); Berg (1926); Zimmerman (1934, pp. 381–4).

[54] Crow (1926). On Crow's advocacy for the Morgan's genetics, see Brush (2002).

published in the *Journal of Genetics*, Crow recognized that the phylogenetic system could no longer be represented by a two-dimensional coordinate system. For Crow, the relationship between genes, and their multifarious interactions and participations, would produce genealogies of great multiplicity, with different characters marking different affinities. Hayata's methodology seemed a way into representing this multifariousness, one which would explode the reified nature of natural categories such as species, individual and the higher categories.

The iconoclastic young Dutch botanist and geneticist, Herman Johannes Lam (1892–1977) first advocated a turn to Hayata in his inaugural address to his students as professor in Batavia [Jakarta] in the early 1930s. Dynamic thinking, Lam would tell his students, 'to use Hayata's metaphor, has been the gush of wind that toppled the classic phylogenetic tree, because it lacked a sufficiently strong root system'.[55] There is something right about Hayata's system, although it has yet to be worked out. (Lam later went on to successfully coin the term 'taxon' to separate the taxonomic collectives from their Linnaean hierarchy.)[56]

The most detailed commentary and enthusiastic support for Hayata's new system came from the eminent Swiss naturalist, G. Einar Du Rietz (1895–1967), later to become the great advocate of the 'Nordic school' of plant sociology. In his lengthy and magisterial survey, 'The Fundamental Units of Taxonomy' (1930), Du Rietz explored the history, status and relationships of systematics in the face of the monumental reforms going on in the life sciences. In the light of the new genetics, and detailed field explorations of colonial science, the old categories were now breaking down. Du Reitz's own work was easily the best indicator of the crisis and the various attempts to overcome the old methodologies. As such, Du Reitz was especially intrigued by Hayata's ties to the new genetics, and especially his ecumenicalism and his sublation of older systems. 'Hayata's excellent idea of the dynamic system certainly ought to make possible for old and new systems to exist in friendship side by side

55 See the discussion of Hayata's influence on Lam in Jacobs (1984).

56 At the International Botanical Congress in Stockholm in 1950 Lam proposed the neutral term 'taxon', coined by Adolf Meyer, to designate a 'taxonomic unit'.

in future taxonomy'.[57] Most importantly, of all his commentators Du Reitz seemed to understand that Hayata was not reducing his taxonomy to a phenomenalist comparison of available characteristics (a common criticism of web-based models in taxonomy), but was ultimately based on the distribution of genes and combinations discovered through intense interaction between field and laboratory work.[58]

That said, Du Reitz had some worry worried about Hayata's deflationary approach to the categories of taxonomy, and his general disregard of discontinuities. For Du Reitz there really *are* such discontinuities in nature, which would bar the widespread (decidedly Buddhist) holistic distribution of Hayata's genes throughout the universe. And in that sense, Du Reitz stood at a crossroad and pointed in the direction that the New Synthesis would take in the next decade, wherein smooth distributions of a-historical genes are closely limited by actual reproductive and geological barriers, producing something like isolated populations (later to be dubbed 'species' in the Mayr sense) that mapped into our classifications.

Perched on the eve of that synthesis, Hayata's system now became the subject of much discussion, if only as a way of visualizing the 'phylogeny' of single characters in what was now a burgeoning new view of the meaning of species.[59] For taxonomists, Hayata formed one of the extreme ends of a possible revolution—one which would re-evaluate the relationships and very entities of taxonomy and the life sciences. Revolution was certainly in the air.[60]

However, Hayata's dynamic system was presented on the eve of two crises, which, in the end, would doom his programme. One was scientific, the other political. As Hayata went mathematical and genetic, he did so in the midst of a science retooling itself, as Du Reitz noted, re-establishing hierarchical relations of nested, exclusive, sets of what would

---

[57] Du Rietz (1930, p. 414). Du Rietz was worried about Hayata's founding his higher units of taxonomy on the distribution of *genes*. Practically, said Du Rietz, we have to be content with the distribution of characters.

[58] However, it is sometimes hard to distinguish between morphological manifestations and genetic grounds in much of what Hayata wrote.

[59] See Zimmerman (1934, p. 384), for Hayata's 'much discussed' advocacy of what Zimmerman claimed was the phylogeny of single characters in species.

[60] On Hayata as one extreme of the new revolution, see Watson (1941).

now be seen as 'populations' in new biology's sense.[61] A new 'modern synthesis', hovering on the horizon, would rework the once discredited Darwinian selection as acting upon the distribution of genetic characters throughout a 'population' under Malthusian pressure. The resurgence of Neo-Darwinist models in the 'Modern Synthesis' of the 1930s and 1940s marked the re-appropriation of the term 'species' and all the hierarchical relationships that went with it. 'Species', as the fundamental ground in a reworked Darwinism would be wholly renegotiated, given new meaning and theoretically loaded into that particular system (and, most strikingly, no longer the exclusive domain of natural history and naturalists).[62] The proper job for a taxonomist, then, would be searching for species as (temporarily) stable islands of potential or actual interbreeding within shifting pools of genetic interaction.

And with the addition of a selectionist turn, old phylogenetic and taxonomic stories began to resurface. Logically nested sets represented the 'true' order of things in a refurbished Neo-Darwinist world. Here logical classification of entailment and hierarchy would be given a new licence. In the new system, there would be something 'right' about the old Linnaean approach, provided it was given meaning in the new population genetics). The culmination would be reached in the extreme logistic system of William Hennig, developed in the second half of the twentieth century.[63] Already by the 1930s, the multifarious connections elaborated by Hayata's web were under attack as but simple artefacts of genetically isolated convergent adaptations to similar environmental challenges, and of no real use to classification and biology.[64]

Of course there would be dissenters. 'Phenetic' classification, would balk at the new synthesis definition of taxonomy and species, and would

---

[61] Mayr (1942). For the strategy of naming populations as 'species', see McOuat (2001).

[62] In the manifesto of the new science, Huxley 1940, species were not some arbitrary, nested, set of characters (or genes) but real entities of isolated, interbreeding, individuals. The literature on the origins (and hardening) of the 'modern synthesis' is enormous. See Gould (1983), Smocovitis (1996), Mayr and Provine (1998), Huneman (2019).

[63] See Hennig (1966).

[64] Gates (1936, pp. 514–15). R. Ruggles Gates, a Batesonian geneticist and eugenicist from King's College, London, led the attack from the side of the new species concept. He would later find himself deeply isolated in his attempt to convince his discipline that the different human races were actually five different species. See Tucker (1994, p. 175).

produce powerful mathematical models of (supposedly non-theoretical and non-historical) distance relations between phenomenological objects in biology. Following the lead of the British philosophical naturalist, J.S. Gilmour, pheneticists such as Sokal, Sneath and Jardine could point to the amassed distribution of 'characters' across a taxonomic space, and reject the rigidity introduced by one so-called perfect 'natural' system (except in the case of expediency). And for a moment, just as their programme began to take flight in the 1950s, they turned again to Hayata's 'web'.[65]

Hayata died in 1934, but could there have been a particularly Eastern way of knowing in biology? While Western biologists until recently constructed the 'true' phylogenetic tree, under the guidance of Neo-Darwinism, Japanese biologists have been more open to explore the movement of genes *across* species, thereby prodding the very historicist and logical sensibilities which gave birth to the new science.[66] 'Pluralism' is now the watchword in our understanding of biology and its entities, even if few have adopted Hayata's full-blown perspectivalism.[67] And now, it is Western scientists (notably Canadian and French scientists) dissatisfied with unilinear trees, discovering the importance of hybridity, of interrelated sets, and the perspectivism of systems theory.[68] We may want

---

[65] See J.S. Gilmour (1940), Winsor (2000), Sneath and Sokal (1973) and Jardine (1969). It wasn't a web that Hayata would recognize, though. Their notion of 'general resemblance' and cluster concepts, although similar to Hayata's (1931) version of his dynamic system, reduced perspectivalism to a unitary notion of overall similarity, thereby negating Hayata's notion of irreducible tracings of multiple possible 'genetic' relations and classifications. Interpersonal, institutional and sometimes philosophical details of the disputes in twentieth-century taxonomy are to be found in Hull (1998).

[66] See, for example, the research by the Japanese biologists K. Ochai and T. Akiba, described in Broda (1979, p. 1). See also Falkow (1975). In the 1960s, Mutoo Kimura could present his 'neutral gene theory' of evolution as a strong alternative to the pan-selectionist model of Neo-Darwinism (Kimura, 1968). It is unclear, however, whether the notion of a particular 'eastern' way of thinking would apply in any of these cases.

[67] See Dupré (1993).

[68] Doolittle (1999). Bapteste and Boucher (2008). The 'Questioning the tree of life' research group at the University of Exeter has begun to explore the historical, philosophical and biological implications of lateral gene transfer and its affect on the unilinear, hierarchical tree of life.

to explore where that metaphor takes us. It certainly should inform how we do the history of science—not as a tree but as a bush.

References

Atran, Scott. 1990. *Cognitive Foundation of Natural History: Towards an Anthropology of Science*. Cambridge: Cambridge University Press.

Bartholomew, James. 1989. *The Formation of Science in Japan*. New Haven: Yale University Press.

Bateson, William. 1886. On the ancestry of the chordata. *Quarterly Journal of Microscopic Science Ns.* 26: 535–571.

Beasley, W.G. 1987. *Japanese Imperialism 1894–1945*. Oxford: Clarendon.

Bentham, George & John Hooker. 1862–1888. *Generum Plantarum ad explaria imprimus in Herbariis kewensisbus servata definite*, 3 vols. London: Reeve.

Bapteste, Eric, and Y. Bouchet. 2008. Lateral Gene Transfer Challenges Principles of Microbial Systematics. *Trends in Microbiology* 16: 200–207.

Berg, L.S. 1926. *Nomogenesis, or Evolution Determined by Law*. London: Constable and Company.

Bessey, Charles E. 1897. Phylogeny and Taxonomy of the Angiosperms. *Botanical Gazette* 24: 155–178.

Bowler, Peter. 1983. *The Eclipse of Darwinism: Anti-Darwinian Evolution Theories in the Decades around 1900*. Baltimore: Johns Hopkins University Press.

Broda, P.M.A. 1979. *Plasmids*. San Francisco: W.H. Freeman.

Brush, Stephen. 2002. How Theories became Knowledge: Morgan's Chromosome Theory of Heredity in America and Britain. *Journal for the History of Biology*, 35: 471–535.

Bureau of Aboriginal Affairs. 1911. *Report on the Control of Aborigines in Formosa*. Taipei: Government of Formosa.

Cain, Arthur J. 1958. Logic and Memory in Linnaeus's System of Taxonomy. *Proceedings of the Linnean Society of London* 169: 144–163.

Carlson, Elof. 1966. *The Gene: A Critical History*. Philadelphia: Saunders.

Churchill, Frederick. 1974. William Johannsen and the Genotype Concept. *Journal for the History of Biology* 7: 5–30.

Cook, Francis. 1977. *Hua-Yen Buddhism: The Jewel Nets of Indra*. Penn State Press.

Crow, William B. 1926. Phylogeny and the Natural System. *Journal of Genetics* 17: 85–155.

Darwin, Charles. 1859. *On the Origin of Species*. London: John Murray.

Doolittle, Ford. 1999. Classification and the Universal Tree. *Science* 284: 2124–2128.

Du Rietz, G.E. 1930. The Fundamental Units of Biological Taxonomy. *Svensk Botanisk Tidrift* 24: 333–428.

Duignan, Peter, and L. Gann. 1972. *The Rulers of German Africa 1884–1914*. Stanford: Stanford University Press.

Dupré, John. 1993. *The Disorder of Things. Metaphysical Foundations of the Disunity of Science*. Cambridge, Mass.: Harvard University Press

Engler, A. 1897. *Erläuterungen zum der Übersicht über die Embryophyta Siphonogama, in den Natürlichen Planzenfamilien, Nachträge zum II-IV.*

Endersby, Jim. 2008. *Imperial Nature: Joseph Hooker and the Practices of Victorian Science*. Chicago: University of Chicago Press.

Falk, Raphael. 2001. The Gene—a Concept in Tension. In *The Concept of the Gene in Development and Evolution: Historical and Epistemological Perspectives*, ed. P. Beurton, R. Falk, and H.-J. Rheinberger, 317–348. Cambridge: Cambridge University Press.

Falkow, S. 1975. *Infectious Multiple Drug Resistance*. London: Pion.

Gates, R. Ruggles. 1935. Mutations and Natural Selection. *American Naturalist* 70: 505–516.

Gilmour, John S.L. 1940. Taxonomy and Philosophy. In *The New Systematics*, ed. J. Huxley, 461–474. Oxford: Oxford University Press.

Gould, Stephen J. 1977. *Ontology and Phylogeny*. Cambridge, Mass.: Belknap Press.

Gould, Stephen J. 1983. The Hardening of the Modern Synthesis. In *Dimensions of Darwinism; Themes and Counterthemes in Twentieth-Century Evolutionary Biology*, ed., Grene, M., 71–934. Cambridge: Cambridge University Press.

Haeckel, Ernst. 1866. *Generelle Morphologie der Organismen*. Berlin: G. Reimer.

Hayata, Bunzo. 1919. Protomarattia, a new genus of Marattiaceae, and Archangiopteris. *Botanical Gazette* 67: 84–92.

Hayata, Bunzo. 1911–21. *Icones Plantarum Formosanarum*, 10 vols. Tokyo: Government of Formosa.

Hayata, Bunzo. 1928a. Succession in the Vegetation of Mt. Fuji and the Formulation of a New Theory, the Succession Theory, in Opposition to the Natural Selection Theory. *Proceedings of the Third Pan-Pacific Science Congress* 2: 1867–1868.

Hayata, Bunzo. 1928b. The Succession and Participation Theories and their Bearings upon the Objects of the Third Pan-Pacific Science Congress. *Proceedings of the Third Pan-Pacific Science Congress* 2: 1869–1875.

Hayata, Bunzo. 1928c. The Relation between the Succession and Participation Theory and their Bearing on the Natural System. *Proceedings of the Third Pan-Pacific Science Congress* 2: 1876–1886.

Hayata, Bunzo. 1931. Über das 'Dynamische System' der Pflanzen. *Berichte der Deutschen Botanische Gesellschaft* 49: 328–348.

Helmholtz, Hermann. 1892. Goethe's Presentiments of Coming Scientific Ideas, in Cahan, D. (ed.), 1995. *Hermann Von Helmholtz, Science and Culture: Popular and Philosophical Essays*. Chicago: University of Chicago Press.

Hennig, William. 1966. *Phylogenetic Systematics*. Urbana: University of Illinois Press.

Henry, Augustine. 1896. A List of Plants from Formosa. *Transactions of the Asiatic Society of Japan*, 24 (supplement).

Hooker, Joseph. 1873. *A General System of Botany. Descriptive and Analytical*. London: Longmans, Green and Company.

Hull, David. 1988. *Science as a Process: An Evolutionary Account of the Social and Conceptual Development of Science*. Chicago: University of Chicago.

Huneman, Philippe. 2019. Editor's Introduction. *Journal of the History of Biology* 52: 509–5118.

Huxley, Julian, ed. 1940. *The New Systematics*. Oxford: Oxford University Press.

I-Te Chen, Edward. 1995. Goto Shimpei, Japan's Colonial Administration in Taiwan 1898–1906: A Critical Examination. *American Asian Review* 13: 29–59.

Jacobs, M. 1984. *Herman Johannes Lamm, 1892–1977*. Amsterdam: Rodopi Press.

Jardine, Nicholas. 1969. A Logical Basis for Biological Classification. *Systematic Zoology* 18: 37–52.

Johannsen, Wilhelm. 1911. The Genotype Conception of Heredity. *American Naturalist* 45: 129–159.

Keller, Evelyn Fox. 2002. *The Century of the Gene*. Cambridge, Mass.: Harvard University Press.

Kimura, Yojira. 1977. Binomial Nomenclature of the Japanese Plant Societies. *Japanese Studies in the History of Science* 16: 105–108.

Kimura, Mooto. 1968. Evolutionary Rate at the Molecular Level. *Nature* 217: 624–626.

Krause, Kurt. 1923. Eine neue Form des Natürlichen Systems. *Die Naturwissenschaften* 11: 60–63.
Lam, H.J. 1936. Phylogenetic Symbols: Past and Present. *Acta Biotheoretica* 2: 153–194.
Lo, Ming-cheng. 2002. *Doctors within Borders: Profession, Ethnicity, and Modernity in Colonial Taiwan*. Berkeley: University of California Press.
Masaharu, Anesaki. 1925. Eastern and Western Civilizations. *Institute of Pacific Relations* 1925: 100–102.
MacLeod, R., and Philip F. Rehbock, eds. 1988. *Nature in its greatest extent: Western science in the Pacific*. Honolulu: University of Hawaii Press.
Masaharu, Anesaki. 1930. *History of Japanese Religion: With Special Reference to the Social and Moral Life of the Nation*. London: Kegan Paul.
Masamune, Genke. 1934. A brief biogiaphy [sic] of Bunzo Hayata; and bibliography of Dr. B. Hayata's work. *Transactions of the Natural History Society of Formosa* 24: 477–485.
Matsumura, Jinzo. (ed.). 1884. *Nippon Shokubutsu Meii or Nomenclature of Japanese Plants in Latin, Japanese, and Chinese*. Tokyo: Maruya & Company.
Mayr, Ernst. 1942. *Systematics and the Origin of Species, from the Viewpoint of a Zoologist*. Cambridge, Mass.: Harvard University Press.
Mayr, Ernst and William. B. Provine (eds.). 1998. *The Evolutionary Synthesis: Perspectives on the Unification of Biology*. Harvard University Press.
McOuat, Gordon. 2001. From Cutting Nature at its Joints to Measuring it: New Kinds and New Kinds of People in Biology. *Studies in History and Philosophy of Biology* 32: 613–645.
Merrill, E.D. and E.H. Walker (eds.). 1938. *A Bibliography of Eastern Asiatic Botany*. Jamaica Plain, Mass.: Arnold Arboretum.
Minoru, W. 1965. Japanese Students Abroad and the Acquisition of Scientific and Technical Knowledge. *Cahiers D'histoire Mondiale, IX* 2: 254–293.
Morgan, Thomas Hunt. 1917. The Theory of the Gene. *American Naturalist* 51: 513–520.
Myers, Ramon. 1972. Taiwan as an Imperial Colony of Japan: 1895–1945. *Journal of the Institute of Chinese Studies, Chinese University of Hong Kong* 6: 425–451.
Myers, Ramon, and Adrienne Chung. 1964. Agricultural Development in Taiwan under Japanese Colonial Rule. *Journal of Asia Studies* 23: 555–570.
Myers, Ramon, and Mark Peattie, eds. 1987. *The Japanese Colonial Empire, 1895–1945*. Princeton: Princeton University Press.
Nagel, Thomas. 1986. *The View From Nowhere*. Oxford: Oxford University Press.

Naik, V.N. 1984. *Taxonomy of Angiosperms*. New Delhi: McGraw Hill.
Naranch, Bradley, and Geoff Eley, eds. 2014. *German Colonialism in a Global Age*. Durham, NC: Duke University Press.
Needham, Joseph. 1954–2008. *Science and Civilization in China*. Cambridge: Cambridge University Press.
Odin, Steve. 1982. *Process Metaphysics and Hua-Yen Buddhism: A Critical Study of Cumulative Penetration vs. Interpenetration*. New York: SUNY Press.
Ohashi, Hiroyoshi. 2009. Bunzo Hayata and His Contributions to the Flora of Taiwan. *Taiwania* 54: 1–27.
Roy, Denny. 2003. *Taiwan, a Political History*. Ithaca: Cornell University Press.
Royal Botanic Gardens. 1907. Miscellaneous Notes. *Bulletin of Miscellaneous Information* 1907: 18–36.
Russell, E.S. 1916. *Form and Function*. London: John Murray.
Shinjo, Shinzo (ed.). 1926. *Scientific Japan, Past and Present*. Tokyo: National Research Council of Japan.
Sarton, George. 1928. Review of Scientific Japan, Past and Present. *Isis* 10: 83–88.
Smocovitis, V. Betty. 1996. *Unifying Biology: The Evolutionary Synthesis and Evolutionary Biology*. Princeton: Princeton University Press.
Sneath, Peter H.A.. 1957. Some thoughts on Bacterial Classification. *Journal of General Microbiology* 17: 184–200.
Sneath, Peter H.A.., and Robert R. Sokal. 1973. *Numerical Taxonomy*. New York: Freeman.
Spang, Christian W., and Rolf-Harald. Wippich, eds. 2006. *Japanese-German Relations, 1895–1945*. London: Routledge.
Stevens, P.F. 1984. Metaphors and Typology in the Development of Botanical Systematics 1690–1960, or the Art of Putting New Wine in Old Bottles. *Taxon* 33: 169–211.
Thompson, D'Arcy. 1917. *On Growth and Form*. Cambridge: Cambridge University Press.
Tsurumi, E. Patricia. 1967. Taiwan under Kodama Gentaro and Goto Shimpei. *Harvard East Asia Research Center, Papers on Japan* 4: 95–146.
Tucker, William H. 1994. *The Science and Politics of Racial Research*. Chicago: University of Illinois Press.
Turrill, William B. 1937. Contacts between Plant Classification and Experimental Botany. *Nature* 137: 563–566.
Virchow, Rudolf. 1861. *Goethe als Naturforscher*. Berlin: Hoheneichen.
Watson, E.V. 1941. The Dynamic Approach to Plant Structure and Its Relation to Modern Taxonomic Botany. *Biological Reviews* 18: 65–77.

Whewell, William. 1858. *Novum Organon Renovatum*. London: Parker.
Winsor, M.P. 2000. Species, Demes, and the Omega Taxonomy: Gilmour and The New Systematics. *Biology and Philosophy* 15: 349–388.
Yamanaka, Chihiro. 2021. *Emergence of Japan Society for the Promotion of Science with a Focus on Overseas Influence*. History of Science Society Annual Meeting, 18–21 November 2021.
Yosaburo, T. 1914. Japan's colonial policy. In *Japan to America: A Symposium of Papers by Political Leaders and Representative Citizens of Japan on Conditions in Japan and on the Relations Between Japan and the United States*, ed. N. Masoaka, 95–99. New York: Putnam.
Yosaburo, Takekoshi. 1907. *Japanese Rule in Formosa*. London: Longmans.
Zeller, Suzanne. 1987. *Inventing Canada: Early Victorian Science and the Idea of a Transcontinental Nation*. Toronto: University of Toronto Press.
Zimmerman, Walter. 1934. Research on Phylogeny of Species and of Single Characters. *American Naturalist* 68: 381–384.
Ziporyn, Brook. 2004. Tendai School. In Robert E. Buswell (ed.). 2004. *Encyclopedia of Buddhism*. New York: Macmillan.

18

Dialogue and Comparison Compared

H. Floris Cohen

Joseph Needham, it is well-known, was occupied over more than half of his long life with his famous Grand Question. He was concerned all the time to make comparisons between the history of science in China and in Europe prior to the rise of modern science, with a view to understanding why that vast rupture in the history of humanity took place in the latter rather than in the former, no less well-developed civilization. He was also concerned to establish cases where Europe was at the receiving end of certain discoveries first made in China. The organizers of the conference the present collection of chapters stems from posited a neat ranking between these two concerns of his:

> ... it [recently] became evident that the impetus of Needham's comparative question had become less urgent. A vast body of literature had emerged since Needham's pioneering work, in part inspired by his

H. F. Cohen (✉)
Utrecht University, Utrecht, The Netherlands
e-mail: h.f.cohen@uu.nl

A. Bala et al. (eds.), *Multicivilizational Exchanges in the Making of Modern Science*,
https://doi.org/10.1007/978-981-97-3541-9_18

> pioneering efforts and inspiration, which showed that what was important was to understand how interconnections of civilisations over the last three thousand years, and the exchanges of cosmological [and many other ...] practices and knowledge, had been woven together to make modern science possible. ... this crucial dimension of science history has hitherto been marginalised by the overemphasis on his comparative question.[1]

And indeed, in several pre-circulated conference papers 'dialogue' became the buzz word and 'comparison' was summarily proclaimed outdated. But does it really make sense to treat 'dialogue' and 'comparison' as hierarchically ordered or even incompatible entities? In this paper I investigate the question by making a thorough-going comparison between the two.

1 Requirements for Making Everyday Comparisons and Dialogues Fruitful

What requirements ought to be fulfilled for setting up a fruitful comparison, and what are the conditions for having a fruitful dialogue?

Conditions for fruitful comparison first. In everyday life there is not much of a problem. When, for instance, we are in a fruitshop comparing apples and pears, we alone are doing the comparing, so we decide for ourselves what specific criteria we find at this moment and in this given situation relevant. What comes out of the comparison is, and cannot but be, fruitful in the sense that we personally benefit from it. We do benefit even if the outcome (for instance, that tasty-looking pear) turns out disappointing—in that case we have learned for the next time to put better criteria to work in our mental balancing of our chosen set of similarities and differences.

With requirements for fruitful dialogue things do not stand so simple. Two major conditions ought to be met in any case. One concerns the alternation of speaking and listening that constitutes the very essence

[1] 'Needham Conference 2023 Background': p. 6 of the program for the conference 'Needham's Dialogical Vision: Understanding Science as a Multi-Civilizational Outcome' held on 23–24 March 2023 at the Hong Kong University of Science and Technology.

of a dialogue. While one speaks, the other makes a genuine effort to pick up and comprehend what it is that is being said—at times, further elucidation may be asked for and given before proceeding any further. A significant component of the art of listening, then, concerns both the active anticipation and the collaborative repair of misunderstandings.[2]

The other major requirement for a dialogue to be fruitful is that it be an open-minded encounter. Both participants genuinely want the dialogue, and both make an effort to ensure that it be a dialogue between equals who perceive each other to be equals. What, then, may prevent a dialogue from being an open-minded one, or even from taking place altogether? Here is a surely incomplete list. Threats or acts of force stand in the way of an open-minded dialogue. So do expressions or even signs of a sense of inherent superiority (be it racial, social, sexual, age-related, or of any other kind) on either side. So does excluding people who are perceived to be outsiders who should mind their own business. So does token rather than full-blooded participation in the dialogue. So does the presence of ulterior motives (for instance, when one or both participants appear to be out to catch the other in the act of making a mistake or of saying something objectionable). So does declaring another person old-fashioned, not in touch with the times any more, or otherwise unfit for dialogue. So does presenting an obvious non-fact as if it were a fact. So does making a statement that the speaker knows to be a lie. There is, further, at least one possibility where both requirements, the open mind and the willingness and capacity to listen, come together: it is when an effort at mutual understanding fails because, in the most literal sense, a common language is lacking, or, if not that, then an at least to some extent shared conceptual equipment.

It appears then that, generally speaking, the making of a fruitful comparison is a good deal less fraught with complications than having a fruitful dialogue. It is true that dialogue may, at its best, yield more creative outcomes than comparison, in that the entities compared are known beforehand, whereas from a dialogue one can learn things never before even heard of. And yet, dialogue may also go wrong in more

---

[2] Speaking comes naturally to us, but listening, really listening in the sense of giving full attention to what the partner in dialogue has to say, is an art, instructively discussed by Kate Murphy in her *You're Not Listening* (New York: Celadon Books, 2021).

numerous and in more incisive ways than may happen with the weighing of similarities and differences that the making of comparisons consists of. Now does this conclusion remain valid when we narrow the situation down to fruitful comparison made between aspects of pre-modern science in different civilizations, and to fruitful dialogue taking place between pre-modern civilizations in regard of science or some specific aspect thereof?

2 To Make Historical Comparison Fruitful; to Think Fruitfully About Past Dialogues

What, so that question comes down to, is needed for a student of the pre-modern past to make fruitful comparisons? And what has in the pre-modern past itself been needed for bringing about a fruitful dialogue on matters scientific? What conditions obtain in the one case; what requirements need to be met in the other?

Please note that these two questions are asymmetrical. With comparisons the actors are, without more ado, people—in our specific case, historians. The requirement of fruitfulness now comes down in effect to 'fruitful in a context of historical scholarship' which, given what scholarship is about in the first place, means in its turn that we are going to investigate what is needed to make our comparisons enhance our understanding. For cross-cultural historical comparison this is, and from here on remains, our fundamental requirement—if some given act of comparison fails to enhance our understanding, it had better be discarded or (a more constructive outcome) be used as a stepping-stone toward doing a better job next time.

In the case of dialogue between civilizations the prime actors are quite different. It is trivially true to say that 'civilizations' do not exist as such—what actually do exist are people reared in and formed by the civilization in which they are rooted. In the case of what I shall for brevity's sake call pre-modern scientific dialogue the actors are people reared in different civilizations who transmit, or exchange, or circulate, certain pre-modern scientific goods. Either they do or they don't, but only if they can be shown or at least reasonably be supposed by historians to have done so in historical reality can we speak of a fruitful dialogue having taken place at some place and time (even if it may not always be possible to pinpoint that place and that time with all desired exactitude).

The comparative investigation of historical cross-cultural comparison and ditto dialogue that is to occupy the remainder of this paper, then, comes down to the following. With the former I shall seek to specify what a present-day historian who may wish to undertake a large-scale comparison would be wise to do in order to make it enhance our understanding. In the latter case I shall seek to specify what was required in the pre-modern past to make fruitful dialogues happen in the first place, and what is needed for such an assumed dialogue to become a suitable object of historical inquiry. I start my investigation with Joseph Needham in the act of comparing, followed by Joseph Needham in the act of reporting on dialogue. And since I actually carried out that investigation 34 years ago, and reported the results in several publications, I confine myself to summing up the principal results then arrived at.[3]

3 Historical Comparison as Handled by Joseph Needham: A Summing-Up

First of all, Needham appeared from one of his numerous essays to the next to have varied the precise formulation of his Grand Question, at times substantially so and with effects enduring in many a later essay. The question famously put to Needham in 1937 was 'why modern science originated only in Europe'. Needham consistently transformed this question into one variant or another of 'the problem of the origin of modern science in Semitic-Occidental (Euro-American) civilization on the one hand, and its failure to arise in Chinese civilization on the other'. Or

[3] In the next two sections I sum up large portions of Chapter 6 of my *The Scientific Revolution. A Historiographical Inquiry* (University of Chicago Press, 1994), especially pages 418–488, which contain all passages by Needham quoted in the main text. An earlier summing-up appeared in my 'Joseph Needham's Grand Question, and How to Make It Productive For Our Understanding of the Scientific Revolution'. In *Science and Technology in East Asia*, (Ed.), A. Arrault and C. Jami (eds.), 21–31, vol. 9: *The Legacy of Joseph Needham* (vol. 51 in series 'De Diversis Artibus'). Turnhout: Brepols, 2001.

rather, that is how he by and large kept phrasing it for a while. But once he got fully immersed in pre-modern Chinese science and technology and became more convinced every day of their richness, an additional question began to take shape in his mind. 'The more you know about Chinese civilisation, the more odd it seems that modern science and technology did not develop there', he wrote, and thus his second problem became: 'Why was it that between the second century BC and the sixteenth century AD East Asian culture was much *more* efficient than the European West in applying human knowledge of Nature to useful purposes?' The history of technology, then, had now also entered his problem definition, from then on to dominate it on some occasions and to be passed over entirely on others. Even so, at the heart of Needham's Grand Question remained always the comparison between two very large entities: ancient China on the one hand and, on the other, the civilization of Europe as it began to take shape in the Carolingian period.

Now on to the answers to his own Grand Question that Needham came up with. Some came down in effect to the schematic projection upon China of certain characteristics often attributed to Europe; for instance, Needham's idea that, unlike China's 'feudal bureaucracy', European capitalism came with a democratic mentality which in its turn favored the rise of modern science. Other, later explanations of his appeared to display a much richer and subtler weighing of elements that he deemed relevant in each of the two civilizations, though always against the background of certain prejudices of his about what had caused modern science to arise in Europe.

All in all I came to identify in Needham's essays six answers that he actually gave to his Grand Question; four answers that he rejected, and one answer that he strenuously avoided. As to some order or hierarchy between those six answers actually given, there proved to be hardly any. His six explanations display much diversity, leaving it unclear how he saw their mutual relationship but also what status he ascribed to each of them, and even whether he regarded one, or two, or even more of them as finally decisive. More than once he grounded an argument in favor of some proposed explanation in source interpretations that are in effect much radicalized summaries of far more balanced considerations developed in the corresponding tome of *Science and Civilisation in China.*

His answers were further pervaded by ideological bias of several kinds, notably his political preferences (Marxist with mildly Communist overtones); his 'organic' view of proper scientific knowledge; his preset ideas about how modern science came about in Europe, and, above all, his touching yet at times quite uncritical love affair with the civilization of ancient China. Further, he often approached the history of world science as if it were a race—which civilization was at some point ahead of the others; which one fell back, and when? Finally, although as a great scholar he was above hiding from his readers indisputable facts that more or less squarely contradicted his conclusions, at times he made it quite difficult for them to hunt such facts down in some subordinate clause of an at first sight perhaps plausible-looking argument.

I drew from all this a dual conclusion. Yes, the vast, comparative question that had got Needham going in 1937 and that kept him occupied for the rest of his many days has vastly enhanced our understanding of the history of science in ancient China, and whatever the shortcomings of the Needham way to handle the question, he deserves our deeply felt admiration for his pioneership and for his overall achievement. But no, Needham did not really succeed in enhancing our understanding of why modern science arose in Europe and not in China. He in effect made a mess both of how he phrased the question and how he answered it. Hence, the question for us is whether making a mess of such historical cross-cultural comparison is inevitable or whether we can learn from his example. Is it possible for later generations, in climbing on his powerful shoulders, to look farther than he did and, thus prepared, do a better job? Before answering, I shall now first sum up how I found Needham to have handled the issue of dialogue between China and the West prior to the Scientific Revolution.

4 Dialogue in History as Handled by Joseph Needham: A Summing-Up

Not, to be sure, that Needham, or anybody else at the time, spoke in this connection of 'dialogue'. The expression then current was 'the transmission of science'. The implication that B learns from what A has found but not the other way round has given occasion to replace the expression 'transmission' with 'exchange', or 'circulation', or, most recently, 'dialogue'. Even so, with Needham transmission was hardly a one-direction affair, or rather, the direction that kept him occupied was from China to Europe.

Excluding from the cases advanced by Needham those that really concern technological rather than scientific findings, I came across five products of Chinese science that, in his view, had moved westward and reached Europe, most of them leaving considerable marks on the pursuit of science there. Whereas Needham spent little time on substantiating the claims he made for alchemy, for observational astronomy, and for cosmology, he waxed eloquent on the claims he made for magnetism and for the measurement of time. With magnetism there was no missing the rhetorical means Needham employed to blow up out of all reasonable proportion the significance of what was, in itself, an impressive find—a note written around 1080, so Needham discovered, by the statesman and polymath Shen Kua about the directive properties of the lodestone. There was also Needham's widely acclaimed discovery of writings and drawings that recorded the construction, by the polymath and statesman Su Sung at the end of the eleventh century, of a huge water clock intrinsically capable of measuring time with an accuracy far exceeding what European clocks were capable of prior to Christiaan Huygens' science-imbued invention of the pendulum clock in 1657. Here, too, for Needham a great deal more was at stake. He claimed now to have found what he called 'the missing-link' between the ancient Greek clepsydra and the appearance in early fourteenth-century Europe of the escapement mechanism. An overland voyage westward of knowledge of the principle underlying Su Sung's water clock, which principle Needham chose to regard as also a kind of escapement, must have been instrumental in this regard. True, he acknowledged that nothing even remotely similar had ever turned up in the Muslim world, which given sheer geography could hardly have been passed by during that purported westward voyage of the design of Su Sung's clock. But no matter, absence of evidence, so Needham claimed, did not count against his reconstruction of what must have happened.

Declaring absence of factual evidence not good enough an argument against actual transmission did not, in Needham's conception of the westward transmission of Chinese scientific knowledge generally, remain confined to the sole case of Su Sung's water clock. To the contrary, he turned it into a methodological rule, which he phrased thus:

> As in all other fields of science and technology the onus of proof lies upon those who wish to maintain fully independent invention, and the longer the period elapsing between the successive appearances of a discovery or invention in two or more cultures concerned, the heavier that onus generally is.

In other words, Needham assumed cross-cultural transmission to be the 'natural' thing, and independent discovery or rediscovery the aberration in need of empirical proof. True, of many an ancient Chinese discovery or invention Needham did feel sure that it had been made independently in Europe. Only, he left it wholly unclear what criterion remained to decide for or against actual transmission, since in more than one case he did not accept the absence of data on actual ways of transmission as a criterion at all.

As always with Needham, the full picture is not nearly so straightforward as it may now seem. To get the full picture in view we must take a step back from specific cases and from the methodological rule that in his view ought to govern them. As he saw it, the transmission of pre-modern scientific goods displayed a global pattern, and it is this:

> While one can easily see that artistic styles and expressions, religious ceremonies and doctrines, or different kinds of music, have tended to be incommensurable; for mathematics, science and technology the case is altered—man has always lived in an environment essentially constant in its properties, and his knowledge of it, if true, must therefore tend towards a constant structure.

Consequently, there is one true, universal (or, as Needham called it, 'ecumenical') science. It has grown out of the modern science that started in Western Europe in the seventeenth century and has accumulated ever since. Into the making of universal science went certain very significant elements of science as cultivated in many pre-modern civilizations. Needham's favorite metaphor for the whole process was, and remained, that of the many rivers of traditional science, all of which have flown into the ocean of universal, modern science.

Needham worked this overall conception out in various directions. The most creative investigation that he derived from it was an arresting piece about what he saw as a heterogeneous yet in the end highly interdependent knowledge system located around the Mediterranean. And he noted that into this system:

> ... East Asian science was not incorporated ... It is not that there was no contact between Arabic civilisation and East Asian science; quite the contrary. But for some reason or other, when the translations were being made from Arabic into Latin, it was always the famous authors of Mediterranean antiquity who were chosen, and not the books of Islamic scholars concerning the science of IndiaIndia and China.

The most striking example here is that of al-Biruni's grand survey of Indian science which, unlike works by Ibn al-Haytham, Ibn Rushd, and so many others, did not come up for translation in twelfth-century Toledo but was first rendered into a European language by the late nineteenth century. All in all, so Needham concluded,

> the examples which have been given show that there was no lack of contacts between Arabic and East Asian science, but it remains true that East Asian science did not filter through to the Franks and Latins, i.e. to precisely that part of the world where ... modern science and technology were later to develop.

No trace here, remarkably, of the alleged superfluousness of empirical data sustaining a claim to actual transmission, as Needham made sure to sustain this grand generalization by means of a whole range of historical facts and circumstances. Whether the generalization itself is still tenable in the light of later scholarship or not I am not qualified to say. The point is rather that, in his views on transmission of scientific goods from

China to Europe, Needham was, as so often in his work, an on occasion brilliant yet hardly a consistent thinker.

In fact, Needham went even further, in that at times he denied not only the *reality* but even the very *possibility* of pre-modern dialogue between civilizations in matters scientific. I end my summary report on my findings at the time about Needham's views on transmission and even dialogue with three passages I encountered on three different spots in his work:

> Western science ... developed on the whole without the benefit of either Indian or Chinese contributions.
>
> The mutual incomprehensibility of the ethnically-bound concept systems did severely restrict possible contacts and transmissions in the realm of scientific ideas. This is why technological elements spread widely through the length and breadth of the Old World, while scientific elements for the most part failed to do so.
>
> The sciences of the mediaeval world were in fact tied closely to their ethnic environment, and it was difficult if not impossible for people of those different environments to find any common basis of discourse. For example, if Chang Hêng had tried to talk to Vitruvius about the Yin and the Yang or the Five Elements, he would not have got very far, even if they could have understood each other at all.

We are now ready for another preliminary conclusion. As with Needham comparing, so also with Needham in the act of considering cases of dialogue between civilizations where scientific goods are concerned we are facing a mess, although in the latter case a mess with some visionary overtones. And of course, when I had completed my mapping of what rather to my surprise had turned out to be a dual mess, I could not help wondering whether a messy outcome could be avoided at all.

As to the comparative approach, was I really bound to conclude that it was pointless to pursue it? Does Needham's widely flung effort conclusively demonstrate that, if you want to become wiser about why modern science arose in Europe, not elsewhere, comparison cannot get you any further? True, some of Needham's smartest and best informed critics at the time, notably A.C. Graham and Nathan Sivin, appeared to have drawn that very conclusion. On the other hand, I encountered at the same time the much more modest yet also vastly more consistent and historically responsible handling of a similar question in a lengthy essay by a Turkish scholar named Aydın Saylı.[4] I further remembered a remark once made by a colleague to the effect that comparison is 'the motor of historical thinking'.[5] So did Needham's example now compel me, or did it not, to conclude that the comparative approach may serve historians of other subjects well but that historians at work on pre-modern science in more than one civilization had better do without the motor?[6] If so, how could we ever hope to get closer to explaining the rise of modern science in seventeenth-century Europe?

When it comes to dialogue, the question raised by Needham's mess is more subtle. After all, there are most certainly some uncontroversial cases, such as the Baghdad-centered translation movement due to which ancient Greek natural philosophical and mathematical insights became known in the world of Islam, just as, centuries later, another, this time Toledo-centered translation movement made these insights, meanwhile appropriated, handled, enriched, and expanded in the Islamic world, known in medieval Europe. So here the question rather became under what conditions such dialogues could take place, and how to establish their historical presence or absence. Was this issue bound likewise to remain enveloped in the part-visionary mess Needham had made of it?[7]

5 Comparison and Dialogue: What Is Needed in Any Case

The prime condition for all work in cross-cultural history of science, so Needham's approach had made unmistakably clear to me, is that how we engage it should remain as free as we can possibly make it of

[4] Discussed at length in my *The Scientific Revolution. A Historiographical Inquiry*, Sect. 6.2.

[5] His name is Dick van Lente.

[6] By 'historians of other subjects' I mean in particular economic historians, for whom comparison belongs to their standard equipment.

[7] In this regard, too, Saylı's approach proved to be much more careful than Needham's.

wishful thinking. Few cognitive errors are quite so devious as the inclination, innate to all of us, to take how we would *like* some given state of affairs to be the *reality* thereof. In the present case the wishes that may lead our thinking astray stem from two sources chiefly. One is the wish that comes so natural to every scholar (myself surely included) to find our own interpretations and conclusions confirmed, not contradicted, by the facts of the matter. The other, more specific yet equally voluminous source of wishful thinking is that veritable cauldron, so familiar to every adult human being, where feelings are brewing of innate superiority to people noticeably different from ourselves in (above all) skin color, cultural habits, and/or social position. Something similar is, conversely, true of feelings of guilt or shame in view of past unjust treatment, and of hosts of related sentiments that may so easily cloud our judgment. A particularly blatant case in point, so even my ultra-brief summaries must have amply shown, is that of Joseph Needham himself. The overriding sentiment behind all those numerous oddities and inconsistencies that detract so needlessly from his formidable overall achievement (without, fortunately, destroying it) is not hard to guess. Needham himself was quite open about it:

> Everyone who undertakes a big inter-cultural job like this must naturally project his own system of beliefs in doing so—it is his opportunity to preach (and I use the word quite advisedly) to his own and later generations. If sometimes we have written like barristers pleading a case, or sometimes over-emphasise the Chinese contributions, it has been consciously to redress a balance which in the past tilted over much too far on the other side. We were out to redress a secular injustice and misunderstanding.

For later generations, not failing to notice that meanwhile twenty-five solid tomes of *Science and Civilisation in China* have effectively redressed that 'secular injustice and misunderstanding', there is not the slightest need (if ever there was a need) to accept that such a projection mechanism would really be 'natural'. To the contrary, it seems to me imperative to adopt as dispassionate a stance as we can possibly muster, guided along in so doing by no other sentiment than that of a genuine *respect* for other civilizations than the one we happen to be rooted in ourselves. Respect in the old-fashioned sense of the word, which is neither that of paying obligatory homage to a demanding autocrat, nor that of abstaining from all quality judgments, but rather of taking our fellow human beings, whatever their cultural background, no less serious in their habits and ways of thinking than we habitually take ourselves.

The condition that I regard as coming next in significance is respect of quite another kind, by which I mean respect for the *facts* of whatever case we find ourselves investigating—a respect, to be sure, in need of being unceasingly refreshed. What may really count as fact (and even more as 'conclusive fact') is, of course, up for further examination always, yet at issue here is something much more elementary. Take your facts seriously; be inquisitive enough to follow them (in particular when they become rebellious) to wherever the central problem you have posed yourself appears to lead you. Another way of phrasing this condition is that you should remain open always to those facts you encounter along the way that, certainly at first sight, do not fit your original interpretation all too well, or even not at all—often these turn out, on closer inspection, to be the most instructive of them all.

Next, it seems to me imperative not to vary the problem that you have posed yourself in the rather light-hearted manner Needham accustomed himself to. If, as your investigation proceeds, you feel bound to alter the gist of your problem in some major way, then it behooves you to reconsider at your ease what the possibly far-reaching consequences may be for such solutions to your original problem as you have in the meantime come up with.

Finally, as to those solutions to your problem, a major condition is that you be clear about what they are and how they stand related to one another. Thinking about causation is not, as a rule, something that historians are particularly good at. The least we should do in my opinion is to spend hours if not days or weeks asking ourselves whether one cause will do (usually not); what kind of causality we are invoking at any given point (for causes come in more than one guise); how far the explanatory power stretches of the various causes that we find it plausible to invoke, and so on.

What, to sum up, is needed in the practice of cross-cultural history of science is respect for other civilizations; respect for the relevant facts; consistency in posing (and in dealing with) the overriding problem that guides the investigation, and, finally, rigorous thinking about what may count as causes and how they stand related.

Moving now from conditions and requirements valid, in my view, to cross-cultural history of science generally to those that more specifically apply to cross-cultural *comparison*, the principal one seems to me to be a need to think long and deep about what you want to compare with what, and at what level of generality. Comparison naturally presupposes the presence of both similarities and differences. So when you set out to define the comparative problem that you want to solve, that is, what the objects of comparison are going to be, then in order to make your problem feasible you have to make choices. In the domain of comparative cross-cultural history of science there are, in principle, three levels to choose from: the macro-level, the micro-level, and the meso-level. The most useful level of comparison is but rarely (if ever) the macro-level of entire civilizations where Needham forever kept situating his problem (pre-modern science in China and in 'the West'). At that lofty level you run two main risks. One is to overshoot your mark, as Needham so amply did. The other is to come up with essentialist explanations, derived from properties allegedly fixed forever (such as an alleged inability of Chinese people to think logically, or of Muslims everywhere to have forever adhered to their faith the fundamentalist way). It is explanations in terms of such purported essences that were quite rightly scorned by Needham but that have been and still are around (perhaps masked nowadays behind euphemisms) for China as well as for the civilization of Islam.

Not that, as a rule, the micro-level of individual practitioners has outcomes of significantly better quality in store. Most often the respective contexts at that level are simply too different to make instructive comparison feasible. Also, we historians have been reared on the admonition to keep our healthy distance from more generalizing disciplines like sociology or philosophy, and always to respect in history the unicity of every individual person or past event. And surely this is by and large true. But if that attitude be reasoned through to its ultimate consequence, it comes to stand squarely in the way of the act of comparison as such—at that level everything becomes incommensurable with everything else. Nonetheless, some rare cases at the micro-level may look so evidently similar and, at the same time, so well-defined, that even a big difference in local context may cease to be a hindrance to instructive comparison. One example here is a comparison made in 2003 by an American historian by the name of Gene Jinsiong Cho between the first exact calculation of an equal division of the octave made around 1605 in Europe by the Dutch polymath, Simon Stevin, and the first, far more detailed and also more accurate calculation of equal temperament made some twenty years earlier in China by the courtier and polymath Zhu Zaiyu.[8]

On the whole, however, the meso-level seemed to me to provide far and away the most promising operational basis for at least potentially fruitful comparison. If you choose the events or developments or states of affairs that you want to compare with sufficient care, there is a good chance that at that level the relevant similarities and the relevant differences will still prove to be manageable.

Now on to *dialogue*, in the sense of cases of fruitful scientific contact in pre-modern times between one civilization and another. Here a first condition rests in the need to be alert all the time to how much may all too easily go wrong in dialogue even between individuals. I categorized above the large collection of possible cases under two headings: poor listening, and misunderstanding. Of course, where collectivities are at stake the picture changes somewhat. One circumstance that we saw Needham pointing at without drawing any further conclusions is a complete incompatibility of conceptual frames. In such a situation no fruitful dialogue can even be imagined to take place between people reared in so different conceptual frames as (to stick to Needham's imaginary example) Chang Heng and Vitruvius. This, however, is an extreme case. While alert to the very real possibility that sincere efforts at dialogue

[8] Gene Jinsiong Cho, *The Discovery of Musical Equal Temperament in China and Europe in the Sixteenth Century* (Lewiston (NY): Edwin Mellen Publishers, 2003).

failed due to profound misunderstandings, we should also keep mental room open for the historical possibility of *creative* misunderstandings.

Needham himself, with that inclination of his to run from one extreme to the other, alternated his somewhat overdone pessimism in this regard—the likelihood of complete absence of what he called 'a common basis of discourse'—with a much more frequent mood of light-hearted optimism. When in that mood he was ready to assume transmission to be the natural state of things unless empirically disproved. In thus shifting the onus of proof from those who posit the transmission of scientific goods in some given case to those who dispute it on the grounds of absence of empirical evidence; that is, in claiming transmission, and *a fortiori* dialogue, without a shred of empirical evidence, Needham demonstrably opened the floodgates for the kind of wholly arbitrary decisions that he reached himself, as for example in the case of Su Sung's water clock.

By the same token the floodgates for wishful thinking are being thrown open as well. And where wishes enter the debate, a scholarly opponent all too easily turns into a person deviously out to prevent your wishes from becoming reality. Indeed, more than once in actual debates on specific cases scholars frame each other as (for instance) 'transmission sceptics' vs. 'transmission believers', as if transmission (or circulation, or dialogue) were just an all-or-nothing, yes-or-no affair.[9]

Also, and even more worrisome in my view, the entire argument about allegedly being absolved from empirical validation of a plausible-looking case of transmission (or dialogue) is squarely opposed to a very basic rule valid for science and scholarship of just about any kind. In all but the rarest disciplines—logic; mathematics, and, perhaps, as a borderline case in its current state, string theory—science and scholarship deal, and ought to deal, beside theories and hypotheses and concepts and many further mental constructs, with facts. Most certainly for any scholarly discipline that deals with human beings facts, while surely disputable in each and every individual case, just cannot be dispensed with. Leaving

[9] Viktor Blåsjö, 'A rebuttal of recent arguments for Maragha influence on Copernicus'. In: *Studia Historiae Scientiarum* 17, 2018; pp. 479–497; p. 481: 'Since Ragep 2017, p. 267, has labelled me a "transmission sceptic," it seems natural for me to refer to these authors [i.e. Barker, Heidarzadeh, Bala, and Ragep] collectively as "transmission believers"'.

them out comes down to speculation, and ought to be called it. To speculate is most assuredly a legitimate business in science and scholarship, a potentially very productive one even—only, it has no demonstrative force whatsoever; in the end, empirical validation is indispensable.

Because it illustrates the point just made so neatly, I shall now spend three easy-to-skip paragraphs on the case mentioned in passing above—the calculation of equal temperament that Zhu Zaiyu made in a memorial for his emperor (that he saw at once rejected at court), and the calculation made some twenty years later by Simon Stevin in a treatise on musical theory written for princely instruction that he eventually withheld from publication (to vanish soon after he died, and to be found back and edited in 1884). Awareness of these bare facts made it tempting to believe that Stevin must have owed the calculation, or at least news about it, to cross-cultural dialogue. After all, Zhu's text was considerably earlier and, speaking in general terms, the route lay open at that time for information on the contents of a courtly memorandum submitted in the Imperial Palace to reach within twenty years the harbors of the Northern Netherlands. Small wonder, then, that Joseph Needham was so tempted. He pointed at the resident Jesuit in Peking, Father Matteo Ricci, as the most likely intermediary. Still, he apparently felt the onus of proof to rest on those who claim transmission, for he acknowledged with some regret that the available evidence for it was, as yet, no more than circumstantial.

Of course it need not stay that way. Gene Jinsiong Cho, even more eager than Needham to solidify the evidence, argued in 2003 in a book wholly dedicated to the matter that he had meanwhile found the hoped-for solid evidence. He pointed at texts showing that Father Ricci had been well acquainted with the calendrical context of Zhu's numerous treatises on music. He further invoked a related case—the huge sailing carriage that Stevin once demonstrated to great effect on Scheveningen beach. He went on to argue that, in all likelihood, Stevin had learned about such a carriage from a hearsay remark about Chinese sailing carriages made in a widely read travel account by an early Dutch explorer. So, given Zhu's precedence, and with a viable route from Peking to the Dutch harbors now established with greater certainty, Cho took transmission in the case of equal temperament to be settled beyond reasonable doubt.

On somewhat closer inspection, however, Cho's unmistakably wishful thinking caused him to deal all too lightly with the very premise of his argument. Rather than taking the trouble to read Stevin's treatise, as also a deservedly authoritative interpretation thereof (published in 1943) in the original Dutch, he relied exclusively on poorly edited and even more poorly translated versions thereof. But when the very heart of a scholarly argument hinges on what its core text has to say, an interpreter should not rest content with confiding in the qualities of the translator but to deal with the text the direct way (if need be, with the assistance of a well-informed local informant). For someone who actually reads Stevin's treatise in the original, some fairly elementary knowledge of musical theory suffices to confirm what Stevin's biographer had already noted in 1943, which is that at issue therein is not musical temperament, let alone the equal variety thereof, but quite another matter.[10] Throughout his treatise Stevin argued that, contrary to what everybody else has always taken for granted, the *natural* division of the octave is not the one that goes with the ratios of the first few integers. Rather, so he sought to prove over the length and breadth of his treatise, the natural division of the octave is the *equal* division (with, for instance, the pure fifth C–E given, not by 2:3, but by the 12th root of ½ to the power of 7). Consequently, calculation of the tonal scale in equal temperament, for Zhu the main business of his treatise, was for Stevin, with his outspoken lack of interest in the very issue of temperament, no more than a side issue. *Numerically* the decimal fractions that he calculated (though not quite so accurately, and computed down to four rather than to twenty decimals) are equivalent to Zhu's, yet *conceptually* equal temperament and the equal division of the octave are miles apart. If, in spite of what Stevin in his treatise went out of his way to argue, we nonetheless consider it as being about equal temperament, we are just projecting our own standard way of tuning our present-day musical instruments back onto a treatise irrelevant to it. With Zhu Zaiyu things stand different—what he was advocating as part of an ongoing debate about the calendar and its

10 The argument refers to ch. 13 of E.J. Dijksterhuis' Dutch-language biography of Stevin, published in 1943, and to Sect. 2.3., on Stevin's musical theory, in my *Quantifying Music. The Science of Music at the First Stage of the Scientific Revolution, 1580–1650* (Dordrecht: Reidel, 1984).

vast contemporary ramifications was, indeed, that musical instruments should henceforth be tempered equally. In short, Zhu Zaiyu and Simon Stevin were addressing different issues, derived from different musical and also local contexts, and the issue of dialogue, however attractive, and however plausible at first sight, vanishes in thin air. It would, of course, be pointless to deny the occurrence of occasional pre-modern dialogue *as such*; only, for establishing its reality in any *specific* case the hard work cannot be dispensed with.

Where, in sum, should we draw the line? Surely it would be unreasonable to insist always and everywhere on hard-written evidence of the act of transmission, or dialogue, itself. My argument about Zhu and Stevin is not about the absence of (so to say) a contemporaneous loan slip. It is, rather, about the need, for us scholars, to go to the attainable bottom of things as an indispensable condition for drawing any fixed conclusions in this regard, especially if those conclusions lie in the direction of where we emotionally prefer them to reside anyway. I do understand, and sympathize with, the wish of East-Asian or Middle-Eastern students and their teachers to see the scientific achievement of their ancestors acknowledged and properly evaluated rather than ignored or belittled as has been done far too often and for far too long by European and American historians of science.[11] And yet, to make the same prejudicial mistake but now in the opposite direction of coming up with insufficiently substantiated claims sanctified by the august term 'dialogue' is not the solution—the sober requirements of sound scholarship come first, always.[12]

So much, then, for the various requirements and conditions regarding both comparison and dialogue that I have come over time to distinguish. The list naturally contains general rules of a somewhat abstract character. Even at their most concrete they are rather like recipes. But a recipe is one thing, a meal prepared according to its prescriptions still quite something else. So I went on to find out how my own recipes would work out in

[11] To mention a personal example: In 1994 still, when my historiographical book on the Scientific Revolution came out, interest of my co-professionals in its 'non-Western' sixth chapter was marginal at best (somewhat unexpectedly, this proved true as well among Sinologists and Arabists, who with rare exceptions like F. Jamil Ragep appeared to regard me as just an intruder).

[12] I made the point most explicitly in what I wrote about Arun Bala's *The Dialogue of Civilizations in the Birth of Modern Science* in my essay review 'From West to East, from East to West? Early Science between Civilizations'. In: *Early Science and Medicine* 17, 2012, pp. 339–350.

practice. This I did in Part I of my 2010 book *How Modern Science Came Into the World. Four Civilizations, One 17th Century Breakthrough.*[13]

6 Comparison and Dialogue Jointly Taken Up in One Analytical Frame

When Needham was active the term in general use was 'transmission', not the 'circulation', or 'exchange', or 'dialogue', that came in use later. It was likewise natural in his time to call his subject 'science' without more ado. Of course he made a distinction between modern and pre-modern science, which after all was the distinction at the very heart of his causal enterprise. Even so, when, for instance, he waxed eloquent about what he called China's 'perennial philosophy of organic materialism' he unhesitatingly categorized it as science, though surely of the pre-modern kind.

It has meanwhile, and for good reason, become customary to use other terms. In my book I rather spoke, with deliberate blandness, of 'the pursuit of nature-knowledge'. I further coined the expression 'modes of nature-knowledge' and set it to work as my basic unit of analysis. That way I gave myself a chance to compare, e.g., China's organic materialism with, e.g., ancient Greece's various natural philosophies without having to call them science or, even worse, 'proto-science'. But I insisted from the very start that those various modes of nature-knowledge were not necessarily static entities. Rather, under certain circumstances and if certain conditions were fulfilled, they were, in principle, open to appropriation, to enrichment, and even to transformation of a more or less radical kind:

> Civilizations may clash; they may also cross-fertilize. Processes of transformation and of what I shall from here on label 'cultural transplantation' offer the single most potent boost to novelty and creativity history knows of. An influx of foreign people, foreign ideas, foreign practices may under

[13] The book came out in 2010 with Amsterdam University Press. A shorter version for a wider public appeared in 2015 with Cambridge University Press as *The Rise of Modern Science Explained. A Comparative History.*

propitious circumstances greatly enhance chances for novel things to happen to ideas or habits worn out or petrified or grown stale in their original setting.[14]

This particular view, then, came to dominate the various cross-cultural comparisons that I subsequently undertook. And sure enough I situated them all at the meso-level. My prime comparison was between the pursuit of nature-knowledge in the only two efforts ever undertaken to face the world of nature afresh and make coherent sense of it on the truly large scale—Warring States and Han China on the one hand, and the cities of Athens and Alexandria on the other. The other three major comparisons that I undertook involved three successive cases of the transplantation of significant portions of the Greek legacy of nature-knowledge from the East Roman Empire and its immediate successors to Islamic civilization, from there to medieval Europe, and from just conquered Byzantium to Renaissance Europe.

I complemented my general point about the chances for renewal inherent in feats of cultural transplantation with two conceptual schemes that I had found elsewhere. The historical sociologist Joseph Ben-David had argued in 1971 that, in the pre-modern world, relatively brief periods of flourishing of the pursuit of nature-knowledge, followed by most often steep decline, were the rule, the natural course of things as it were, rather than facts of history in need, as such, of explanation. This brilliant insight accomplished at one stroke three major feats. It put an end to the very notion of *failure*, as if Chinese (or, for that matter, Islamic) civilization had been charged by some nebulous authority to discover the basics of modern science but, alas, for some reason or other failed to do so. Ben-David's insight further turned Europe into the world-historical exception, in that the science it began to produce in the seventeenth century never went through a period of decline but, all the while spreading worldwide, kept flourishing (and even expanding) to the present day. Consequently, the big riddle in need of explanation by historians is not the many cases of decline but this unique case of forever ongoing advance. Also, this way the notion of 'Eurocentrism' was

[14] The passage appears on p. 45/46 of my *How Modern Science Came into the World.*

robbed at one healthy stroke of what made it so deservedly objectionable to Needham and to many later scholars who came to repeat his objections. What remained of the notion of Europe at the center of events was the indisputable historical fact that that is where modern science originally came into being. But the notion of failure, or the profoundly unhistorical idea that there was some kind of race between civilizations going on, with one among them being the first to cross the finish line, went at one stroke overboard as well. So did a whole range of similar prejudices. Or at least it did so for those who were prepared to listen to what Ben-David had to say.

In the conceptual equipment just outlined I combined a conception of the possibility of transformative cross-fertilization between civilizations (hence, 'dialogue' of a special kind) with laying the groundwork for viable comparison. The bridge that enabled me actually to forge the combination was the concept of 'latent developmental potential' that I owed to the economic historian David Landes. He introduced it in 1983 when discussing Needham's claim of Su Sung's water clock being the missing-link half-way the escapement mechanism by means of which around 1320 large tower clocks began to be constructed all over Europe. Landes argued instead that the Chinese water clock and the European mechanical clock embodied two quite different construction principles, one of which—the escapement—happened to possess a far greater developmental potential. The Chinese water clock, with its capacity for vastly superior accuracy, had already reached the summit of its constructive potential, whereas the European clock proved over time to lend itself to improvement of many kinds. In 1657 Huygens' invention of how in the clock mechanism to couple the escapement to a freely swinging pendulum enhanced at one stroke the accuracy of the mechanical clock far beyond what the water clock was still inherently capable of. Convincing as I found this style of historical reasoning, one major implication is that the presence or absence of a latent developmental potential can only be established in retrospect. Contemporaries cannot know beforehand with any degree of certainty whether some major, freshly made invention will prove to contain inside itself any developmental possibilities at all—at best, they may intuit it.

As a result of all this, for the 'modes of nature-knowledge' that constituted my basic unit of analysis, I now came to distinguish between three historical possibilities. A given mode of nature-knowledge, such as the Chinese philosophy of organic materialism or the ensemble of ancient Greek natural philosophies, may or may not possess a latent, innate potential for being adopted in another civilization and appropriated there, enriched perhaps, or even, just possibly, transformed in more or less radical ways. To be sure, for such a feat of cultural transplantation to become historical reality some major outside event was necessarily required—paradoxically, in all cases that I investigated warfare, in so many ways the very opposite of dialogue of any kind, proved to have served as the crucial incentive.

No such major outside event, so I went on to argue, ever happened to Chinese civilization. Though overrun by neighboring foreigners twice, the conquerors were quickly taken up in a civilization stable enough to remain basically unchanged by the event. As a result, and just like so much else in China until the onset of modernization, the philosophy of organic materialism survived by and large unscathed (while gradually growing stale) until way into the nineteenth century. With Greco-Roman civilization things went quite different—unlike with its Chinese counterpart a whole range of foreign invasions brought it to an end. In the seventh century AD the establishment by military means of an Islamic polity around large portions of the Mediterranean created an environment suitable for a first large-scale feat of cultural transplantation. The chance for a host of preserved documents in ancient Greek natural philosophy and mathematical science to be translated, studied, expanded, and enriched came when, as the outcome of civil war, caliph Abbas' descendants came to power and founded the city of Baghdad—site of the first cultural transplantation to befall the Greek legacy. Two more followed in their turn, first (in the wake of the Reconquista) from the Islamic world to medieval Europe and then (upon sultan Mehmet II's conquest of Byzantium) from that city to Renaissance Europe. In all three cases transplantation brought with it a range of enrichments, of a kind to lend themselves well to mutual comparison (directed, for instance, at how Ptolemy's original approach and his specific findings fared in the civilization of Islam and then in Renaissance Europe).

By 1600 Western Europe had by and large completed its assimilation of the Greek heritage, meanwhile enriched by a sizable number of well-ascertainable scientific goods stemming from Islamic civilization but also by some locally developed, empiricist elements of nature-knowledge. At that point a threefold, revolutionary transformation of the resulting knowledge took place that historians came in due time to characterize as the onset of the Scientific Revolution. Part of my analysis of those revolutionary transformations consisted of surveying (by means of such serious literature as proved available) what its various pioneers owed to insights and practices picked up in the meantime from other civilizations. (In Chapter 16 of this book Arun Bala makes an effort to show that, in the cases of Descartes' natural philosophy; of the advent of mathematical realism with Kepler and Galileo, and of the rise of Bacon-style experiment, 'dialogue' played a much larger role than I attributed to it; there is no space left to demonstrate here how for each of these three cases a deeper investigation that goes into indispensable detail shows otherwise).[15]

So much, then, for the manner in which I sought to weld the method of comparison and the occasional historical reality of dialogue together

[15] Frankly, I have some difficulty recognizing the 'Floris Cohen' who figures as the main person in Arun Bala's Chapter 16—he strikes me as reasoning in a significantly different manner than the Floris Cohen who, in the present section of this chapter, has summed up his own views on dialogue, comparison, and how the two may harmoniously go together in one analytical frame. This is not the place to set up a full-scale debate, so I just invite the reader to form his own judgment by comparing the present section (or, even better, the argument developed in the opening chapters of *How Modern Science Came Into the World*) with Arun Bala's chapter. As to, in particular, Arun Bala's claims about Descartes and Ibn al-Haytham, I refer the reader to the following specific passages in my book: on Descartes' almost fully *non-mathematical* philosophy of nature: pp. 228–238 and pp. 378–379; on Descartes' work in mathematical analysis (without any connection to his natural philosophy): pp. 348–352; on Ibn al-Haytham: p. 59/60 and pp. 295–299. Further, the passage on p. 137/138 regarding the Chinese approach to nature stresses above all the far-going *parallelism* with what took place in Renaissance Europe at the interface of nature-knowledge and the crafts, whereas the lengthy passage on China's 'worldview of organic materialism' that Arun Bala does quote represents Joseph Needham's, hardly my, way of capturing what difference between the two he believed to remain. Arun Bala's argument about Francis Bacon seems partly to state the obvious and for the rest to be a bit of a fight against windmills. Finally, and perhaps most importantly, he seems to regard every observation I have made about any specific, well-defined and empirically well-established case of dialogue as a grudging concession on my side rather than as an integral part of my views on these matters (see also, in this regard, the argument on specifics that I have unfolded in my book on pages 203–208).

in one comprehensive frame—a frame meant to be sufficiently ample but also sharply delineated enough to be put to analytical work in explaining how, in early seventeenth-century Europe, some basic elements of unmistakably modern science could emerge, and what historical trends and events had come together at that place and at that time to make such an event at all possible. 'Dialogue' and 'comparison', then, are neither superior/inferior to one another, nor are they opposite or even incompatible entities. Instead, they may go peacefully and, with some luck, fruitfully together.

References

Blåsjö, Viktor. 2018. A Rebuttal of Recent Arguments for Maragha Influence on Copernicus. *Studia Historiae Scientiarum* 17: 479–497.

Cho, Gene Jinsiong. 2003. *The Discovery of Musical Equal Temperament in China and Europe in the Sixteenth Century*. Lewiston (NY): Edwin Mellen Publishers.

Cohen, H. Floris. 1984. *Quantifying Music. The Science of Music at the First Stage of the Scientific Revolution, 1580–1650*. Reidel.

Cohen, H. Floris. 1994. *The Scientific Revolution. A Historiographical Inquiry*. Chicago: University of Chicago Press.

Cohen, H. Floris. 2001. Joseph Needham's Grand Question, and How to Make It Productive For Our Understanding of the Scientific Revolution. In *Science and Technology in East Asia*, ed. A. Arrault and C. Jami, 21–31, vol. 9: *The Legacy of Joseph Needham* (vol. 51 in series 'De Diversis Artibus'). Turnhout: Brepols.

Cohen, H. Floris. 2010. *How Modern Science Came into the World: Four Civilizations, One 17th-Century Breakthrough*. Amsterdam University Press.

Cohen, H. Floris. 2012. From West to East, from East to West? Early Science between Civilizations. *Early Science and Medicine* 17: 339–350.

Cohen, H. Floris. 2015. *The Rise of Modern Science Explained. A Comparative History*. Cambridge: Cambridge University Press.

Murphy, Kate. 2021. *You're not Listening*. New York: Celadon Books.

Index

A. Bala et al. (eds.), *Multicivilizational Exchanges in the Making of Modern Science*, https://doi.org/10.1007/978-981-97-3541-9

The manufacturer's authorised representative in the EU is Springer Nature Customer Service Centre GmbH, Europaplatz 3, 69115 Heidelberg, Germany. If you have any concerns regarding our products, please contact ProductSafety@springernature.com

Printed and bound by CPI Group (UK) Ltd, Croydon, CR0 4YY
01/10/2026
02225913-0007